COMPUTATIONAL MATERIALS CHEMISTRY

COMPUTATIONAL MATERIALS CHEMISTRY

METHODS AND APPLICATIONS

Edited by

L.A. Curtiss
Argonne National Laboratory,
Argonne, IL, U.S.A.

and

M.S. Gordon
Iowa State University,
Ames Laboratory, Ames IA, U.S.A.

KLUWER ACADEMIC PUBLISHERS
DORDRECHT / BOSTON / LONDON

A C.I.P. Catalogue record for this book is available from the Library of Congress.

ISBN 978-90-481-6497-4 (PB)
ISBN 978-1-4020-2117-6 (e-book)

Published by Kluwer Academic Publishers,
P.O. Box 17, 3300 AA Dordrecht, The Netherlands.

Sold and distributed in North, Central and South America
by Kluwer Academic Publishers,
101 Philip Drive, Norwell, MA 02061, U.S.A.

In all other countries, sold and distributed
by Kluwer Academic Publishers,
P.O. Box 322, 3300 AH Dordrecht, The Netherlands.

Printed on acid-free paper

Printed in the Netherlands.

TABLE OF CONTENTS

COMPUTATIONAL MATERIALS CHEMISTRY: METHODS AND APPLICATIONS

PREFACE

Computational materials chemistry is the study of chemical properties of materials using theoretical approaches. Computational methods used include ab initio molecular orbital theory, density functional theory, quantum chemical/classical hybrids, quantum dynamics methods. These methods are used to investigate structures, energetics, and dynamics of materials. Recent advances in computational techniques and computer technology have significantly enhanced the potential contributions that simulation and modeling can make to materials chemistry research.

The knowledge gained from these types of computational studies is used to help understand the effect of chemical reactivity on the behavior of materials and to help design new materials with novel or improved properties. Computational materials chemistry research encompasses surface chemistry, heterogeneous catalysis, polymer chemistry, and solid state chemistry. It underpins technological applications in areas such as fuel cells, batteries, catalysis, friction and lubrication, membranes, environmental chemistry, and electronics. Recently, computational chemistry has been playing an important role in the emerging field of nanoscale materials such as nanoparticles, carbon nanotubes, nanocrystalline diamond, and molecular wires.

This book provides a survey of topics in computational materials chemistry that are representative of different computational approaches and different areas of applications including 1) surface chemistry, 2) solid state chemistry, 3) polymer chemistry, and 4) nanoscale materials. In addition, methodologies used in computational materials chemistry are reviewed.

L. A. Curtiss
Argonne National Laboratory
Argonne, Illinois U.S.A.

M. S. Gordon
Iowa State University
Ames, Iowa U.S.A

Chapter 1

USING QUANTUM CALCULATIONS OF NMR PROPERTIES TO ELUCIDATE LOCAL AND MID-RANGE STRUCTURES IN AMORPHOUS OXIDES, NITRIDES AND ALUMINOSILICATES

J. A. Tossell

Dept. of Chemistry and Biochemistry University of Maryland College Park, MD 20742

1. INTRODUCTION

Determination of local and mid-range structure is an important problem in glass science. Many experimental techniques can be used to ascertain local structure, including x-ray diffraction, EXAFS, IR-Raman and NMR spectroscopies.[1] In most cases a single spectral technique is not adequate to definitively assign the local species present. Particularly valuable techniques for the characterization of local and mid-range environments in glasses are Raman spectroscopy and solid-state NMR spectroscopy.[2] The success of these techniques for the characterization of local order in glasses is partly a result of the existence of numerous crystalline materials of related chemical composition and structure, which makes possible a "finger-printing" approach to the assignment of the spectrum of a glass. It is also partly a result of an often rather simple and systematic dependence of NMR shieldings on the coordination number and the degree and type of polymerizationof the magnetic atom.[3] Trends in Raman spectral energies also exist, but they tend to be more difficult to interpret qualitatively. However, there are numerous cases in which glass structures are different from those observed in crystalline solids of the same (or any) stoichiometry. For such cases quantum mechanical calculations of properties can be very valuable in assessing the local

L.A. Curtiss and M.S. Gordon (eds.),
Computational Materials Chemistry: Methods and Applications, 1–34.

structures of the species present[4]. Knowledge of local and mid-range structure is very important both in understanding the macroscopic thermodynamic and transport properties of natural glasses and in designing synthetic glasses with specific desirable characteristics, such as low thermal expansion or chemical inertness.

We have applied quantum mechanical techniques to the calculation of properties for a number of different species occurring in borate, silicate, aluminosilicate and nitride glasses. There is a continuing controversy about the presence and properties of planar boroxol rings $B_3O_3(O^-)_3$ in glassy B_2O_3. Dynamic angle spinning NMR studies have recently yielded NMR shieldings and quadrupole coupling constants for boroxol ring and non-ring sites in B_2O_3 glass[5]. The boroxol ring B's were found to be significantly deshielded compared to the non-ring B atoms. We have been able to reproduce this shielding difference as well as to interpret the persistence of boroxol ring NMR signatures despite changes in the Raman spectrum of B_2O_3 glass[6]. Although early molecular dynamics studies failed to find boroxol rings, recent MD studies[7] are finally reproducing significant numbers of such boroxol rings, after incorportating polarization effects in the O atoms. Defect species existing in small concentration in aluminosilicate glasses, such as O atoms in Al-O-Al linkages[8] or in O[3] (O tricluster sites, where the O is bonded to three tetrahedrally coordinated atoms) have been characterized by NMR[9]. Calculations of the spectral properties associated with Al-O-Al linkages have been quite successful[10,11] while those on O[3] species[12,13] have so far not been able to produce properties in agreement with experiment. Addition of P_2O_5 to aluminosilicate glasses produces great changes in phase relations and viscosity and solid-state ^{31}P NMR spectroscopy shows a number of different species which we have assigned to a number of distinct complexes[14] using NMR calculations. Spinning side-band intensities have yielded chemical shift anisotropies which have been very helpful in confirming assignments. Calculations of the C, Si and N shieldings in models for C_3N_4, Si_3N_4 and P_3N_5 established that N shieldings show strong non-local effects, consistent with large solvent effects seen in ^{14}N NMR [15,16]. Such long-range effects (observed to a lesser extent for O) make the calculation of N NMR shieldings with local cluster models difficult[17], but can potentially give additional information on mid-range order

In this manuscript we utilize molecular quantum mechanical methods to calculate the equilibrium structures, NMR properties and in some cases the vibrational spectra for a number of molecular models for local species occurring in the various glasses described above. Such molecular cluster approaches have been used for some time[18], particularly for the analysis of vibrational spectra and NMR spectra, which are generally determined by the local atomic environment. Our goals are: (1) to reproduce shielding trends caused by variations in composition and structure in the first and second coordination shells about the magnetic atom, (2) to establish the connection between different spectroscopic signatures for a given species , e.g. to interrelate features seen in vibrational and NMR spectra, (3) to test the cause and effect relationships thought to underly empirical correlations of shielding and structure and (4) to assist in the identification of species existing in glasses and melts.

Several problems to be surmounted in the accurate calculation of NMR properties of condensed phase species are: (1) the n^3 problem - the fact that the computational resources needed for an ab initio Hartree-Fock calculation scales as about the third power of the size of the system or the size of the basis set used to expand the MO's, tempting us to use model systems which are too small and/or basis sets which are too small, (2) the correlation problem[19] - the neglect of the instantaneous correlation of electron motions in the Hartree-Fock method leading (in general) in calculations of NMR shieldings to undershielding and exaggerated shielding anisotropies, and (3) the problem of truncating or terminating the model cluster without introducing (unacceptably large) computational artifacts. Previous problems arising from the gauge dependence of the shielding constant have been virtually eliminated by modern methods like the gauge-including AO[20] (GIAO) method, although some other methods give comparable results. For small molecules the first two problems can essentially be overcome by employing large basis sets, correlated methods and fast present day computers, but for large molecules (as models for mid-range order in solids will necessarily be), such constraints will seriously limit the accuracy of the results.

A good general procedure for modeling the NMR shielding in a condensed phase species is: (1) choose a particular set of model molecules, which have the same local geometries as those expected in the solids, (2) choose a quantum mechanical method rigorous enough to accurately reproduce the geometric structures of such model molecules, (3) establish the stability of the calculated NMR shielding trends (although not necessarily the absolute values of the shieldings) with respect to basis set expansion and method selection by study of a related set of gas-phase molecules (if such a series exists), (5) establish the stability of the calculated shielding trends vs. enlargements of the model molecules, and (6) test the calculated shielding trends vs. those assigned experimentally to particular species in the glasses. However, this procedure can fail at a number of points. The model of local structure suggested (usually by experimentalists) may be poorly or incompletely specified, there may be no comparable set of gas-phase molecules on which to test the computational methods, it may be impossible to include or even identify further shells of atoms in the glass and experimental trends may actually be affected by other structural parameters than those which the study focuses on.

One difference between our approach and that of many chemists and materials scientists is that we try to make contact with the crystal-chemical concepts typically employed as interpretive tools by mineralogists and glass scientists. These include standard concepts such as coordination number but other concepts little used in current (molecular) chemistry such as bond strength[21], which within an ionic model is simply cation charge divided by coordination number and within a covalent model relates to bond overlap population. We will show for Na NMR shieldings and for O EFG's how such simple concepts can be used to organize and explain the data.

2. COMPUTATIONAL METHODS

The basic Hartree-Fock self-consistent-field MO theory used in this work has been thoroughly reviewed and the errors in bond distances and vibrational frequencies expected at various computational levels for small gas-phase molecules are well understood Calculation of the equilibrium structures was done with the programs GAMESS[22] and GAUSSIAN[23] while the calculations of NMR shieldings were done mainly with the the GIAO implementation in GAUSSIAN. Descriptions of the major methods for calculating NMR shieldings and reviews of their applications can be found in the proceedings of a NATO workshop on the calculation of NMR shieldings edited by the author[24]. Jameson has recently discussed advances in NMR theory and computational methods[25].

The standard expansion basis sets employed in our geometry and vibrational spectra calculations are the 6-31G* polarized split-valence bases designed by Pople and coworkers[26]. For the most accurate of the NMR calculations presented here doubly polarized bases were used with a 6-311G s, p basis, two d polarization functions on the heavy atoms and single p polarization functions on the H atoms. Thus, even our best basis sets are necessarily modest by NMR shielding standards for gas-phase molecules, since the molecular clusters considered are often very large. Remember that our goal is to accurately evaluate shielding trends and differences in shielding between related molecules, not to obtain highly accurate absolute shieldings.

Our work has much the same goal as the recent comprehensive studies carried out on the structures and NMR spectra of numerous boranes and carboranes using the earlier IGLO method for calculating NMR shieldings by Schleyer and coworkers[27]. In both cases NMR shieldings are calculated for energy optimized structures and matched against experimental data to support (or contradict) the structural assignment. An important difference between the Schleyer, et al. work and the present work is that for the boranes and carboranes interactions with counterions or solvent can often be quite small while in glasses such interactions are typically stronger and computational artifacts due to truncation of the model cluster present a more serious problem.

It has recently become possible to calculate NMR shieldings using density functional band theory[28] but the method has so far been applied only to fairly simple crystalline materials with light atoms[29]. We will not discuss this method in detail but will later describe some results using this approach for crystalline C_3N_4.

For quadrupole nuclides such as ^{17}O and ^{27}Al in addition to NMR shieldings it is important to calculate the electric field gradient at the nucleus (EFG) which can be used to evaluate the quadrupole coupling constant (QCC). However, the QCC also depends upon the nuclear quadruple moment, which can be determined only by correlating EFG's from very high level calculations on small molecules with their experimental QCC values[30]. The absolute value calculated for the EFG depends upon the quality of the basis set used and upon the treatment of correlation, but trends in EFG's are often reproduced well at fairly low computational levels. It is therefore common to evaluate scaling factors, appropriate to different computational levels, which give the best fit of calculation to experiment for a set of test molecules, and then employ those factors in calculating EFG's for other related molecules[31]. One can also calculate EFG's using band theory, and agreement with experiment is quite good, using either Hartre-Fock[32] or density functional methods[33]. A recent study comparing periodic band theory results with cluster results for the simple ionic compound $NaNO_2$ showed that choosing the "right" cluster was rather difficult and that even fairly distant atoms could significantly affect the results[34].

3. RESULTS

3.1 B_2O_3 Glass

Although crystalline B_2O_3 does not contain boroxol rings, $B_3O_3(O^-)_3$, the existence of such rings in glassy B_2O_3 is supported by B and O NMR, Raman and x-ray diffraction data[35]. The calculated symmetric oxygen breathing frequency[36] for the gas-phase boroxine molecule $B_3O_3H_3$ has also been shown to agree well with the frequency of this mode in glassy B_2O_3. For the simplest possible boroxol ring model $B_2O_3(OH)_3$ and the corner sharing model $[B(OH)_2]_2O$ we first calculated equilibrium geometries, vibrational spectra, electric field gradients and B and O NMR shieldings[6(a)], all with 3-21G* basis sets, using the older RPA-LORG method [37] to evaluate the NMR shieldings. The calculated values of the quadrupole couplings and the trends in shieldings agreed well with the experimental data shown in Table 1, establishing that the different local environments can indeed be distinguished computationally (note that according to the modern definition of the the NMR chemical shift, δ, and the NMR shielding constant, σ, a more negative (or less positive) value of δ corresponds to a more positive value of σ, i.e. to increased shielding). Thus, the experimentally observed more positive chemical shift for the boroxol ring borons in B_2O_3 glass is consistent with their smaller calculated values of shielding constant. We actually compare experimental and calculated EFG values using the quadrupole product, which is given by $QCC[1 + \eta^2/3]^{1/2}$ (in MHz).

Table 1. Comparison of calculated and experimental NMR properties for B_2O_3 glass (from ref. 6(a)).

	Experiment		Calculated	
property	boroxol ring	corner-sharing	boroxol ring, $B_3O_3(OH)_3$	corner-sharing, $[B(OH)_2]_2O$
^{17}O NQCC (MHz)	4.69	5.75	4.49	5.28
^{11}B quadrupole product, NQCCx $\sqrt{1 + \eta^2/3)}$ (MHz)	2.68	2.56	2.41	2.34
	δ		σ	
^{11}B NMR shielding (ppm)	18.1 ±1 (deshielded)	13 ±1	117.0 (deshielded)	122.0

The deshielding of B in the boroxol ring was also found in other ring structures studied, such as the $B_3O_6^{3-}$ anion. The calculated values for symmetric breathing motions of $B_3O_3(OH)_3$ and $[B(OH)_2]_2O$ (after standard corrections for correlation and anharmonicity) were also consistent with the experimental assignments.

^{11}B and ^{17}O NMR shieldings and nuclear quadrupole coupling constants were later calculated using larger models for B_2O_3 and borates, and using the more accurate GIAO method to obtain the NMR shieldings[6(b)] Those calculations gave qualitatively similar results to the earlier ones, reproducing the experimental distinction between B atoms in boroxol rings compared to those in non-ring environments, thus giving further support to a boroxol ring model. However, the differences in shift calculated between the simplest boroxol ring and non-ring sites were smaller than in the earlier calculations. The new calculations accurately reproduced the difference in ^{11}B shift seen experimentally between boroxol ring and non-ring sites only if larger clusters containing all boroxol ring B or all non-ring B were compared, not using small clusters containing some of each type of B. This supported the experimentalists's interpretation[5(a)] that there are boroxol ring rich and poor regions within the glass, with relatively few B atoms at the boundaries between regions.

In Figure 1 we reproduce a figure modified from ref. 6(b) showing calculated ^{11}B and ^{17}O shieldings and ^{17}O QCC values for the $B_8O_{15}H_6$ cluster, which serves as a model for the boroxol ring B's. The boroxol ring B's are systematically deshielded, while the O shieldings are unexpectedly lowest for the central atoms. In Fig. 2 we show the same quantitities for $B_6O_{13}H_8$, a model for several connected non-ring sites. The characteristic B shielding was calculated to be about 106 ppm for the boroxol ring sites and about 112 ppm for the non-ring sites, consistent with the experimental difference of about 5 ppm. In this case larger clusters gave qualitatively similar results to the minimal ones, but also provided support for a new physical interpretation of clustering of different regions within the glass.

The calculations also established that the characteristic Raman signature of the boroxol ring is lost if one of the B's of the ring is converted to 4-coordination, while the boroxol ring ^{11}B NMR signature is unaffected, explaining how the boroxol ring Raman signature can disappear while the number of boroxol ring B's indentified by NMR signatures remains the same.

These calculations were less successful in interpreting the ^{17}O NMR of the borates (Fig. 1(b)), since they showed that the O NMR shieldings of chemically equivalent atoms at different positions within the cluster could have systematically different values, with deshielding for the central O atoms compared to the peripheral ones. As discussed later, similar long-range effects on NMR shieldings have also been seen for N, using a number of different theoretical approaches.

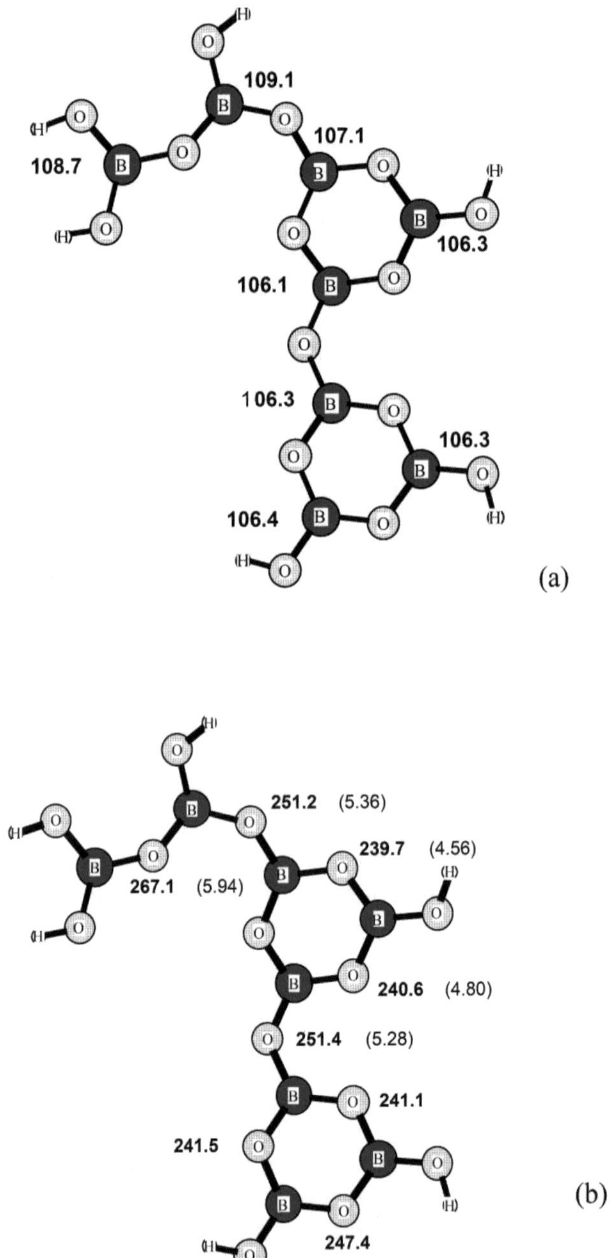

Figure 1. Calculated (a)[11]B (bold) and (b) [17]O shieldings (bold) and QCC's (in parentheses) for the $B_8O_{15}H_6$ cluster model of the boroxol ring sites.

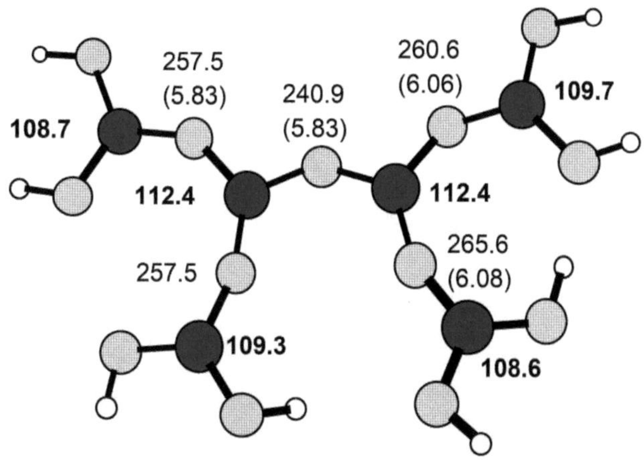

Figure 2. Calculated ^{11}B (bold) and ^{17}O shieldings and QCC's (in parentheses) for the $B_6O_{15}H_8$ cluster model of the non-ring B sites

3.2 Al-O-Al and O[3] species in aluminosilicate glasses

In aluminosilicate glasses, e.g. albite glass with composition $NaAlSi_3O_8$, the oxygens are predominantly either "bridging" oxygens (BO) between Si and/or Al, i.e. in Si-O-Si or Si-O-Al linkages, or "nonbridging" oxygens (NBO), bonded covalently to a single Si or Al and ionically to one or more network modifying cations like Na^+ or Ca^{+2}. The BO and NBO are usually readily distinguished by their values of QCC and often by their NMR shifts as well. Al-O-Al linkages are expected to be unfavorable since the resulting O's are underbonded in the Pauling [21(b)] sense, receiving an insufficient bond strength from the two Al^{+3} cations. However, it is well establihsed that Al-O-Al linkages do occur in glass and the energetics for their formation have been detemined calorimetrically[38]. Structures, energies and NMR properties for the O atoms in the two different isomers of the 4-ring molecule $Si_2Al_2O_4H_8$ shown in Fig. 3 were calculated[10] and the isomeric energies were found to be in semiquantitative agreement with the experimental data (which showed rather wide ranges). Previous theoretical studies on systems which were of varying total charges had yielded highly erratic results. The isomer energy difference yielded an energy penalty for formation of the Al-O-Al bond that was intermediate between the calorimetric results obtained for Na^+ and Ca^{2+} aluminosilicate glasses. The calculated differences in isomer energy were essentially unchanged when the -H termination was later replaced by -OH termination[39], with values of 17.0 and 17.5 kcal respectively for the energy penalty of the "paired" isomer vs. the "alternating" one.

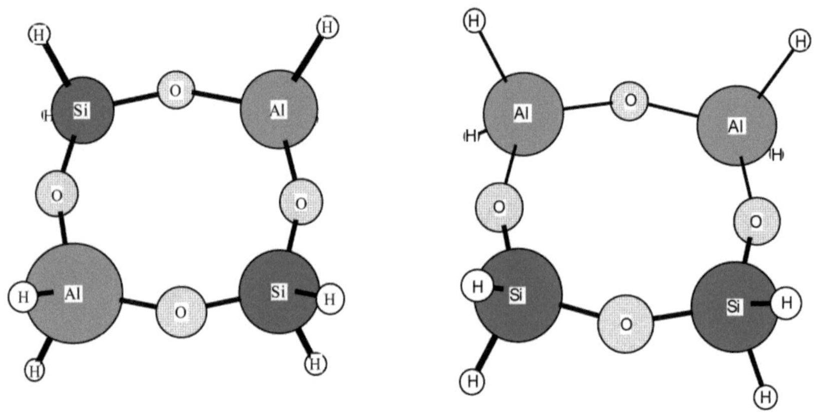

Figure 3. Paired and alternating isomers for the 4-ring model molecule $Si_2Al_2O_4H_8$(adapted from ref. 10).

The NMR shieldings and QCC values for the O in these isomers also clearly distinguished between Al-O-Al, Al-O-Si and Si-O-Si linkages. . QCC values were subsequently determined experimentally in crystals containg Al-O-Al linkages[8], which were in agreement with the calculations Calculated QCC values (using the scaling factors from ref. 31) for bare Al-O-Al were around 1.8 MHz, while experimental values were around 2.0 -2.2 MHz Several different models of the Al-O-Al linkage (of varying size) yielded similar QCC values. Thus even simple calculations were able to accurately predict the energetics and the EFG of the Al-O-Al species.

Experimental studies on other aluminosiliicate glasses, expected to have entirely BO oxygens, instead showed small populations of NBO, which were consistent with the overall glass stoichiometry only if there were also O[3] species, in which the O is coordinated by three tetrahedrally coordinated cations[9]. O[3] or "tricluster" species had been previously suggested to explain some viscosity data[40], In ref. 9 a peak was identified with a ^{17}O NMR shift of +20 ppm and a quadrupole coupling constant of 2.3 MHz, only slightly larger than that observed for Al-O-Al. Subsequent calculations by

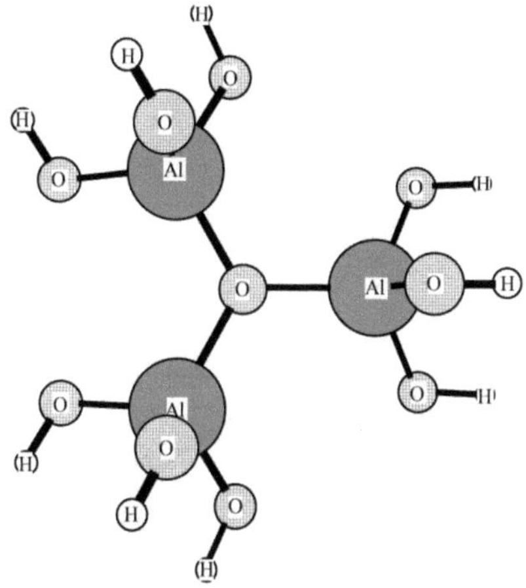

$Al_3O(OH)_9{}^{-3}$

Figure 4. Structure calculated for the O[3] molecular cluster $Al_3O(OH)_9{}^{-3}$, with a calculated QCC of around 3. 9 MHz

Xue and Kanzaki[12] using Hartree-Fock molecular cluster techniques cast doubt on this assignment since they predicted quadrupole coupling constants at the O for several plausible O[3] species that were 50% larger than those of the identified peak. Later calculations by Tossell and Cohen[13] for molecular cluster species like those studied by Xue and Kanzaki yielded similar results. One of the species considered by both groups is shown in Fig. 4.

This O[3] model was qualitatively like that suggested by the experimentalists. Only the local coordination of the three tetrahedral Al's about the O was really specified in the original formulation and it was assumed that the O environment was planar (and for this model system geometry optimization in fact yielded a planar structure). Xue and Kanzaki also showed that their results were stable toward basis set expansion and incorporation of correlation. However, they failed to consider other tricluster geometries analogous with those already characterized in inorganic systems (e.g. in alumoxanes[41]). In Fig. 5 we show a tricluster species having O-O shared edges, which has a substantially reduced value of ^{17}O NQCC.

Band theoretical calculations by Iglesias, et al.[33] and Tossell and Cohen[13] on the various O sites in the andalusite, sillimanite and kyanite polymorphs of crystalline Al_2SiO_5, some of which had tricluster-like sites (three bonds to O, but with the Al cations in 5- or 6- coordination), yielded QCC values larger than for a Al-O-Si linkage, except for the andalusite O(C) site, shown in Fig. 6, which gave a considerably smaller value of the QCC, around 2.4 -

2.5 MHz (calibration factors are not so well established for the band theoretical methods).

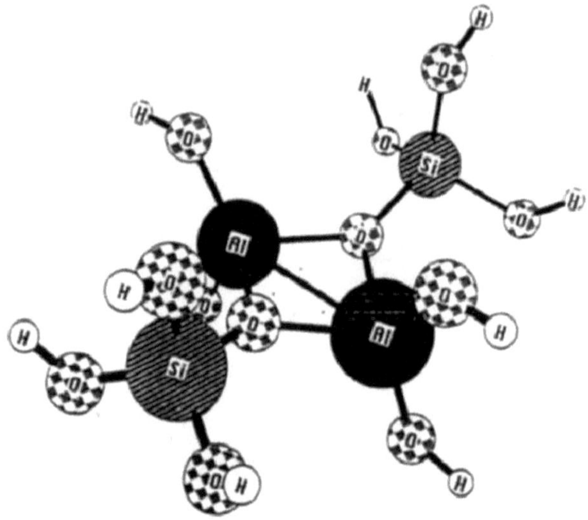

Figure 5. Calculated structure of $Al_2O_2(OH)_4[Al(OH)_3]_2^{-2}$, with QCC at central O[3] of 2.8 MHz (compared to 3.9 MHz for $Al_3O(OH)_9^{-2}$).

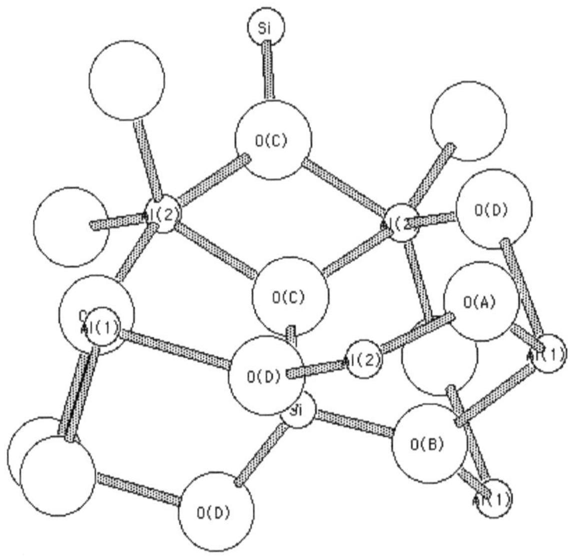

Figure 6. O(C) site in andalusite polymorph of Al_2SiO_5, obtained from the crystal structure using CRYSTALMAKER software[42]

In this site the O is coordinated to two Al[5] and one Si[4]. More importantly, two O(C) share an edge. The calculated QCC parameters for the various sites in the Al_2SiO_5 polymorphs were found to be well correlated with the sum of bond strengths received at the O, weighted by covalency, except for O(C) in andalusite, which falls well below the correlation line shown in Fig.7. The purpose of correlating quantum mechanically calculated EFG's with simple measures of bond strength and covalency was to provide a semiquantitative approach to aid experimentalists with assignments, in the absence of quantitative calculations. The small QCC value of O(C) was associated with its edge-sharing geometry, as confirmed by cluster

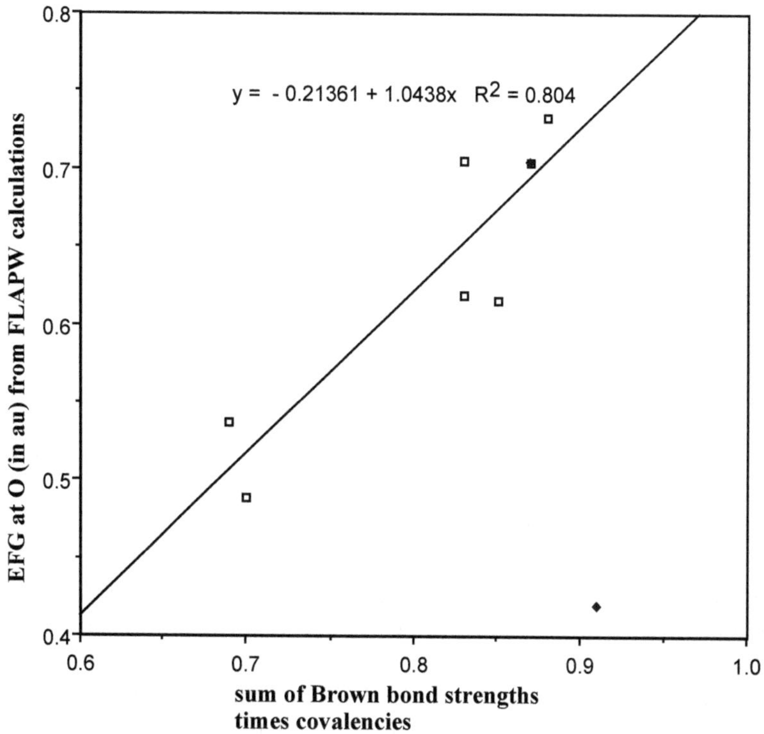

Figure 7. Correlation plot of EFG calculated with band theory vs. the sum of bond strengths times covalencies for andalusite and sillimanite (from ref. 13)

MO calculations on other species, with and without shared edges.

Stebbins et al[43] later measured an O QCC value of 2.5 MHz for a O[3] in the crystalline compound $CaAl_4O_7$. Calculations by Toplis, et al. (mentioned in ref. 43) have generated QCC values in agreement with this experimental result. We have used the $CaAl_4O_7$ crystal structure of Goodwin and Lindop[44] to generate clusters with formulas $Al_3Ca_3O_{10}^{-5}$ and $Al_3O_{10}^{-11}$, the larger of which is shown in Fig. 8.

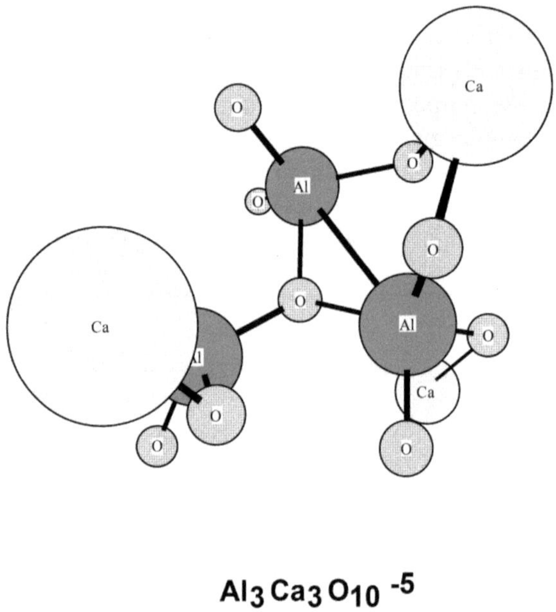

Figure 8. Molecular cluster obtained as a fragment from the crystal structure of $CaAl_4O_7$ (ref. 44).

These two clusters give EFG values at the Hartree-Fock 6-31G* level of 0.410 and 0.419 au, respectively, which using the calibration of OCC and electric field gradient developed by Ludwig, et al,[31] gives QCC values of 2.39 and 2.44 MHz, compared to an experimental value of about 2.5 MHz. These calculated values are much smaller than the value of 3.86 MHz calculated using the same method for the (energy-optimized) planar, corner-sharing tricluster $Al_3O(OH)_9^{-2}$ The $Al_3O(OH)_9^{-2}$ molecular cluster certainly resembles the O[3] site in $CaAl_4O_7$ qualitatively, since it has 3-coordinate O in essentially planar geometry and there is no actual edge-sharing, although some of the other O atoms are close to the central O[3] in the $CaAl_4O_7$ fragment. However, the calculated Al-O distance of 1.843 Å in the cluster is somewhat longer than the experimental values averaging about 1.79 Å in $CaAl_4O_7$. In general, one would expect an overestimate of the bond distance to give an EFG which is too small, (since in a point charge model the EFG scales as charge/distance^3) but in $T_3O(OH)_9$ systems, where T=Al, Si, increasing the O-T distance actually increases the O EFG, as shown in Fig. 9.

The largest component of the EFG is in the direction perpendicular to the central 4 atom plane (e.g. OT_3) and its magnitude arises from an imbalance of contributions from the lone pair orbital perpendicular to the plane and the sigma bonding orbitals. The sigma bonding contribution becomes larger as the bond distance decreases and this gives a smaller imbalance and a smaller EFG.

Thus the problem in obtaining an accurate QCC value for such a tricluster does not appear to be in the method for calculating the EFG itself from a given cluster geometry, but in the accurate determination of the local geometry. This shows the limit of local models for calculating properties like the EFG. It is well known that distant atoms can sometimes strongly affect EFG values[34], but in this case it appears to be small discrepancies in the local geometry that cause the problem.

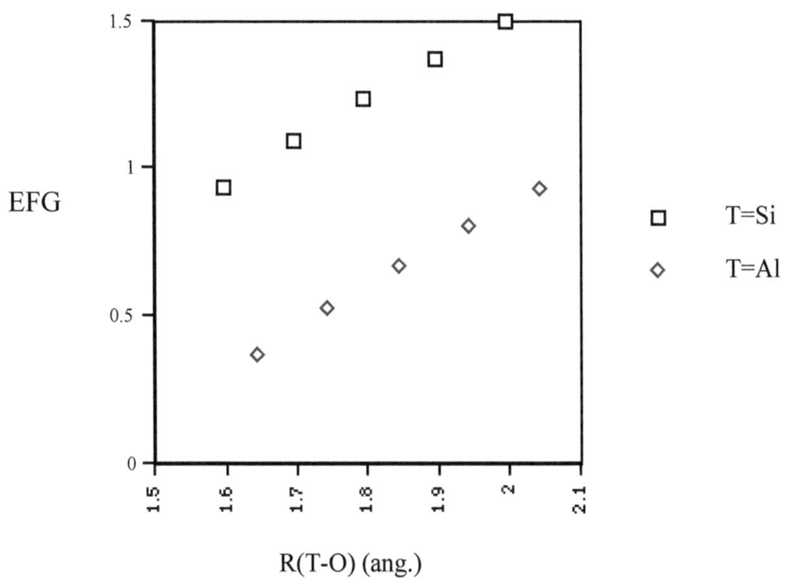

Figure. 9. Calculated O EFG in $T_3O(OH)_9$ clusters vs. T-O distance

3.3 Water in aluminosilicate glasses

The addition of small amounts of water to aluminosilicate glasses drastically lowers their viscosity. The molecular mechanism for this important property has been the subject of numerous studies. For probing the local structure of such a "wet" aluminosilicate glass solid-state NMR seems ideal. However, the large changes in physical properties are accompanied by relatively small and ambiguous changes in the Si, Al and O NMR. Only the ^1H and ^{23}Na NMR show significant changes. Two qualitative local models have been developed to explain the interaction of water with the aluminosilicate melt. In one the H_2O depolymerizes the predominant Si-O-Al bonds[45], giving so-called T^3 species (tetrahedrally coordinated Si or Al, with three of their O atoms shared with other T cations) which have Si-OH or Al-OH bonds. In the second[46], H^+ from the water displaces the Na^+ expected to be closely associated with a Si-O-Al linkages, giving a Si-OH-Al protonated species and the OH^- from the H_2O is then associated in some way with Na^+. For the case of water hydration of Si-O-Si linkages, e.g. in SiO_2 glass, the depolymerization mechanism is in fact well established. It was first thought that since Al-O bonds are expected to be weaker than Si-O, the Si-O-Al linkage would be more susceptible to hydrolysis than the Si-O-Si. It now appears that the Na^+ associated with a Si-O-Sl linkage may in a sense shield it from attack

Several different molecular models have been used to describe the local geometries expected in the glass, for which QCC and NMR shift values have then been calculated with standard quantum mechanical methods. A simple

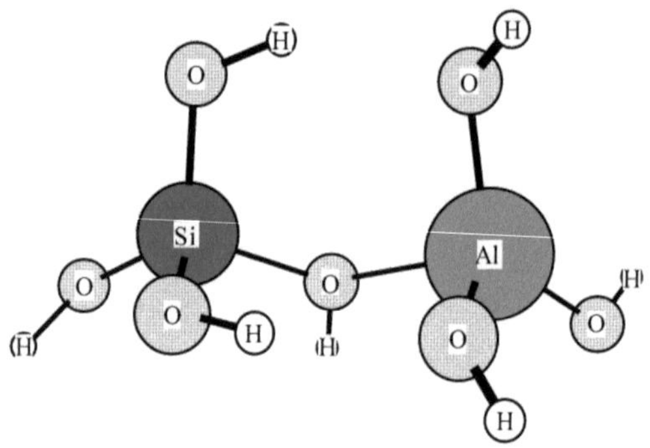

Figure 10. The $SiAl(OH)(OH)_6$ model for the protonated Si-O-Al bond.

model for the protonated Si-O-Al linkage is shown in Fig. 10.

However, most of the models actually used by Liu, et al [47] were considerably larger, as shown in Fig. 11 for the models of albite. Such larger models are particularly important in the accurate modeling of the ^{17}O NMR parameters. With respect to O as the central atom these clusters are 3NN, atomically correct through the 3rd nearest-neighbor shell. Comparison of calculated ^{17}O, ^{27}Al and ^{29}Si NMR shifts with experiment for crystalline albite gave very good agreement using these clusters.

On the basis of comparison of calculated and experimental ^1H, ^{17}O, ^{27}Al and ^{29}Si NMR shifts the conclusion now emerging seems to be that the calculated properties of any T^3 Al-OH species (formed by hydrolysis of a Si-O-Al linkage) are inconsistent with the experimental ^{17}O NMR data. Hydrolysis of only the Si-O-Si bonds in the structure, possibly with replace-

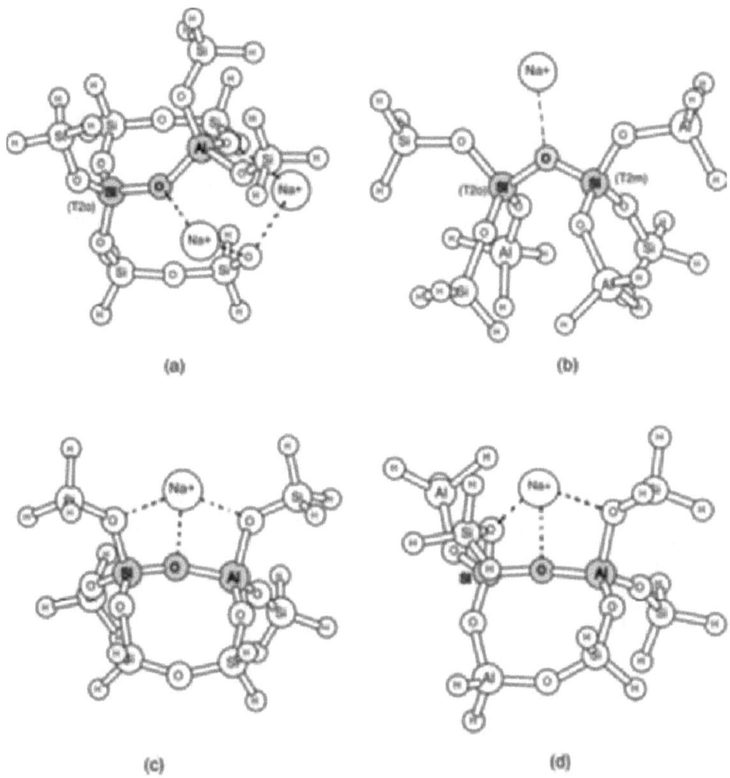

Figure 11. Cluster models for (a) Si-O-Al fragment of crystalline low albite; (b) Si-O-Si fragment of crystalline low albite; (c) Si-O-Al fragment of albite glass and (d) Si-O-Al fragment of albite glass (from ref. 47).

ment of Na^+ by H^+ adjacent to the Si-O-Al bonds could explain the results. However, the nature of the Na^+- OH^- interaction, if there is one, is unclear.

There are also still a number of problematic aspects of these calculations. Most of the attention in recent studies has been focused upon the NMR shifts since several researchers have shown that calculating QCC values using cluster models is rather tricky, sometimes giving good agreement with experiment and sometimes giving misleading results. Therefore a large QCC calculated for Al in the $SiAl(OH)(OH)_6$ model system was somewhat discounted. However, experimental studies on protonated zeolites give very large QCC values for Al in such a Si-(OH)-Al linkage[48] (around 13-14 MHz), consistent with the calculations. But no such large QCC Al NMR signature is seen in the hydrated aluminosilicate glasses. All the geometries have been calculated at the Hartree-Fock level with modest basis sets. This gives only modest accuracy in the description of H-bonding, which may well be important in such systems. All the calculations of molecular clusters with water have also considered only one water molecule. Krossner and Sauer[49] have shown in computational studies of water interaction with zeolites that both correlation and the presence of multiple waters favors ion-pair species over neutrals, i.e. it facilitates H_2O dissociation. It is now clear that water dissociation will usually occur only in environments in which several additional waters are available to solvate the ionic species produced. This has been shown for example in a number of studies of water adsorbed on metal oxide and metal surfaces [50].

Table 2. Experimental ^{23}Na NMR shifts, calculated deshieldings and calculated bond strength sums at Na for crystalline silicates and aluminosilicates. Deshieldings calculated using average Na-O distances and those directly calculated from Na-centered clusters are given in bold (from ref. 51).

crystal	**site**	**expt. shift**	**deshielding using Si_2H_6O**	**deshielding using SiH_3ONa**	**bond strength sum**
Na_2SiO_5-α			51.7 (50.0)	62.0	1.039
Na_2SiO_5-β	Na(1)	9.4	49.0 (47.2)	59.9	0.985
	Na(2)	15.6	52.5 (52.5)	65.5	1.046
Na_2SiO_3		23.0	52.7 (52.1)	63.1 **50.4**	1.06
$Na_2BaSi_2O_6$	Na(1)	25.0	56.5 (56.2)	66.9	1.148
	Na(2)	5.4	40.5 (33.8)	52.7	0.791
$NaAlSi_3O_8$ low albite		-8.5	46.5 (36.4)	-	0.911
$(NaK)Si_3O_8$ microcline		-24.3	24.0 (21.5)	-	0.424
anhydrous sodalite		-1.7	34.3 (33.4) **29.8**	-	0.66
nepheline	Na site	-5.5	40.0 (37.8)	-	0.765
	K site	-19.5	23.1 (22.7)	-	0.417

Also, the ^{23}Na NMR data, which shows the biggest change with hydration, has not yet been explained. Since one must normally work with clusters centered on the atom of interest in order to obtain accurate shieldings, we would need to consider Na centered clusters. There are two problems with constructing such clusters: (1) they tend to be quite large if even a couple of nearest-neighbour shells are considered and (2) calculating their geometries using cluster models is problematic since Na^+ ions are "network-modifiers", which basically sit in "holes" in the network structure defined by the network-forming cations and the O atoms.

3.4 Calculation of ^{23}Na NMR shieldings in crystals and glasses

On a more positive note, we earlier showed[51] that calculating ^{23}Na NMR shifts reliably is not difficult if accurate geometries are available. GIAO calculations at the HF level with 6-31G* bases seem generally sufficient to give reliable shifts, as shown in Table 2 for several Na silicates and aluminosilcates.

When the experimental chemical shifts are plotted against the calculated deshieldings the correlation line gives a slope of 1.34 and a correlation coefficient of 0.869, indicating that most of the experimental variation is accounted for by the model.

Choosing the right Na-centered cluster for geometry optimization can be difficult but results generally match experimental structures well if we retain the correct atoms through 3NN and incorporate any constraints arising

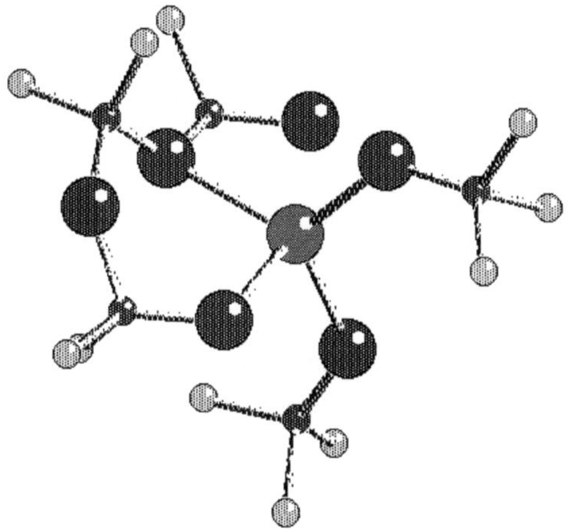

Figure 12. Na-centered cluster model for Na_2SiO_3, of composition $NaSi_5O_6H_{12}^{-3}$.

from the presence of aluminosilicate rings in the structure. A simple cluster representing the 5-coordinate Na site in Na_2SiO_3 is shown in Fig. 12.

It is also possible to correlate the deshielding of the Na with the sum of the bond strengths received by the Na^+ as in the model of Brown and Altermatt[21(b)], using either accurate Na-O distances from the crystal structure of the material, averaged Na-O distances for the site or (in some cases) geometries optimized for a cluster model. Increased bond strength received at the Na from surrounding ligands leads to increased charge transfer from these donor ligands into the Na3p orbital, which in term increases the magnitude of the deshielding. The correlation of calculated absolute deshielding with bond strenth sum is shown in Fig. 13.

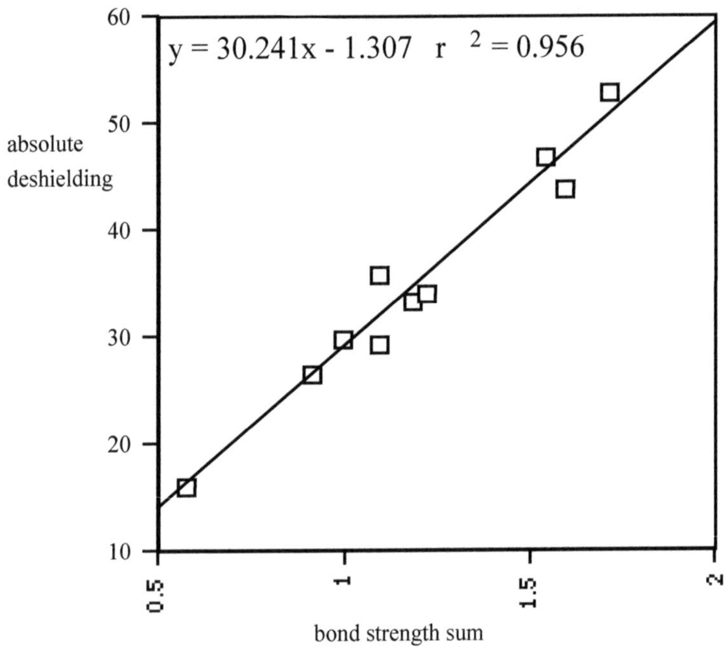

Figure 13. Calculated Na^+ absolute deshieldings vs. bond strength sum (adapted from ref. 51)

One might think that given this success in calculating ^{23}Na NMR shifts in crystalline materials and correlating those shifts with simple crystal chemical concepts such as bond strengths, interpreting the ^{23}Na NMR of aluminosilicate glasses would be easy. However, the problem in interpreting the aluminosilicate glass results is that each Na^+ has a number of different possible nearest-neighbor ligands, which can differ in both Na-O bond distance and bond strength, e.g. H_2O, OH^-, O in Si-O-Si and O in Si-O-Al are not equivalent in their desheilding of Na^+. Using the crystal structure of albite, as in Fig. 14, we can calculate the ^{23}Na NMR shielding, obtaining a shift of -3.4 ppm vs. Na^+ (aq), somewhat deshielded with repect to the experimental value of about -14 ppm [46(b)].. A model using only Si atoms (no Al's) actually gives a better result, with a calculated shielding of about -10 ppm. This suggests that in the finite cluster of Fig. 14 the Na^+ approaches the O in the Si-O-Al linkages too closely.

If we now simply allow a single H_2O to approach Na^+, in the region

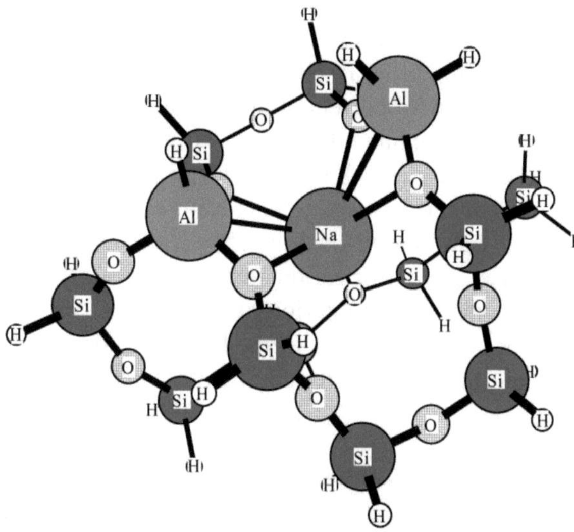

$NaSi_{10}\,Al_3\,O_{13}\,H_{26}\,^{-2}$

Figure 14. Model for the Na^+ site in albite

between the aluminosilicate ring structures enclosing it, we find that the shift becomes more negative (while the experimental result gives it as more positive) and the QCC goes up (while experimentally it remains nearly constant). The changes occur because bonding of Na^+ to the added water reduces its interaction with the O atoms of the aluminosilicate network. It is probable that any study of water interaction with this system must include more than one water.

3.5 P_2O_5 dissolved in aluminosilicate glasses

Small amounts of P_2O_5 dissolved in aluminosilicate melts can influence not only their viscosities but the activities of silica and the redox ratio of Fe. P has a strong effect upon the degree and type of polymerization of aluminosilicate melts. When P is dissolved in aluminosililcate melts there are basically two different types of species produced: (1) various isolated orthophosphate and polyphosphate groups, presumably closely associated with Na^+, and (2) species in which P is linked by bridging O to the Si,Al lattice. Within each group the detailed composition varies with the mole % Al_2O_3. In our recent study at least 13 different P-containing species were identified experimentally[14] For these different species it was possible to determine both ^{31}P isotropic NMR shifts and spinning side-band intensities, which correlate with chemical shift anisotropy. Shifts and anisotropies for a large number of species were then calculated using the GIAO method and a 6-31G* basis set at optimized 6-31G* HF geometries. Generally the clusters considered are fairly small, with correct atom identities only through second nearest neighbors, although we have done some larger calculations to test the stability the results to cluster size. A $P_3O_{10}^{-5}$ cluster model of a triphosphate and a $PO_4(AlH_3)_4^{-2}$ cluster model used to represent a P with 4 P-O-Al linkages (as in solid $AlPO_4$) are shown in Fig. 15. These are only two of the many clusters considered. Simplifications such as termination by -H and neglect or use of only one Na^+ counterions were necessary because of the large number of species considered and seemed justified since we were basically trying to elucidate trends in shift and anisotropy along several related series of molecules.

Table 3. ^{31}P Shielding Calculations of Na-P Species (from ref. 14).

Species	σ(Calc.)	δ (Calc.)	δ(Exp.)	% Side Band (Calc.)	% Side Band (Exp.)
H_3PO_4	419.5*	0	0	-	-
PO_4^{3-}	404.6	15.1	NM	-	-
Na_3PO_4	402.9	16.6	14.1	0.09	0.11
$P_2O_7^{4-}$	417.2	2.3	NM	-	-
$Na_4P_2O_7$	418.2	1.3	2.0	0.32	0.37
$P_3O_{10}^{5-}$	412.1	7.4	0.6	0.22	0.31
	430.9	-11.4	-8.0	0.40	0.46
$P_3O_9^{3-}$	438.0	-18.5	-20.0	0.65	-
P_4O_{10}	463.6	-44.1	-46.0	-	-

*This value is used throughout as the shielding reference.

For the anionic species, such as $P_3O_{10}^{-5}$ the proper positioning of counterions such as Na^+ within the glass is unknown. Nonetheless the agreement of calculated and experimental shifts and sideband intensities for such species, as given in Table 3, was reasonably good, particularly for the more compact cyclic structures, for which interaction with the Na^+ counterions is expected to be smaller. This indicates that even at this modest level it is possible to calculate ^{31}P shielding trends semiquantitatively.

In general, anisotropies are considered to be more difficult to calculate accurately than isotropic shifts but in this case the trends in anisotropies were determined primarily by the symmetry of the bonds about P, so that changes from one species to another are quite pronounced and easy to calculate. Based on such calculated shift systematics it was possible to assign all the species identified experimentally.

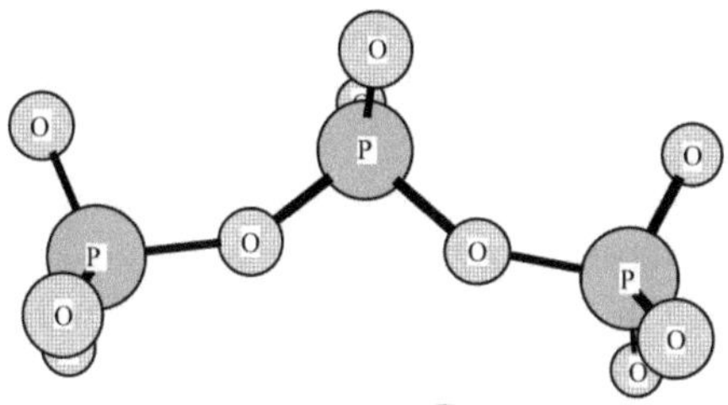

$P_3O_{10}^{-5}$

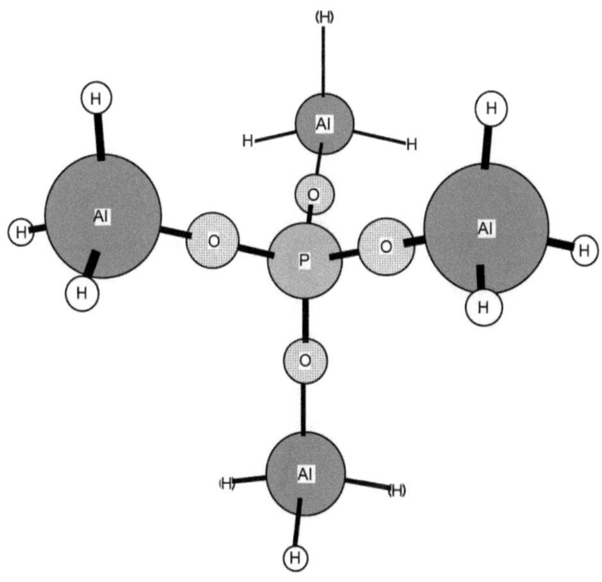

$PO_4(AlH_3)_4^{-2}$

Figure 15. Cluster models for the triphosphate anion and for a P site with 4 P-O-Al linkages

3.6 Ti in aluminosilicate glasses

We have recently studied the properties of Ti as a substituent in polyhedral oligomeric silsesquioxanes as models for its incorporation in zeolites, gels and glasses. We have examined the NMR properties (Si,Ti, O shieldings, O EFG) for the silsesquioxane model compounds $Si_4Ti_4O_{12}H_8$ and $Si_4Ti_4O_{20}H_8$. We also studied the effect of increasing the coordination number of Ti from 4- to 5- to 6- in : $Si_4Ti_4O_{20}H_8$, $Si_4Ti_4O_{21}H_9^{-1}$ and $Si_4Ti_4O_{20}H_{10}^{-2}$.. Both HF and B3LYP GIAO calculations were performed using a 6-31G* basis. The $Si_4Ti_4O_{12}H_8$ molecule is shown in Fig. 16. There are a number of isomers of this molecule, depending upon the relative positioning of the 4 Si and 4 Ti, but the difference in energy between the different isomers is

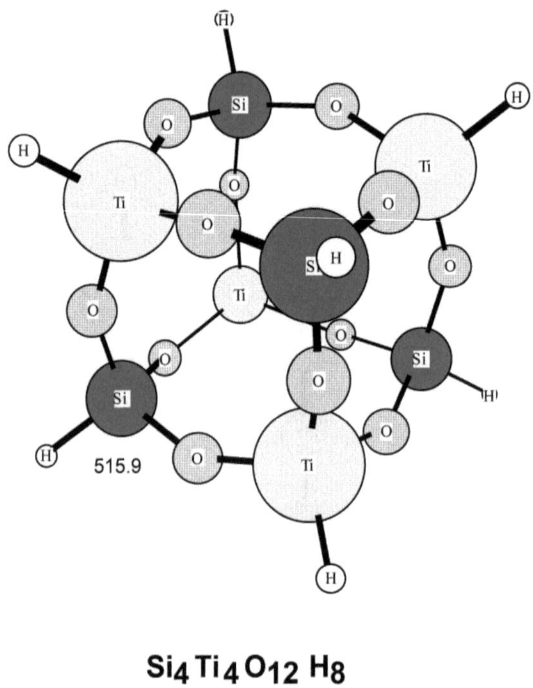

Figure 16. Alternating isomer of the Ti-substituted silsesquioxane

very small, indicating that Ti and Si would be effectively disordered in such a material The geometry we show is the "alternating" geometry, with only Ti-O-Si linakges. The maximally "paired" structure, with 4 Ti on one side of the cage and 4 Si on the other (4 Ti-O-Ti, 4 Si-O-Si and 4 Ti-O-Si) is only 0.6 kcal/mol higher in energy. This is very different from the Si,Al analog where the completely "alternating" isomer, with only Si-O-Al linkages, is strongly preferred.

Some basic questions we hoped to answer concerning Ti in silsesquioxanes (or gels, or glasses, or zeolites) were: (1) what are the spectroscopic signatures for the different Ti local geometries? and (2) how are the spectroscopic signatures of Si and O changed by proximity to Ti's with different local geometries? Ricchiardi and Sauer[52] had previously studied ^{29}Si NMR shifts in related models for Ti in zeolites and had found that the ^{29}Si NMR shifts were only slightly changed by the presence of next-nearest-neighbour Ti. Gervais, et al.[53] had identified Ti-O-Si and Ti-O-Ti linkages in other Si, Ti containing materials using ^{17}O NMR.

As shown in Table 4 we also found that Si shieldings were affected only weakly and apparently unsystematically by the presence of Ti in a next-nearest-neighbour position. By contrast the calculated shieldings and EFG's at O were substantially smaller in Si-O-Ti compared to Si-O-Si linkages. For O in Ti-O-Ti, e.g. in the paired form of $Si_4Ti_4O_{12}H_8$, the shieldings were even more negative and the EFG's even smaller. Thus O NMR, rather than Si NMR, would be the preferred method to characterize these species. We must be cautious, however, because it is well known that calculating ^{17}O NMR shifts is difficult, due to correlation effects and long-range effects. The silsesquioxane molecular models considered are too large for MP2 or other HF-based correlated calculations so we must employ DFT approaches, with the accompanying uncertainty in accuracy of results[54]. In the Table below we see that B3LYP[55] and HF results are quite similar (in both absolute shielding and trend) for Si, show larger differences for O and extreme differences for Ti (hundreds of ppm and different trends). Unfortunately, there have been few experimental studies of simple Ti compounds for which we might do calculations to establish the adequacy of the various methods, partly because Ti is a quadrupolar nuclide and its shielding is therefore difficult to measure unless it is in a highly symmetric environment.

The limited experimental data[53] on ^{17}O NMR shifts lies between the HF and B3LYP values, as shown in Table 5 and 6.

It is also apparent that we can distinguish 4-, 5- and 6- coordinate Ti, with terminal -OH groups by virtue of a systematic increase in shielding, but a deprotonated 5-coordinate speces (titanyl) is strongly deshielded. Our calculated values for 5- and 6- coordinate Ti, formed by adding $-OH^-$ ligands to the parent silsesquioxanes (with 4-coordinate Ti) are given in Table 7. These trends in shielding with change in coordination number are qualitatively the same as those seen for Al, Si or P.

Table 4. Calculated shieldings (in ppm) and EFG's (in au) for $Si_8O_{12}H_8$ and related molecules, using GIAO HF and B3LYP methods.

molecule	$Si_8O_{12}H_8$ HF B3LYP	$Si_7TiO_{12}H_8$ HF B3LYP	$Si_4Ti_4O_{12}H_8$ Alternating HF B3LYP	$Si_4Ti_4O_{12}H_8$ Paired HF B3LYP
σ^{Si}	516.5 485.0	514.8 481.5	515.9 488.6	516.9 484.6
σ^{Obr}	257.8 226.3	26.4 -54.0 (Si-O-Ti)	28.8 -52.3	-244.4 -367.1 38.0 -49.0 257.9 224.9
σ^{Ti}	-	27.4 -345.1	44.1 -338.4	35.0 -333.5
q^{Obr}	0.930 0.877	0.559 0.475 (Si-O-Ti)	0.567 0.487	0.250 0.208 0.567 0.486 0.932 0.887

Table 5. Comparison of calculated ^{17}O NMR shifts for Si-O-Si, Si-O-Ti and Ti-O-Ti linkages in $Si_4Ti_4O_{12}H_8$ (paired) with average experimental values from ref. 53.

	HF		B3LYP		Exp.
linkage	σ	δ	σ	δ	average δ
Si-O-Si	258	34	225	61	20
Si-O-Ti	38 (81,64[b]	254 211,228)	-49	335	300
Ti-O-Ti (all Ti^{IV})	-244	536	-367	653	400 - 600

Table 6 Comparison of calculated ^{17}O EFG's (atomic units) and QCC's (e^2qQ/h, MHz) for Si-O-Si, Si-O-Ti and Ti-O-Ti linkages in $Si_4Ti_4O_{12}H_8$ (paired) with average experimental values from ref. 53.

	q		NQCC	
linkage	HF	B3LYP	HF	Exp.
Si-O-Si	0.932	0.887	5.0	5.1
Si-O-Ti	0.567	0.486	2.9	≈3.0, ≈2.0
Ti-O-Ti (all Ti^{IV})	0.250	0.208	1.3	Small (OTi_3,OTi_4)

Table 7. Trends in shielding of Ti and of O bonded to Ti, with change in Ti C.N. (using GIAO HF method).

molecule	σ^{Ti}	σ^{O}
$Si_4Ti_4O_{20}H_8$ (Ti^{IV})	474	64, 81
$Si_4Ti_4O_{21}H_9^{-1}$ (Ti^{V})	510	133
$Si_4Ti_4O_{21}H_8^{-2}$ (Ti^{V})	204	204, 241, 248
$Si_4Ti_4O_{22}H_{10}^{-2}$ (Ti^{VI})	575	215,222

3.7 Long-range effects on N NMR shieldings in solids

In crystalline ß-Si_3N_4 there are two inequivalent N sites which have N shieldings differing by almost 20 ppm. Comparing the intensity ratio of these peaks with the crystal structure establishes that the more numerous N(1) sites, locally distorted from C_{3v} symmetry, have a higher shielding than the N(2), locally C_{3v} symmetric sites. But NMR calculations at both singly and doubly polarized basis set levels on optimized $(SiH_3)_3N$ and on $(SiH_3)_3N$ structures with local geometries identical to those in the N(1) and N(2) sites showed that locally non- C_{3v} sites in fact have lower shieldings[4(a)]. This indicated a problem with either the method used to calculate the N NMR shieldings or with the size of the cluster model.

We have now calculated the N NMR shieldings in larger molecular models for the N2 site of β-Si_3N_4 and have found that the N2 shielding is greatly reduced (by as much as 60 ppm) when additional N1 atoms (2nd-nearest-neighbors to the central N2 are included[56]. By contrast the EFG at N2 is only slightly affected by changes in the size of the molecular cluster model. A series of calculations on cluster models for the analog compound C_3N_4 yields similar results. These effects are seen in Table 8.

Similar effects were seen in a later study[57] of the N NMR of P_3N_5, using molecular models of different varying size and the GIAO method. Large molecular models gave a systematic deshielding of the central N atoms, with relatively little change in the value of the N EFG. For example the NMR shieldings of the central N atoms in the molecules in Fig. 17 are 268 ppm for the smaller model and 140 ppm for the larger model, a difference of 128 ppm, arising from the replacement of -H by -NH_2 in the 2nd-neighbour position. By contrast the calculated EFG at the central N atom changes only from 1.15 au to 0.99 au.

Table 8. ^{14}N NMR shieldings evaluated at the GIAO SCF level using 6-31G* basis sets and 6-31G* geometres for Si compounds, 6-311(2d) geometries for C compounds.

molecule	σ^N (ppm)
$(SiH_3)_3N$	288.1
$(SiH_3)_2N(SiH_2NH_2)$	256.8
$(SiH_2NH_2)_3N$	247.7
$Si_6N_5H_{15}$	244,7
$Si_9N_9H_{21}$	206.1 (inner) 237.1 (outer)
$(CH_3)_3N$ nonplanar	263.2
planar	307.2
$C_6N_5H_{15}$	259.3
$C_9H_9H_{21}$	235.9 (inner) 248.7 (outer)

This suggests that the main change in electronic structure between the different molecular cluster models occurs in the unoccupied orbital space, which influences the paramagnetic contributions to the NMR shieldings. Calculations by Yoon, et al.[29] on C_3N_4 using a very different method (band theoretical DFT with a plane wave pseudopotential basis set) also

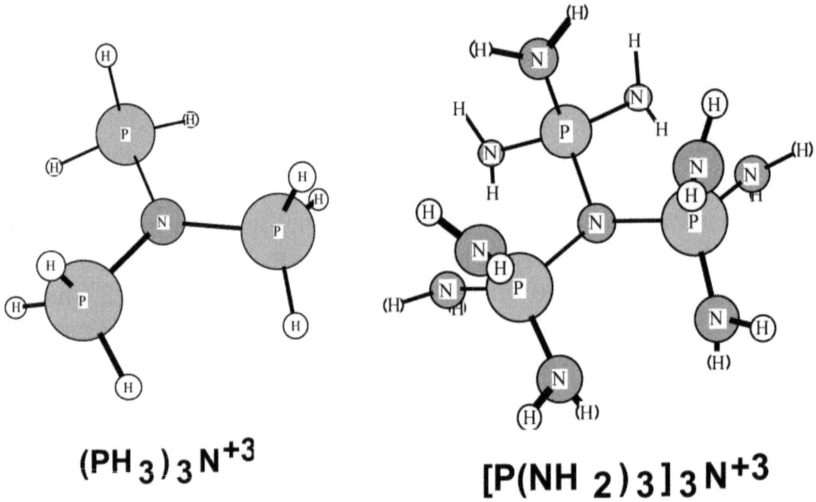

Figure 17. Two different cluster model for the central 3-coordinate N in P_3N_5.

showed a strong deshielding of N in their larger cluster models. Their calculated shielding of 226 ppm for 3-coordinate N in $(CH_3)_3N$ was from 30 to 130 ppm larger than that calculated in any of the crystalline forms of C_3N_4 and they noted "other researchers had noticed the extraordinary sensitivity of N shifts to the environment before". Large solvation effects on ^{14}N NMR shieldings have been described for example by Witanoski, et al.[58], but the effects seen in these cluster calculations on solids are much larger.

By contrast, the size of the molecular cluster had only a modest effect upon the calculated ^{13}C shieldings. It thus appears that N (and possibly other electronegative atoms) show long-range effects upon their NMR shieldings For example, the calculated shieldings of the central O's differ by about 60 ppm between $(SiH_3)_2O$ and $[Si(OH)_3]_2O$, a similar but less extreme effect than in the N case. In this case the use of a cluster approach has actually revealed some interesting physics - the importance of non-local effects upon the NMR shieldings of electronegative atoms.

3.8 Recent developments for SiO_2 glass

Since SiO_2 is probably the most important glass-forming oxide it is often the first target for new, state-of-the-art methods, even though its properties have been studied for a long time. The crystalline forms of SiO_2 also serve as test cases for new quantum mechanical calculations of properties, using either band theoretical or molecular cluster methods.

For example, the ^{17}O and ^{29}Si spectra of the siliceous zeolites ferrierite and faujasite, with composition SiO_2, have recently been determined by Bull, et al[59], using experimental NMR and a molecular cluster computational approach. The zeolites have a large number of chemically inequivalent Si and O sites, which differ in experimental shifts and (for the O) QCC values by small amounts, thus providing a demanding test for theory. A systematic approach was used in which crystal structure data was employed to generate molecular clusters of varying size, which the authors labeled in terms of the number of shells of atoms included about the central magnetic nucleus. In this way the convergence of the cluster model calculations with size of the cluster could be tested. This is a clearly a desirable approach but it has seldom been applied so systematically. Some testing of different basis sets and methods was also performed. It was found that 4NN shell models (4 shells of nearest neighbours about the central atom) gave excellent ^{29}Si NMR shifts compared to experiment. For ^{17}O NMR, for which differences in shifts and quadrupole couplings between the different sites were often very small, even the 4NN model was not accurate enough to assign the (rather complicated) spectra. In general, convergence with cluster size was slower for the ^{17}O shifts and coupling constants than for the ^{29}Si shifts. This sobering result indicates that even when the structure is known to long distance from the magnetic nucleus precise identification of species with similar NMR parameters will be difficult.

Mixed quantum mechanical-molecular mechanical (QM/MM) and embedded cluster methods have also been used to study properties (although not yet NMR shieldings) in SiO^2. As implemented by Eichler, et al[60] a scheme combining ab initio calculations with analytical potential methods allows geometry optimizations for very large systems, thus potentially reducing artifacts associated with the use of small clusters. The ONIOM method developed by Morokuma and coworkers[61a] and later applied to the calculation of NMR shieldings[61b] has also been used to optimize geometries of very large fragments in a study of the peroxy bridge defect center in amorphous SiO_2 by Ricci, et al.[62] In this method different parts of the molecule are treated with different quantum mechanical methods and/or different basis sets and those different parts self-consistently linked. In covalently bonded glasses like SiO_2 defect structures and energies are strongly influenced by the mid-range structure of the material so that many atoms must be included with a cluster approach to obtained converged results. Treating the more distant atoms at a lower level greatly reduces the computational cost. Once the structure of the defect has been accurately determined one could construct a molecular cluster, possibly considerably smaller than the one used for optimization, for the NMR calculation. Alternatively we could use the ONIOM method to do the NMR calculation. From studies on molecular systems, Karadakov and Morokuma concluded that treating the region near the magnetic nucleus at the GIAO MP2 level with a large basis set and the rest of the molecule at the GIAO HF with a smaller basis would provide a feasible scheme for accurate calculations on very large systems.

Finally, ab initio molecular dynamics (MD) methods based on density functional theory have been used to study local structure[63] in amorphous SiO_2 and to study H_2O diffusion[64] in SiO_2. In each case it was necessary to utilize a super-cell approach to the periodicity of the solid, which forces assumptions about the symmetry and thus potentially reduces the generality of the results. The ab initio MD calculations on liquid and glassy SiO_2 obtained significant numbers of small rings in the bulk material (with three or four Si atoms, e.g. 3-rings or 4-rings) which had been earlier hypothesized and supported by molecular cluster calculations of Raman frequencies and ^{29}Si NMR shifts[65]. The ab initio MD results also gave frequencies for the D1 and D2 Raman "defect" lines of SiO_2 glass, associated with the 3- and 4- rings, in excellent agreement with experiment[66]. However, the initial studies gave populations of small rings which were too large, resulting in a distribution of <Si-O-Si angles which was strongly peaked at low angles (characteristic of the 3-rings) but was in poor agreement with experimental <Si-O-Si distributions. Later results indicated that the ab initio MD was probably not fully equilibrated, a problem endemic to MD, whether pair-potential or ab initio.

The studies on H_2O diffusion in SiO_2 glass[64] were performed using plane-wave pseudopotential DFT band theory and yielded both interesting qualitative and quantitative results. This study basically showed that H_2O diffused easily through large silicate rings (6-rings and higher) and that the calculated energetics for an interconnected large ring diffusion path were consistent with experiment. By contrast, when the water molecule entered a 5-ring or smaller it was very likely to react with the ring network, hydrolyzing a Si-O-Si bond to form two silanols (Si-OH) units and later reforming on the same or other side of the ring. Clearly it would be very desirable to extend such calculations to the interaction of H_2O with an aluminosilicate glass like albite.

4. CONCLUSIONS

Local and mid-range order in glasses can often be determined by comparing calculated properties for candidate structures with the available experimental data from x-ray diffraction, vibrational and NMR specctroscopy.

Generally, no single experimental technique provides an unambiguous assignment, so the more experimental probes of glass structure, the better! But for the purpose of ascertaining local geometry NMR seems now to be the most powerful single technique.

Since molecular models for mid-range units in glasses are too large for very accurate quantum calculations and since artifacts due to cluster termination can be significant, considerable caution and common sense must be used to assess agreement of calculation and experiment. In general, geometries and vibrational spectra calculated at the polarized split valence basis set Hartree-Fock level seem to describe trends in the experimental data reasonably well while more flexible basis sets, e.g. 6-311G(2d,p) are desirable for calculations of the NMR shielding trends. DFT techniques which have enjoyed considerable success in reproducing geometries of gas-phase molecules have been applied to molecular cluster modeling of solids in a few cases and seem to be giving somewhat improved results compared to HF.

While shielding trends can usually be understood in terms of first and second neighbour bond distances and angles we have established the importance of second-neighbour lone-pair orientation for the N shieldings of nitrides . The clusters considered should have correct atom identifies to at least 3rd nearest-neighbour if possible. This is more important for electronegative atoms like N and O, but even for electropositive atoms the local and mid-range geometries can be strongly affected by cluster termination effects. Most of the studies described herein have been done with quite conventional approaches, in which molecular quantum chemical codes are applied to cluster models treated simply as free molecules. More sophisicated approaches are possible, in which the cluster is embedded in the potential from the rest of the atoms or different parts of the cluster are treated at different quantum mechanical levels and we can anticipate these will be used more in the future.

ACKNOWLEDGEMENTS

This work was supported by the National Science Foundation, grant number EAR0001031.

REFERENCES

1. (a) S. R. Elliott, *Physics of Amorphous Materials*, 2nd ed., Longman Press, London, 1990 (b) R. H. Doremus, *Glass Science*, 2nd. Ed., Wiley, New York, 1994
2. C. A. Fyfe, *Solid State NMR for Chemists,* C. E. C. Press, Guelph, 1983
3. R. J. Kirkpatrick., "MAS NMR spectroscopy of minerals and glasses", in *Spectroscopic Methods in Mineralogy and Geology*, ed. F. C. Hawthorne, Rev. Mineral., 18, 341, 1988

4. (a) J. A. Tossell, *Int. J. Quantum Chem., Quantum Chem. Symp.*, **29**, 443 (1995); (b) J. A. Tossell, "Calculating the NMR properties of minerals, glasses and aqusous species", in *Molecular Modeling Theory: Applications in the Geosciences*, ed. R. T. Cygan and J. D. Kubicki, Rev. Mineral. Geochem., **42**, 437 (2001).
5. (a) R. E. Youngman and J. S. Zwanziger, *J. Non-Cryst. Sol.*, **168**, 293 1994 (b) C. Joo, U. Werner-Zwanziger, J. W. Zwanziger, *J. Non-Cryst. Sol.,* **261**, 282 (2000).
6. (a) J. A. Tossell, *J. Non-Cryst. Sol.*, **183**, 307 (1995) (b) J. A. Tossell, *J. Non-Cryst. Sol.*, **215**, 236, (1997).
7. J. K. Maranas, Y. Chen, D. K. Stillinger and F. H. Stillinger, *J. Chem. Phys.*, **115**, 6578 (2001)
8. J. F. Stebbins, S. K. Lee, and J. V. Oglesby, *Amer. Mineral.,* **84**, 983 (1999).
9. J. F. Stebbins and Z. Liu, *Nature*, *290*, 60, 1997
10. J. A. Tossell, Amer. Mineral., 78, 911 (1993).
11. X. Xue and K. Kanzaki, Phys. *Chem. Mineral.,* **26**, 14, (1998).
12. X. Xue and K. Kanzaki, *J. Phys. Chem. B* **103**, 10816 (1999).
13. J. A. Tossell and R. E. Cohen, *J. Non-Cryst. Sol.*, **286**, 187 (2001).
14. G. D. Cody, B. Mysen, G. Saghi-Szabo and J. A. Tossell, *Geochim. Cosmochim. Acta,* **65**, 2395 (2001).
15. J. A. Tossell, *J. Magn. Reson.*, **127**, 49 (1997).
16. J. A. Tossell, *Chem. Phys. Lett.*, **303**, 435 (1999).
17. J. A. Tossell, "Local and long-range effects on NMR shieldings in main-group metal oxides and nitrides:, pp. 304-319, in *Modeling NMR Chemical Shifts" Gaining Insights into Structure and Environment*, ed. J. C. Facelli and A. C. DeDios, ACS Symp Series, no. 732, ACS, Washington, DC,1999
18. J. Sauer, *Chem. Rev.,* **89**, 199 (1989).
19. (a)J. A. Pople, *Ber. Bunsen-Ges. Phys. Chem.,* **86**, 806 (1982) (b) J. Simons, *J. Phys. Chem.*, **95**, 1017 (1991).
20. (a) R. Ditchfield, *Mol. Phys.*, **8**, 397 (1974) (b) K. Wolinski, J. F. Hinton and P. Pulay,, *J. Am. Chem. Soc.*, **112**, 8251 (1990) (c) J. F. Hinton, P. L. Guthrie, P. Pulay and K. Wolinski, *J. Magn. Reson. A*, **103**, 188 (1993).
21. (a) L. Pauling, *J. Am. Chem. Soc.*, **51**, 1010, (1929) (b) I. D. Brown and D. Altermatt, *Acta Cryst.* **B41**, 244 (1985).
22. M. W. Schmidt, K. K. Baldrige, J. A. Boatz, S. T. Elbert, M. S. Gordon, J. H. Jensen, S. Koseki, M. Matsunaga, K. A. Nguyen, S. J. Su, T. L.Windus, M. Dupuis,and J. A. Montgomery, *J. Comput. Chem.*, **14**, 1347 (1993).
23. M. J. Frisch, G. W. Trucks, H. B. Schlegel, P. M. W. Gill, B. G. Johnson, M. A. Robb, J. R. Cheeseman, T. Keith, G. Petersson, J. A. Montgomery, K. Raghavachari, M. A. Al-Laham, V. G. Zakrzewski J. V. Ortiz, J. B. Foresman, C. Y. Peng, P. Y. Ayala, W. Chen, M. W. Wong, J. L. Andres, E. S. Peplogle, R. Gompers, R. L. Martin, D. J. Fox, J. S. Binkely, D. J. Defrees, J. Baker, J. P. Stewart, M. Head-Gordon, C. Gpnzalez and J. A. Pople, GAUSSIAM94, Rev. B.3, Gaussian, Inc., Pittsburgh, PA, 1994
24. J. A. Tossell, ed. *Nuclear Magnetic Shieldings and Molecular Structure*, Kluwer, Dordrecht, 1993.
25. C. J. Jameson "Recent advances in nuclear magnetic shielding theory and computational methods", pp. 1-23, in *Modeling NMR Chemical Shifts" Gaining In-*

sights into Structure and Environment, ed. J. C. Facelli and A. C. DeDios, ACS Symp Series, no. 732, ACS, Washington, DC,1999
26. W. A. Hehre, L. Radom, P. v. R. Schleyer, and J. A. Pople, *Ab Initio Molecular Orbital Theory*, Wiley, New York, 1986
27. P. v. R.Schleyer, M. Buhl, U. Flesicher and W. Koch, *Inorg. Chem.,* **29**,153, (1990).
28. F. Mauri, B. G. Pfrommer and S. G. Louie, *Phys. Rev. Lett*., **77**, 5300 (1996).
29. Y.G. Yoon, B. G. Pfrommer, F. Mauri and S. G. Louie, *Phys. Rev. Lett.,* **80**, 3388 (1998).
30. P. Pyykko, *Z. Naturforsch*, **471**, 189 (1992).
31. R. Ludwig, F. Weinhold and T. C. Farrar, *J. Chem. Phys.,* **105**, 8223 (1996).
32. M. H. Palmer and J. A. Blair-Fish, *Z. Naturforsch.* **49a**, 137 (1994).
33. M. Iglesias, K. Schwarz, P. Blaha and D. Baldomir, *Phys. Chem. Mineral*., **28**, 67, (2001).
34. E. A. Moore, C. Johnson, M. Mortimer and C. Wigglesworth, *Phys. Chem. Chem. Phys*., **2**, 1325 (2000).
35. (a) G. E. Jellison,. Jr., L. W. Panek,., P. J. Bray and G. B. Rouse, Jr., *J Chem. Phys.,* **66**, 802, (1977) (b) F. L. Galeener, *Solid State Comm.,* **44**,1037 (1982) (c) P. A. V Johnson, A. C. Wright and R. N. Sinclair, *J. Non-cryst. Sol*, **50**, 281, (1982).
36. J. A. Tossell and P. Lazzeretti, *J. Phys. Chem.,* **94**, 1723 (1990).
37. A. E. Hansen and T. D. Bouman, J. Chem. Phys., 82, 5033, 1985
38. A. Navrotsky, G. Peraudeau, P. McMilland and J.-P. Coutures, *Geochim. Cosmochim. Acta*, **46**, 2039, (1982).
39. J. A. Tossell and G. Saghi-Szabo, *Geochim. Cosmochim*. Acta, **61**, 1171 (1997).
40. M. J. Toplis, D. B. Dingwell and T. Lenci, *Geochim. Cosmochim. Acta*, **61**, 2605 (1997).
41. J. A. Tossell, *Inorg. Chem*., **37**, 2223 (1998)
42. D. Palmer. CRYSTALMAKER3, 1998.
43. J. F. Stebbins, J. V. Oglesby and S. Kroeker, *Amer. Mineral*., **86**, 1307 (2001).
44. D. W. Goodwin and A. J. Lindop, *Acta Cryst.* **B26**, 1230 (1970).
45. D. Sykes and J. D. Kubicki, Geochim. *Cosmochim. Acta*, **57**, 1039 (1993).
46. (a) S. C. Kohn, R. Dupree, M. G. Mortuza and C. M. B. Henderson, *Amer. Mineral*., **76**, 309 (1991); (b) S. C. Kohn, M. E. Smith, P. J. Dirken, E. R. H. van Eck, A. P. M. Kentgens and R. Dupree, *Geochim. Cosmochim. Acta*, **62**, 79, (1998).
47. Y. Liu, H. Nekvasil and H. Long, *Geochim. Cosmochim. Acta*, 66, 4199 (2002).
48. M. Huner and T. Horvath, *Ber. Bunsenges. Phys. Chem*., **99**, 1316 (1995).
49. M. Krossner and J. Sauer, *J. Phys. Chem*.**100**, 6199 (1996).
50. (a) R. M. Lynden-Bell, L Delle Site and A. Alavi, *Surf. Sci*., **496**, L1 (2002) (b) P. J. Feibelman, *Science*, 295, 99 (2002).
51. J. A. Tossell, *Phys. Chem. Mineral*., **27**, 70 (1999).
52. G. Ricchiardi and J. Sauer, *Zeit. Phys.. Chem*., **209**, 21 (1999).
53. C. Gervais, F. Babonneau and M. E. Smith, *J. Phys. Chem. B,* **105**, 1971 (2001)
54. P. J. Wilson, R. D. Amos and N. C. Handy, *Molec. Phys.,* **97**, 757 (1999).
55. A. D Becke, *J. Chem. Phys.,* **96**, 2155 (1992).
56. J. A. Tossell, *J. Magn. Reson*., **127**, 49 (1997).
57. J. A. Tossell, *Chem. Phys. Lett.*, **303**, 435 (1999).

58. M. Witanowski, Z. Biedrzycka, W. Sicinska, Z. Grabowski and G. A. Webb, *J. Mag. Resn.,* **124**, 127 (1991).
59. (a) L. M. Bull, A. K. Cheetham, T. Anupold, A. Reinhold, A. Samoson, J. Sauer, B. Bussemer, Y. Lee, S. Gann, J. Shore, A. Pines and R. Dupree, *J. Am. Chem. Soc.*, **120**, 3510 (1998); (b) L. M. Bull, B. Bussemer, T. Anupold, A. Reinhold, A. Samoson, J. Sauer, A. K. Cheetham and R. Dupree, *J. Am. Chem. Soc.*, **122**, 4948 (2000).
60. U. Eichler, C. M. Kolmel and J. Sauer, *J. Comput. Chem.*, **18**, 463 (1996).
61. (a) M. Svensson, S. Humbel, R. D. J. Froese, T. Matsubara, S. Sieber, and K. Morokuma, *J. Phys. Chem.*, **100**, 19357 (1996) (b) P. B. Karadakov and K. Morokuma, *Chem. Phys. Lett.*, **317**, 589 (2000).
62. D. Ricci, G. Pacchioni, M. A. Szymanski, A. L. Shugar, and A. M. Stoneham, *Phys. Rev. B*, **64**, 224104 (2001)
63. (a) J. Sarnthein, A. Pasquarello and R. Car, *Phys. Rev. B*, **52**, 12690 (1995) (b) A.Pasquarello and R. Car, *Phys. Rev. Lett.*, **80**, 5145, (1998)
64. T. Bakos, S. N. Rashkeev and S. T. Pantelides, *Phys. Rev. Lett.*, **88**, 055508 (2002).
65. (a) J. A. Tossell, *J. Non-Cryst. Sol*, **120**, 13 (1990) (b) C. J. Brinker, R. K. Brow, D. R. Tallant and R. J. Kirkpatrick, *J. Non-Cryst. Solids*, **120**, 26 (1990)`
66. F. L. Galeener, *Solid State Comm.*, **44**, 1037 (1982)

Chapter 2

MOLECULAR MODELING OF POLY(ETHYLENE OXIDE) MELTS AND POLY(ETHYLENE OXIDE)-BASED POLYMER ELECTROLYTES

O. Borodin and G. D. Smith
Department of Materials Science and Engineering, 122 S. Central Campus Drive, Rm. 304, University of Utah, Salt Lake City, Utah 84112, USA

1. INTRODUCTION

Property structure relationships in materials can be studied by a number of computational approaches such as ab initio quantum chemistry calculations, ab initio molecular dynamics (MD) simulations, classical MD and Monte-Carlo simulations, finite element modelling, etc. A choice of the computational method depends on the time and length scales and computational resources available. At the current stage of method and hardware development, ab initio quantum chemistry calculations are best suited for studying energy-structure relationships in relatively small systems consisting of tens of atoms. Ab initio MD simulations[1] allow one to study dynamics of systems on a picosecond time scale for systems consisting of hundreds of atoms. Energy and forces in ab initio MD simulations are obtained from solving the electronic structure problem "on the fly". Parameterization of the energy a system as a function of the relative atom positions, e. g., development of a classical force field, significantly speeds up calculations of the energies and forces in MD simulations, positioning classical MD simulations as the most suitable tool to obtain properties of the systems containing 10^3-10^5 atoms on the time scales from 10^{-15} to 10^{-6} s. However, the value of the property predictions using classical MD simulations is limited to the accuracy of the force field used, making a consistent derivation of a high quality classical force fields central to accurate prediction of the property-structure relationship from MD simulations.

L.A. Curtiss and M.S. Gordon (eds.),
Computational Materials Chemistry: Methods and Applications, 35–90.

The purpose of this chapter is to provide a consistent methodology for development of the many-body polarizable classical force fields for MD simulations polymers and polymer electrolytes from ab initio quantum chemistry and density functional calculations. Using oligomers of poly(ethylene oxide) (PEO) and their clusters with Li^+ and BF_4^- as examples, we establish adequate levels of theory and basis sets for the determination of the relative conformational energies, repulsion and dispersion parameters, dipole moments, and molecular polarizability. Then, we demonstrate through extensive comparison of MD simulation results with experimental data that the force fields derived from model ethers and their interactions with the Li^+ and BF_4^- ions are able to accurately predict structure, dynamics, and thermodynamic properties of PEO, its oligomers, and PEO/$LiBF_4$. The possibility of deriving a set of two-body parameters representing the polarization in the system in the mean-field sense[2,3] is also examined. Finally, MD simulations of PEO/$LiBF_4$ polymer electrolytes are used to examine the timescales to reach the equilibrium ion aggregation state for the polymer electrolytes. The structure of the PEO/$LiBF_4$ polymer electrolytes, and a mechanism of the Li^+ cation transport are also discussed.

2. TECHNOLOGICAL IMPORTANCE OF POLY(ETHYLENE OXIDE)

Poly(ethylene oxide) (PEO) is a ubiquitous polymer. Lightly cross-linked PEO is the main constituent of many hydrogels that are used for drug delivery applications.[4] The grafting of PEO on the surface or adsorption of PEO-poly(propylene oxide)-PEO block copolymers (Pluronic™) prevents protein adsorption and denaturation, thus making surfaces more biocompatible.[5-10] Attaching PEO to proteins prevents them from being adsorbed on surfaces,[7] whereas aqueous solutions of PEO with dextran, starch, and poly(vinyl alcohol) are used for industrial protein partitioning.[7] Solutions of PEO with alkali-metal salts have been widely studied as potential polymer electrolytes.[11] Other important applications of PEO include novel surfactants,[12] modification of natural and artificial membranes,[13,14] and aqueous biphasic separations.[15] In all these applications, PEO interacts with the salts and/or with waters, indicating a need for accurate and transferable PEO/water[16] and PEO/salt potentials that can be applied for the study of a broad range of potential applications. Development of the PEO/water potential is described elsewhere[16] together with the results of molecular dynamics (MD) simulations of PEO in aqueous solutions,[16-25] whereas the development of the many-body polarizable PEO, PEO/$LiBF_4$ force fields and MD simulations of them are presented in this chapter.

3. MOLECULAR MODELING OF PEO OLIGOMERS AND PEO MELTS

3.1 Previous MD Simulations of PEO

Several nonpolarizable PEO force fields have been employed in MD simulations of PEO, its oligomers, and PEO-based polymer electrolytes.[26-33] None of these models included many-body polarizable interactions in a consistent way. The nonbonded parameters used in the PEO force fields were taken from fitting crystal structures or generic force fields, with an exception of the work by Halley et al.[26] where ab initio quantum chemistry-based nonbonded parameters were used. The torsional parameters in the PEO force fields were usually obtained by fitting ab initio quantum chemistry calculations at various levels or were taken from previous force fields: Smith et al.[27] used MP2/D95+(2df,p)//HF/D95** 1,2-dimethoxyethane (DME) energies, whereas Neyertz et al.[28] used the MP2/6-311++G**//HF/3-21G diglyme relative conformational energies calculated by Gejji et al.[29] Halley at el.[30] used a torsional potential from previous work that predicted the energies of the major conformers within 0.5 kcal/mol from the MP2/D95+(2df,p)//HF/D95** DME quantum chemistry data of Smith el al.[27] with the $tg^{+}g^{-}$ conformer being 1.6 kcal/mol below the quantum chemistry values. Müller-Plathe et al.[31,32] modified the empirical potential to reproduce the gauche effect, whereas the AMBER force field was used by the Wheeler group without any validation.[33]

Our group used a quantum chemistry based PEO force field with the empirical repulsion and dispersion parameters and quantum chemistry based torsional parameters.[27] The MD simulations with this force field yielded the correct temperature dependence of the PEO characteristic ratio and the characteristic ratio itself,[34,35] structure factor,[36] position of the maximum of the dielectric loss spectra,[37] and spin-lattice relaxation times[37] (T_1) in good agreement with the experimental data. However, the recent ^{13}C spin-lattice relaxation times from NMR experiments[38] differ by a factor of 2-3 from the previously reported values[37] indicating that the PEO dynamics predicted by that force field is somewhat slow. This issue is addressed in below.

The Neyertz et al.[28,39] force field was validated against the available crystal PEO data. They found PEO characteristic ratio of 5.5 in good agreement with experiments[35] but their abundant populations of $g^{+}g^{-}g^{+}$ is contrary to our quantum chemistry results casting doubt on the accuracy of their force field as discussed previously.[34] Halley et al.[26] compared the structure factor for the PEO from MD simulations with the experimental one achieving only limited success. The de Leeuw's group used a modified version of the Neyertz et al.[28] force field in their MD simulations of PEO oligomers and found good agreement for the intermediate incoherent structure factor between MD simulations and neutron spin echo (NSE) experiments.[40] The other groups limited validation of their PEO simulations to comparing density from MD simulations and experiments making it difficult to judge the ability of the force fields used to reproduce polymer structural and dynamic properties.

3.2 Quantum Chemistry Studies of Model Ethers

3.2.1 Relative Conformational Energies and Geometries of 1,2-Dimethoxyethane and Diglyme

Structural properties of polymers, such as the characteristic ratio, are crucially dependent on the relative conformational energies and conformational geometries, where accurate representation of the barriers between the most important conformers is important for realistic representation of polymer dynamics. Therefore, conformational properties are central to the investigation of the influence of chemical structure on polymer dynamic and structural properties. Conformational characteristics of 1,2-dimethoxyethane as a model compound for PEO have been investigated for the last 40 years because of the differences between the crystal-, liquid-, and gas-phase conformations. Analysis of infrared (IR) and Raman spectra indicate that the -C-C- bonds adopt *gauche* (g) conformation in the crystal phase,[41] whereas a mixture of *trans* (t) and *gauche* conformers is observed in the gas and liquid phases.[42-44] In an IR study of DME trapped in an argon matrix, Yoshider et al.[42] suggested that the *ttt* conformer around the -O-C-C-O- bonds is the most stable. A subsequent IR study of gaseous DME under reduced pressure allowed authors to estimate the $tg^{+}g^{-}$ ($tg^{-}g^{+}$) energy being 0.31 (±0.04) kcal/mol relative to the *ttt* conformer from the ratio of the adsorption peaks.[44] Analysis of NMR vicinal coupling constants of DME in the gas phase indicated a *trans* fraction of ~0.25 around the -C-C- bond at 130-170 °C.[44] Electron diffraction experiments[45] indicated that the energy difference between *ttt* and *tgg* or $tg^{+}g^{-}$ conformers is small, which lead to large population of $tg^{+}g^{-}$ or *tgg* conformers along with *ttt* conformers, in the gas phase.

A number of *ab inito* quantum chemistry studies of DME have been reported.[46,,48,49] The effect of basis set size and importance of correlation effects have been investigated by studying the energy difference between the *tgt* and *ttt* conformers. Inclusion of the electron correlation effects, using Møller-Plesset second order perturbation theory (MP2), reduced the *tgt* energy by 0.6-1.1 kcal/mol.[48] Inclusion of the third-order perturbations increased the *tgt* energy by about 0.2-0.25 kcal/mol, whereas inclusion of the forth order terms results in a reduction of the *tgt* energy, by 0.1 kcal/mol, relative to the MP3 energies.[47] Coupled-cluster calculations that included single and double excitations (ΔCCSD) yielded energies that were similar to the MP4 energies. Inclusion of triple excitations (ΔCCSD(T)) reduced the energy by about 0.1 kcal/mol for the largest basis sets investigated (6-311+G**), bringing ΔCCSD(T) energies within 0.1 kcal/mol of the MP2 energies for the 6-311G* - 6-311+G** basis sets and suggesting that for a large basis set inclusion of the electron correlation beyond the MP2 level had an insignificant effect on the conformational energy of *tgt* (relative to *ttt*).[48] Jaffe and Smith[48] have extensively investigated the effect of basis set size on the *tgt* conformational energy. They have found that inclusion of the first set of diffuse functions into D95** and 6-311G** basis sets reduced the gauche energy by 0.23-0.35 kcal/mol at the MP2 or CCSD(T) levels, whereas additional sets of diffuse *sp*-functions did not have much effect on the gauche energy. They also found that it was important to include polarization functions beyond the minimum representation. It was concluded that the MP2/D95+(2df,p)//HF/D95** level was capable of predicting DME conformational energies within an accuracy of 0.2-0.3 kcal/mol.

The conformational energies for all DME conformers was calculated at the MP2/D95+(2df,p)//HF/D95** level.[48] Table 1 lists only the lowest energy conformers. Abe et al.[50] after careful comparison of the existing RIS models against available experimental data, concluded that the RIS model based on MP2/D95+(2df,p)//HF/D95** conformational energies provided a reasonable depiction of DME in the gas phase. Conformational populations obtained at the MP2/D95+(2df,p)//HF/D95** level also were found to be in good agreement with the electron diffraction data.[45]

Despite the MP2/D95+(2df,p)//HF/D95** level conformational energies being successful in reproducing electron diffraction data, NMR vicinal coupling data,[48,50] and tg^+g^- conformer energy from IR spectra, we wish to continue ab initio studies of DME conformational energies in order to answer three additional questions: a) What is the influence of the level of theory (HF, MP2, DFT-methods) that is used in geometry optimization on the conformational energies? b) Does an additional significant increase of the basis set size and augmentations change the relative conformational energies of DME at the MP2 level? c) Are much less computationally expensive DFT methods capable of adequately predicting the conformational energies of DME, when compared to MP2 methods?

We have chosen augmented correlation consistent polarized valence basis sets (aug-cc-pvXz, where X=D,T for double, triple zeta basis sets)[51] for our further quantum chemistry studies that are performed with the Gaussian 98 package.[52] These basis sets are constructed by grouping together all the functions that reduce the atomic correlation energy by the same amount and then adding these groups to the atomic Hartree-Fock orbitals. Thus, for a given accuracy in the correlation energy, these sets are as compact as possible. The conformational energies of the most important DME conformers were calculated at the MP2/aug-cc-pvDz//HF/aug-cc-pvDz level and are compared in Table 1 with the previous energies obtained at the MP2/D95+(2df,p)//HF/D95** level. The relative energies for the tg^+g^- and *ttt* conformers at the MP2/aug-cc-pvDz//HF/aug-cc-pvDz level differ from those at the MP2/D95+(2df,p)//HF/D95** level by ~0.1 kcal/mol, whereas the relative energies of the other most important conformers (tgt, tgg, ggg, ttg) differ by less than 0.04 kcal/mol. The similarity of the MP2/aug-cc-pvDz//HF/aug-cc-pvDz and MP2/D95+(2df,p)//HF/D95** energies is expected, because the aug-cc-pvDz and D95+(2df,p) basis sets are similar in size containing 236 and 228 basis functions respectively.

Table 1. Calculated relative conformational energies (in kcal/mol) of the most important DME conformers and the following force fields: many-body force fields FF-1, FF-2 and FF-3 and the two-body FF-1 force field. (Reproduced with permission from Ref. 49)

energy[a]	MP2/ D95+ (2df,p)	MP2/ Dz	MP2/ Dz	MP2/ Dz	MP2/ Tz	Force fields FF-1 MB, (FF-1 TB), FF-2 MB, FF-3 MB	B3LYP /Dz	HF/ Dz
geom	HF/ D95**	HF/ Dz	B3LYP/ Dz	MP2/ Dz	MP2/ Dz		B3LYP /Dz	HF/ Dz
ttt	0	0.00	0.0	0.00	0.00	0.00, (0.0), 0.0, 0.0	0	0.00
tgt	0.14	0.16	0.18	0.15	0.21	0.16, (0.20), 0.19, 0.21	0.19	0.92
tg^+g^-	0.23	0.35	0.24	0.19	0.39	0.38 (0.37), 0.41, 0.46	0.74	1.66
ttg	1.43	1.35	1.28	1.26	1.41	1.24 (1.17), 1.22, 1.53	1.54	1.85
tgg	1.51	1.47	1.44	1.27	1.52	1.41 (1.40), 1.49, 1.46	1.73	2.80
ggg	1.64	1.55	1.54	1.20	1.60	2.00 (1.85) 2.23, 2.10	2.57	3.79

[a] The aug-cc-pvDz and aug-cc-pvTz basis set are denoted as Dz and Tz, respectively.

Table 2. Backbone torsional angles (degrees) for most important DME conformers calculated using the aug-cc-pv-Dz basis set at the HF, MP2, and B3LYP levels of theory. Geometries for the following force fields: many-body (MB) force fields FF-1, FF-2 and FF-3 and the two-body (TB) FF-1 force fields are also shown. (Reproduced with permission from Ref. 49.)

conf.	level of theory, FF	ϕ_1	ϕ_2	ϕ_3
ttt	All	180.0, 180.0,180.0	180.0, 180.0, 180.0	180.0, 180.0, 180.0
tgt	HF, B3LYP, MP2	184.1, 183.4,183.8	72.2, 74.7, 72.8	184.1, 183.4, 183.8
	MB FF-1,-2,-3	183.0, 183.1,186.3	68.4, 68.5, 67.3	183.0, 183.1, 186.3
	TB FF-1	183.0	68.8	183.0
tgg	HF, B3LYP, MP2	180.7, 181.0,179.8	65.0, 67.5, 58.5	75.9, 74.8, 59.9
	MB FF-1,-2,-3	181.1, 181.3,181.7	60.0, 59.6, 56.0	69.2, 70.2, 58.0
	TB FF-1	180.8	59.3	68.3
tg⁺g⁻	HF, B3LYP, MP2	182.1,180.6,180.7	71.7, 76.6, 75.1	268.2, 274.6, 278.8
	MB FF-1,-2,-3	181.5, 181.8,182.5	81.7, 82.3, 87.0	288.2, 287.3, 293.1
	TB FF-1	181.5	82.2	288.5
ttg	HF, B3LYP, MP2	181.7, 181.4,180.8	179.3, 179.7, 178.7	89.3, 83.0, 81.2
	MB FF-1,-2,-3	181.2, 181.1, 82.0	168.8, 166.8, 168.2	78.5, 78.3, 71.7
	TB FF-1	181.5	168.5	77.5
ggg	HF, B3LYP, MP2	64.3, 63.9, 58.1	49.9, 51.4, 45.6	64.0, 63.9, 58.1
	MB FF-1,-2,-3	63.8, 63.4, 56.9	50.0, 50.4, 49.7	63.8, 63.4, 56.9
	TB FF-1	63.2	50.6	63.2

We proceed by investigating the effect of the level of theory on the geometries (dihedral angles) of the most important conformers. Torsional angles for the most important DME conformers are summarized in Table 2 for the MP2, HF and B3LYP[53] calculations using the aug-cc-pvDz basis set. Torsional angles of the *tgt* conformer are within 2° for all levels of theory, whereas more-significant deviations were seen for the other conformers. The HF level yields torsional angles that differ by up to 10-15° from the MP2 torsional angles for the *tgg* and tg^+g^- conformers. In general, the B3LYP DFT calculations yield dihedral angles more similar to those from the MP2 calculations than those from the HF calculations.

The energies of DME conformers that have been calculated for the MP2, HF and B3LYP geometries, using the MP2/aug-cc-pvDz single-point energy calculations are presented in Table 1. The *tgt* energy for the HF and MP2 geometries are almost identical, whereas the energies for the other conformers differ by 0.1-0.35 kcal/mol, with the effect being the largest for the *ggg* conformer. The MP2/aug-cc-pvDz conformational energies calculated for the B3LYP geometries tend to agree better with those calculated for the MP2 geometries than the conformational energies calculated for the HF geometries. The average deviation of the backbone DME dihedral angles of the B3LYP/aug-cc-pvDz geometry from those at the MP2/aug-cc-pvDz geometry is 3.6°, which is slightly better than the average deviation of the DME backbone dihedrals at HF/aug-cc-pvDz from the MP2/aug-cc-pvDz level (4.4°), suggesting that it is preferential to use the B3LYP geometries rather than the HF geometries if the MP2 geometries are too expensive to compute.

Table 3. Correlation contribution (MP2 energy – HF energy) to relative conformational energy for most important DME conformers (kcal/mol) using MP2/aug-cc-pvDz geometries. (Reproduced with permission from Ref. 49).

conf.	(MP2-HF)/aug-cc-pvDz	(MP2-HF)/aug-cc-pvTz
ttt	0.0	0
tgt	-0.78	-0.72
tg⁺g⁻	-1.64	-1.63
ttg	-0.69	-0.67
tgg	-1.81	-1.71
ggg	-2.98	-2.93

Given the increase in computational power over the past decade, it is now feasible to perform calculations at the MP2 level with much larger basis sets than the D95+(2df,p) basis set used in the previous study of DME in 1993.[48] The conformational energies of DME were calculated at the MP2/aug-cc-pvTz level for MP2/aug-cc-pvDz geometries and are presented in Table 1. The increase of the basis set size from the aug-cc-pvDz to aug-cc-pvTz at the MP2/aug-cc-pvDz geometries increases the conformational energy of *tgt* only by 0.06 kcal/mol, the energies of the tg^+g^-, *ttg, tgg* conformers are increased by 0.15-0.25 kcal/mol, whereas a significant increase in the *ggg* energy by 0.4 kcal/mol is observed.

The correlation energy (the MP2 energy minus the HF energy) of the DME conformers is summarized in Table 3. The *ggg* conformer has the largest correlation contribution (~2.9 kca/mol) to the relative conformational energy, whereas the *tgg* and tg^+g^- have the second largest correlation contribution of about 1.6-1.7 kcal/mol. The large correlation energy of these conformers is responsible for the largest deviations of the HF geometries from the MP2 geometries for these conformers. The smallest correlation contribution for the *ttg* and *tgt* conformers, on the other hand, results in the smallest differences between the HF and MP2 geometries. It is also interesting that the correlation energy contribution to the relative conformational energies is weakly dependent on the basis set size (triple zeta vs. double zeta) and is less than 0.02 kcal/mol for the *ttg* and tg^+g^- conformers and less then 0.1 kcal/mol for the other conformers.

The next issue that we address is the ability of the density functional theory to predict relative conformational energies. Yoshida et al.[46] performed DFT and *ab initio* quantum chemistry calculation of the DME conformers at the 6-31G* level and found that the B3LYP functional yielded a conformational energy of the tg^+g^- conformer of 0.31 kcal/mol in good agreement with the estimates from IR experiments, whereas the B3LYP energy of the *tgt* conformer of 0.51 kcal/mol agreed nicely with the MP3/6-311+G*//HF6-311+G* energy of 0.51 kcal/mol. These results suggest that the B3LYP functional is able to provide reliable conformational energies for DME. To check this supposition, we calculated the conformational energies of DME using the same B3LYP functional with the much larger aug-cc-pvDz basis set, compared to the 6-31+G* basis set used by Yoshida et al.[46] Table 1 indicates that the B3LYP energy for the tg^+g^- conformer with the aug-cc-pvDz basis set is 0.74 kcal/mol, which is 0.4 kcal/mol higher than the B3LYP/6-31+G* energy. This indicates that the good agreement between the B3LYP/6-31+G* energy of the tg^+g^- conformer (0.31 kcal/mol) and the experimental value of 0.31 kcal/mol and our best estimate of 0.39 kcal/mol is serendipitous and is caused by the cancellation of errors that are due to basis set incompleteness and deficiencies in the B3LYP density functional. The B3LYP energy of the *ggg* conformer is also off by ~1 kcal/mol from our best estimate, whereas the energy of the *tgt* conformer is predicted within 0.05 kcal/mol of our best estimate. After comparing the HF, MP2 and B3LYP energies for the most important DME conformers, we conclude that despite the B3LYP energies being in better agreement with the MP2 energies compared to the HF energies, they are still unreliable (causing errors up to 1 kcal/mol in conformational energies) and, therefore, should not be used for parametrization of the potential energy function, which requires accuracies of 0.2 kcal/mol for the relative energies of the most important conformers. However, the B3LYP functional, even with a small 6-31+G* basis set yielded vibrational frequencies in excellent agreement with the experiment after uniform scaling by a factor of 0.97[46] indicating that a set of reliable force constants could be obtained at the B3LYP/6-31+G* level.

The MP2/aug-cc-pvTz//MP2/aug-cc-pvDz energies are our best estimate, and we believe that a further increase in basis set size will change the relative conformational energies by less than 0.1 kcal/mol for the most important DME conformers. We estimate the uncertainty in the energies due to incomplete treatment of electron correlation to be 0.1 kcal/mol and the uncertainty due to neglect of zero-point energies and thermal-vibrational contributions to be 0.1 kcal/mol. We expect that the use of the conformer geometries obtained at the level higher than MP2/aug-cc-pvDz will make a difference of less than 0.1 kcal/mol. Therefore, the uncertainty of the relative conformer energies obtained at the MP2/aug-cc-pvTz//MP2/aug-cc-pvDz level is 0.3 kcal/mol or even less because of cancellation of errors introduced by various approximations. In fact, because of error cancellation, the DME conformational energies at the MP2/aug-cc-pvTz//MP2/aug-cc-pvDz level are within 0.16 kcal/mol of the previous calculations, using a much smaller basis set and the HF geometries (MP2/D95+(2df,p)//HF/D95**), with an average absolute deviation of only 0.06 kcal/mol.

3.2.2 Dipole Moment and Polarizability of Dimethyl Ether and 1,2-Dimethoxyethane

Electrostatic interactions such as Coulomb and polarization interactions yield a major contribution to the PEO-cation complexation energy[2] and a significant contribution of the PEO-water binding,[16] indicating that care must be taken to use an adequate level of theory and a basis set to ensure accurate prediction of the electrostatic potential, dipole moments, and polarizability from quantum chemistry during force field parameterization. Dipole moment and dipole polarizability of a dimethyl ether are known from experimental studies,[54] therefore, we investigate the effect of basis set and level of theory on the ability of quantum chemistry calculations to represent these properties of dimethyl ether. The results of this investigation are presented in Table 4. Commonly used small basis sets such as 6-31+G* overestimate the dimethyl ether dipole moment by 12%-17 %, whereas its polarizability is underestimated by 16%-25 %. A basis set increase from the split-valence double-zeta to split-valence triple zeta results in marginal changes of the dipole moment and polarizability. The addition of a set of *d*- and *f*- polarization functions on heavy atoms and a set of *p*-functions on hydrogen atoms decreases dipole moment by 4-10 % and brings it in within 8% of the experimental values, whereas the dipole polarizability improves only by 0.2 $Å^3$. The D95+(2df,p) basis set yields results similar to those of the 6-311+G(2df,p) basis set. The aug-cc-pvDz basis set, however, yields results that best agree with the experimental values. Increasing of the basis set size to aug-cc-pvTz does not result in a significant improvement of the dipole moment and polarizability, whereas the removal of the augmentation leads to significant underestimation of the polarizability for all levels of theory and significant underestimation of dipole moment at the B3LYP level.

In summary, the B3LYP/aug-cc-pvDz level is the most accurate for combined prediction of the dipole moment (electrostatic potential) and polarizability and therefore is used in the future studies of electrostatic properties of PEO oligomers.

Table 4. Dipole moment and dipolar polarizability of dimethyl ether as a function of basis set and level of theory. (Reproduced with permission from Ref. 49).

Basis set	Basis func.	Dipole moment (Debye)			Polarizability (Å^3)		
		HF	MP2	B3LYP	HF	MP2	B3LYP
6-31+G*[a]	69	1.53	1.61	1.46	3.96	4.27	4.44
6-311G*[a]	72	1.44	1.38	1.29	3.81	4.03	4.19
6-311+G*[a]	84	1.53	1.56	1.45	4.04	4.36	4.51
6-311+ G(2df,p)[a]	138	1.40	1.41	1.33	4.28	4.56	4.74
D95+(2df,p)[a]	129	1.41	1.43	1.36	4.20	4.47	4.68
cc-pvDz[a]	72	1.37	1.24	1.14	3.87	4.07	4.21
aug-cc-pvDz[a]	123	1.37	1.36	1.27	4.60	5.01	5.14
aug-cc-pvDz[b]	123	1.51	1.36	1.32	4.71	5.01	5.15
aug-cc-pvTz[b]	276	1.51		1.32	4.73		5.17
experiment[54]		1.3 (±0.01)			5.29[c]		

[a] The geometry optimization was performed at the level of theory corresponding to the one used in the dipole moment and polarizability calculations.
[b] MP2/aug-cc-pvDz geometry was used.
[c] uncertainty of 0.3-8 % was reported for this data compilation

Table 5. Molecular dipole moment and polarizability of the most important DME conformers from quantum chemistry calculations and force fields. Dipole moments from three force fields presented below correspond to the fits to the electrostatic potential using the following objective functions: (a) $\sum (\phi_i^{QC}-\phi_i^{FF})^2$ or no weighting function; (b) $\sum|\phi_i^{QC}|(\phi_i^{QC}-\phi_i^{FF})^2$, or the absolute value of the electrostatic potential ϕ-weighting used in the FF-1 force field; and (c) $\sum(\phi_i^{QC})^2(\phi_i^{QC}-\phi_i^{FF})^2$ objective function or the ϕ^2-weighting used in the FF-2 and FF-3 force fields. (Reproduced with permission from Ref. 49.)

	dipole moment (Debye)			polarizability (Å^3)	
prop. calc.	HF/D95+ (2df,p)	B3LYP/aug -cc-pvDz	Force fields with no weighting, $\|\phi\|$-weighting (FF-1), ϕ^2-weighting (FF-2 and FF-3)	B3LYP/aug -cc-pvDz	force field
geom.	HF/D95**	B3LYP/aug -cc-pvDz		B3LYP/aug -cc-pvDz	
ttt	0.0	0.0	0.0, 0.0, 0.0	9.30	9.98
tgt	1.52	1.37	1.42, 1.60, 1.61	9.84	10.2
tg$^+$g$^-$	1.65	1.52	1.59, 1.76, 1.80	9.55	10.4
ttg	1.93	1.63	1.57,1.74, 1.77	9.43	10.3
tgg	2.67	2.39	2.37, 2.64, 2.67	9.46	10.4
ggg	1.49	1.41	1.51, 1.67, 1.71	9.14	10.4

The dipole moment and dipole polarizability were calculated for the most important DME conformers and are summarized in Table 5. The *ttt* conformer has a zero dipole moment, because the dipole moments of the monomers are antiparallel. The *tgg* conformer has the largest dipole moment. The dipole moments of the other most important conformers are significant, ranging from 1.37 to 1.63 Debye, suggesting that we can expect strong electrostatic interactions of most of the conformers with ions and polar liquids like water. The B3LYP/aug-cc-pvDz level yields DME dipole moments ~5-20 % lower than the HF/D95+(2df,p)//HF/D95** level[48] in accord with the aforementioned investigation of the basis set and a level of theory on the dimethyl ether dipole moment. Polarizability for the most important DME conformers was calculated at the B3LYP/aug-cc-pvDz level and is summarized at Table 5. The polarizability variation of <10 % between all conformers indicates that the DME polarizability is only weakly dependent on the DME geometry. It is also only slightly less than the polarizability of two non-interacting dimethyl ethers of 10.3 $Å^3$ at the B3LYP/aug-cc-pvDz level suggesting that the polarizability is approximately additive on the monomer length scale.

3.3 Force Field Development

3.3.1 Force Field Development Methodology

In the classical force field developed below, the total potential energy of the ensemble of atoms, represented by the coordinate vector **r**, is denoted as $U^{tot}(\mathbf{r})$. The latter is represented as a sum of nonbonded interactions $U^{NB}(r_{ij})$ as well as energy contributions due to bonds $U^{BOND}(r_{ij})$, bends $U^{BEND}(\theta_{ijk})$, and dihedrals $U^{TORS}(\phi_{ijkl})$ and is given by eq. 3.1,

$$U^{tot}(\mathbf{r})=U^{NB}(\mathbf{r})+\sum_{ij}U^{BOND}(r_{ij})+\sum_{ijk}U^{BEND}(\theta_{ijk})+\sum_{ijkl}U^{TORS}(\phi_{ijkl}) \tag{3.1}$$

The nonbonded energy $U^{NB}(\mathbf{r})$ (eq. 3.2) consists of a sum of the two-body repulsion and dispersion energy terms between atoms i and j and is represented by the Buckingham (exp-6) potential, and the energy due to charge-charge interactions (Coulomb), and the energy due to many-body polarization $U^{pol}(\mathbf{r})$

$$U^{NB}(\mathbf{r})=\frac{1}{2}\left[\sum_{i}\sum_{j}A_{ij}\exp(-B_{ij}r_{ij})-C_{ij}/r_{ij}^{6}+\frac{q_i q_i}{4\pi\varepsilon_0 r_{ij}}\right]+U^{pol}(\mathbf{r}) \tag{3.2}$$

The potential energy due to dipole polarization is not pairwise additive and is given by a sum of the interaction energy between the induced dipoles (μ_i) and the electric field $\mathbf{E}_i^0$ at atom i generated by the permanent charges in the system (the first term in eq 3.3), the interaction energy between the induced dipoles (the second term in eq 3.3) and the energy required to induce the dipole moments μ_i (the third term in eq 3.3)[55]

$$U^{pol}(\mathbf{r}) = -\sum_i \mu_i \bullet \mathbf{E}_i^0 - 0.5\sum_i \sum_{j\neq i} \mu_i \bullet \underline{\mathbf{T}} \bullet \mu_j + \sum_i (\mu_i \bullet \mu_i / 2\alpha_i) \quad (3.3)$$

where $\mu_i = \alpha_i \mathbf{E}_{tot}$, α_i is the isotropic atomic polarizability, $\mathbf{E}_{tot}$ is the total electrostatic field at the atomic site i due to permanent charges and induced dipoles, the second order dipole tensor $\underline{\mathbf{T}}_{ij}$ is given by

$$\underline{\mathbf{T}}_{ij} = \nabla_i \nabla_j \frac{1}{4\pi\varepsilon_0 r_{ij}} = \frac{1}{4\pi\varepsilon_0 r_{ij}^3}\left[\frac{3\,\mathbf{r}_{ij}\,\mathbf{r}_{ij}}{r_{ij}^2} - 1\right] \quad (3.4)$$

The nonbonded contributions including the potential energy due to polarization were calculated for all intermolecular atom pairs (ij) and for the intramolecular atom pairs separated by more than two bonds.

The contributions due to bonds $U^{BOND}(r_{ij})$, bends $U^{BEND}(\theta_{ijk})$ and torsions $U^{TORS}(\phi_{ijkl})$ for atoms i, j, k, l are given by eq 3.5-3.7,

$$U^{BOND}(r_{ij}) = \tfrac{1}{2} k_{ij}^{BOND}(r_{ij} - r_{ij}^0)^2 \quad (3.5)$$

$$U^{BEND}(\theta_{ijk}) = \tfrac{1}{2} k_{ijk}^{BEND}(\theta_{ijk} - \theta_{ijk}^0)^2 \quad (3.6)$$

$$U^{TORS}(\phi_{ijkl}) = \sum_n \tfrac{1}{2} k_{ijkl}^t(n)\,[1 - \cos(n\phi_{ijkl})], \quad (3.7)$$

where r_{ij}^0 is an equilibrium bond length; θ_{ijk}^0 is an equilibrium bend angle; ϕ_{ijkl} is a dihedral. The k_{ij}^{BOND}, k_{ijk}^{BEND}, $k_{ijkl}^t(n)$ are the bond, bend force constants and torsional parameters.

We adopt the following methodology for the many-body polarizable force field development. First, the partial charges are estimated by fitting electrostatic potential around PEO oligomers assuming zero polarizabilities. Then, the atomic polarizability is obtained by fitting polarization energies around DME using the previously obtained set of partial charges to account for intramolecular charge-induced dipole interaction. The final set of partial charges is obtained by fitting electrostatic grid around the ethylene oxide oligomers and using atomic polarizabilities to account for induced dipoles due to intramolecular interaction. A set of the repulsion and dispersion parameters then is determined. Bond and bend force constant could be obtained by fitting frequencies of the model compounds; however, in this work, we simply used the parameters from the previous MD simulations of the nonpolarizable PEO.[27] The equilibrium bond length and bend angles (eqs 3.5-3.6) are fit to obtain the best representation of the conformational geometries, and the torsional parameters (eq 3.7) are fitted to obtain the best representation of the conformational energies of DME most important conformers and barriers between them.

3.3.2 Partial Charges and Atomic Polarizabilities

In our model, which includes the many-body induced dipole polarizability energy, the electrostatic potential around a molecule is no longer only a function of partial charges. Indeed, the partial charges on the atoms separated by more than three bonds from a given polarizable atom create an electric field at that atom, resulting in an induction of an atomic dipole moment, which contributes to the electrostatic field around the molecule and should be considered during partial charge fitting. Thus, the atomic polarizabilities are required before fitting partial charges.

The molecular dipole polarizability of dimethyl ether and DME calculated at the B3LYP/aug-cc-pvDz level is given in Tables 4 and 5, respectively. Allinger's group[56] and Dykstra's group[57] suggested that the interaction between induced dipoles has a minor effect on molecular polarizability, and a sum of atomic polarizabilities could be used to approximate the molecular polarizability tensor with an average error of 10 % and the isotropic polarizability with an average error of 3%.[57] Applequist[58,59] and Thole[60] have developed interacting undamped and damped induced dipole model for description of anisotropy of molecular polarizability tensor of small molecules using only isotropic atomic polarizabilities.

Our observation of components of the molecular dipole polarizability tensor of the most important DME conformers and dimethyl ether being the similar suggests that interaction between the induced dipoles is week and we do not need to use the interacting atomic polarizability model for the closest induced dipoles, i. e., the induced dipoles connected by bonds and bends thus ensuring that polarization catastrophe[60] does not occur during MD simulations. Another consequence of weak interaction between the atomic polarizabilities is that one cannot use the molecular polarizability tensor to partition molecular polarizability into the atomic polarizabilities. A least-squares fit of the atomic polarizabilities to molecular polarizability for a large set of compounds is often used to partition the molecular polarizability to its atomic contributions. Dykstra's group[57] reported values of the following isotropic polarizabilities from such a fit: $\alpha_{C(sp3)}$=1.874 Å^3 and α_{-O-}=0.748 Å^3 with the hydrogen polarizability included into the polarizabilities of heavy atoms. Summation of the atomic polarizabilities yields molecular polarizabilities of 4.5 Å^3 for dimethyl ether and 9.0 Å^3 for DME approximately 10-20 % below the quantum chemistry values of 5.1 Å^3 for 9.1-9.8 Å^3 for dimethyl ether and DME, respectively, as reported in Tables 4 and 5. We wish to investigate the ability of these atomic polarizabilities to represent the polarization energy due to a unit charge

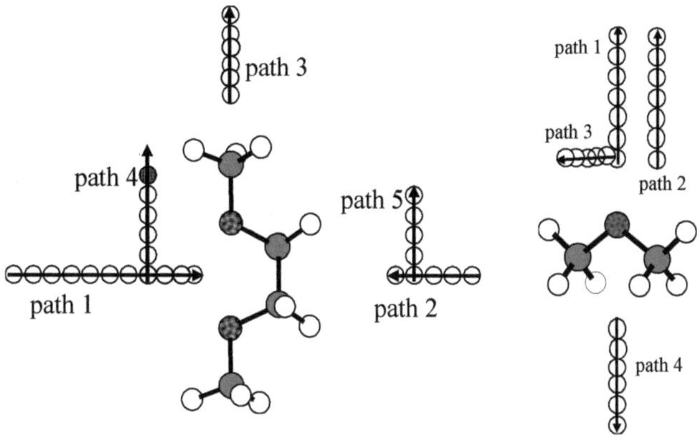

Figure 1. Paths for calculation of the electrostatic potential and polarization energy around the *tgt* conformer of DME and the dimethyl ether. Reproduced with permission from Ref. 49.

(+1e) around a DME molecule in tgt conformation along the paths shown in Figure 1 because this polarization energy is expected to contribute significantly to the Li^+ or another cation complexation energy with PEO and DME.

The polarization energy is calculated as the energy of the DME with the (+1e) charge, minus the self-energy of DME at the same geometry, minus the electrostatic potential at the point location of the unit charge. Figure 2 demonstrates that polarizabilities from Ref. 57 noticeably underestimate polarization energy along paths 1, 3 and 4. For example, for the most probable Li^+ position r(O-Q)=2.4 Å along path1, Dykstra's model yields a polarization energy of -27 kcal/mol, whereas the B3LYP/aug-cc-pvDz level quantum chemistry yields a value of -35 kcal/mol indicating an underestimation by >20% of the polarization energy by this model.

To obtain an improved description of the polarization energy around the most important PEO oligomers, we suggest partitioning the atomic polarizabilities by fitting the polarizable energy around a molecule obtained at the B3LYP/aug-cc-pvDz level quantum chemistry calculations. The atomic polarizabilities of similar atoms types were constrained to be equal during the fit. The ability of the atomic polarizable dipoles from the fit to reproduce polarization contribution to the potential energy is shown in Figure 2(a). This figure also indicates that the atomic polarizabilities from the fit are better able to represent polarization energy around DME than atomic polarizabilities from Dykstra's model[57] and, therefore, will be used for modeling many-body polarization effects in PEO and PEO/salt systems.

The values of the atomic polarizabilities from the current fit are shown in Table 6. The molecular polarizability tensor was calculated using these atomic polarizabilities. The average molecular polarizability tensor from quantum chemistry calculations is compared with that from the force field in

Table 5. The force field yields a molecular polarizability that is, on average, 8 % higher than the quantum chemistry values. Indeed, the atomic polarizabilities in the force field are fitted to reproduce the total polarization energy, not just dipole polarization, and, therefore, are representing the "effective" atom dipole polarizabilities that effectively include higher order polarizabilities whereas the molecular dipole polarizability from quantum chemistry does not incorporate hyperpolarizabilities. Therefore, the force field molecular polarizability is expected to be slightly higher than the quantum chemistry values.

Table 6. Charges and polarizabilities for the PEO force field. (C_m denotes methyl carbon of the CH_3 group, C_e denotes all other carbon atoms). Reproduced with permission from Ref. 49.

atom	Polarizability ($Å^3$)	Charge (e) for FF-1 force field.[a]	Charge (e) for FF-2 and FF-3 force fields.[b]
C_m	1.67	0.0200,	-0.1187
C_e	1.40	0.0661,	-0.0326
O	1.13	-0.3166,	-0.2792
H	0.18	0.0461,	0.0861

[a] from electrostatic potential fit using absolute value of the electrostatic potential (ϕ) weighting and the following WDR radii $r^{VDW}(O)$=1.7 Å, r^{VDW} (H)=1.8 Å, r^{VDW} (C)=2.1 Å
[b] from electrostatic potential fit using the square of the electrostatic potential (ϕ^2) weighting and the following WDR radii $r^{VDW}(O)$=2.0 Å, r^{VDW} (H)=1.8 Å, r^{VDW} (C)=2.5 Å

At the next stage of the force field development, we determine partial charges by fitting an electrostatic grid around a dimethyl ether molecule, five conformers of DME (*ttt, tgt, ttg, tgg, ggg*) and the *tttttt* conformer of

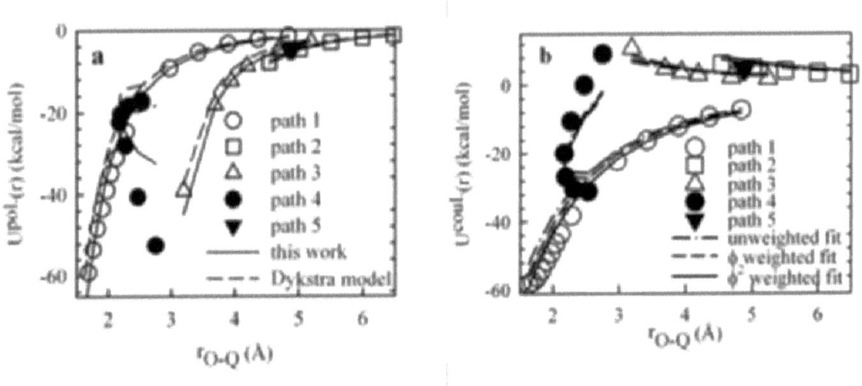

Figure 2. Polarization energy (a) and electrostatic potential (b) as a function of ether oxygen – charge separation ($r_{O\text{-}Q}$) along paths 1-5 around the *tgt* conformer from the B3LYP/aug-cc-pvDz quantum chemistry calculations (symbols) and force fields (solid lines).

diglyme. During the fit, the mean-square deviation of the force field electrostatic potential from that obtained at the B3LYP/aug-cc-pvDz level is minimized. Following Breneman et al.[61] we exclude electrostatic potential grid points inside molecular VDW radii from the fit, because the approach of any atom inside VDW radii is highly unlikely in MD simulations. Previous MD simulations of PEO melts, PEO/LiI[62] and PEO/$LiPF_6$[2] are used to define the VDW radii. The oxygen atom VDW radius of $r^{VDW}(O)=1.75\pm0.1$ Å is based on the closest approach of the Li^+ cation to an ether oxygen from the $g^{OLi}(r)$ radial distribution function for the PEO/Li-salts MD simulations; the hydrogen VDW radius $r^{VDW}(H)=1.8\pm0.15$ is based on $g^{HH}(r)$, $g^{HO}(r)$ and $g^{HF}(r)$ radial distribution functions, whereas $r^{VDW}(C)=2.2\pm0.2$ is based on $g^{CLi}(r)$, $g^{CH}(r)$ and $g^{CF}(r)$. The electrostatic grid points that are farther than 3.5 Å from any atom are also excluded from the fit, to obtain a relatively homogeneous layer of electrostatic grid points around a molecule. In the initial fit, all points around the DME and diglyme molecules are weighted equally; i.e., no position-dependent weighting is used. The partial charges from this fit are given in Table 6. The dipole moments of the most important DME conformers calculated using the resulting charges are shown in Table 5, and indicate good agreement between the force field and quantum chemistry results. The mean-square average deviation of the force field electrostatic potential from the quantum chemistry electrostatic potential was 0.7-1.2 kcal/mol also indicates that the partial charges can reproduce electrostatic potential around DME and diglyme molecules well.

A closer examination of the electrostatic potential along the paths 1 and 4 around the *tgt* conformer shown in Figure 2b indicates that the force field severely (≈10 kcal/mol) underestimates the electrostatic potential in the proximity of the ether oxygen atoms (r(O-Q) ≈2.0 Å), where the strongest binding of the Li^+ cation is expected.[62] This underestimation occurs because the number of points farther from the atom scales as a square of the distance, resulting in more grid points being included in the fit at longer distances, effectively giving larger weight to the more distant points in the electrostatic potential ϕ_i^{QC} fit. In order to increase the contribution to the objective function from the points in the proximity of the oxygen atom during the partial charge fitting, we examined two approaches: (a) weighting contributions of the grid points during the charge fitting by multiplying the $(\phi_i^{QC}-\phi_i^{FF})^2$ terms of the objective function by the electrostatic potential or a square of it; or (b) reducing the VDW radii of the atoms. Weighting of the grid points by the absolute value of the electrostatic potential, and especially the square of the electrostatic potential, results in a much better description of the electrostatic potential in the proximity of the ether oxygen atom, as seen in Figure 2b, at the expense of overestimating the DME dipole moments, as seen in Table 5. On the other hand, reducing the VDW radii by 0.2-0.3 Å, results in little change in the description of the electrostatic potential but changed the absolute values of the charges by as much as 0.04-0.06 e for oxygen and 0.04e for hydrogen. The partial charges from the fits are presented in Table 6. The set of partial charges obtained using the least-

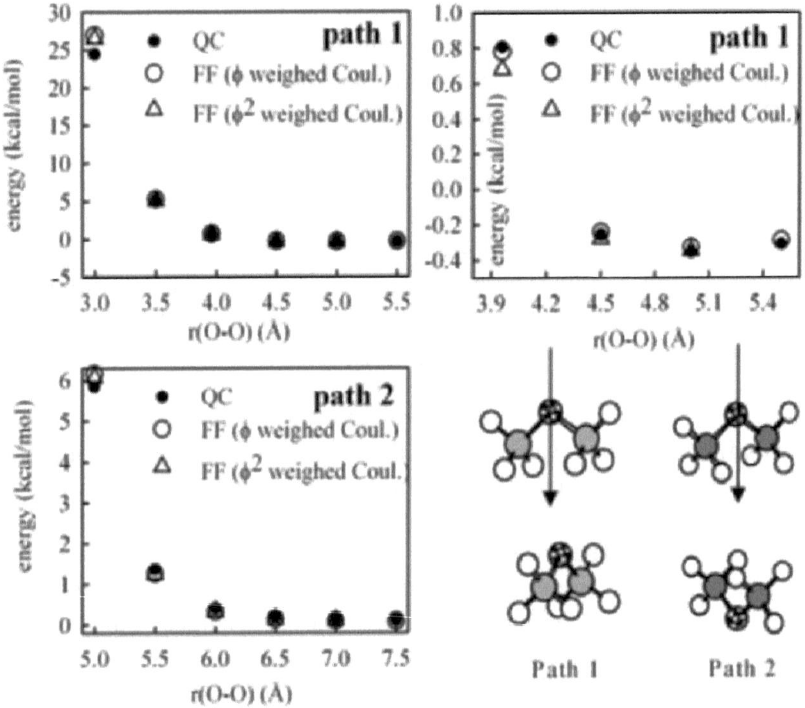

Figure 3. Complex energy of the dimethyl ether dimer at HF/aug-cc-pvDz level as a function of the distance between ether oxygen atoms r(O-O) along the two paths and from molecular mechanics using the developed force fields with the dispersion set to zero. Reproduced with permission from Ref. 49.

squares minimization of the $\sum(\phi_i^{QC})^2(\phi_i^{QC}-\phi_i^{FF})^2$ objective function, i.e., the $(\phi_i^{QC})^2$ weighting function, provides the most satisfactory description of the electrostatic potential for the purpose of modeling PEO/Li^+salts. To improve the description of the electrostatic potential close to the molecule, as well as far from it, one must include a set of the permanent dipoles, quadrupoles, etc. and possibly lone pairs in the model in addition to partial charges.[63-65] In this contribution, we limit ourselves to the simple point-charge model plus induced dipole many-body polarization, because it yields an adequate, albeit not excellent, description of the electrostatic potential and keeps computational cost of simulations down.

3.3.3 Repulsion and Dispersion Parameters

The repulsion and dispersion parameters are typically obtained from experiments by fitting densities and heats of vaporization of oligomers,[66] fitting crystal structures,[67] or using vapor-liquid equilibrium data.[68] Obtaining the repulsion and dispersion parameters by fitting the complex energies of two representative model compounds using the HF level and correlated quantum chemistry methods is less popular. Here, we would like to examine the ability of quantum chemistry calculations to predict the repulsion and dispersion parameters accurately. This can be done by calculating the repulsion and dispersion contribution to the binding energy separately, using quantum chemistry, and comparing the results to the empirically determined repulsion and dispersion parameters.[67]

Previous work on noble gases[69] indicated that the basis set superposition error (BSSE) corrected HF complex energies converge rather quickly as the basis set size increases, making calculation of repulsion contribution relatively easy. The accurate dispersion contribution to the complex energy is calculated using the energy from the correlated method minus the energy at the HF level. The dispersion (correlation) contribution to the complex energy is much more difficult to obtain than the HF energy and it requires the MP4 or coupled-cluster calculations with the triple zeta and larger basis sets. We start with an easier task of determining the repulsion parameters from quantum chemistry calculations and then discuss the ways to obtain the dispersion parameters. The complex energy of a dimethyl ether dimer at the HF level can be approximated as a sum of the Coulomb, many-body polarization and repulsion interactions. The first two contributions are adequately described by the previously determined charges and polarizabilities, leaving only the repulsion parameters as being unknown.

The complex energy with the exception of dispersion is calculated the following way. The geometry of the dimethyl ether dimer is optimized at the MP2/aug-cc-pvDz level. One of the dimethyl ether molecules is shifted along path 1, as shown in Figure 3. The second path is generated by rotating a dimethyl ether by 180° and shifting it along path 2 as shown in Figure 3. The complexation energy is defined as the energy of the complex minus the energy of the individual molecules frozen at the complex geometry. The BSSE correction is performed using the counterpoise method.[70] The BSSE-corrected dimethyl ether dimer complex energy at the HF/aug-cc-pvDz level is similar to the HF/aug-cc-pvTz energy, within 0.01 kcal/mol, for the minimum configuration (path 1) indicating that the HF energy is essentially converged at the aug-cc-pvDz basis set.

The complex energies at the HF/aug-cc-pvDz level for the two paths are shown in Figure 3 along with the complex energies from the force field using the empirical repulsion parameters obtained by fitting poly(oxymethylene) (POM) crystal structures[67] together with the quantum chemistry-based charges and polarizabilities. The force field with the empirical repulsion parameters shown in Table 7 and charges from the ϕ^2-weighted electrostatic potential fit describes the most-important part of the dimethyl ether-dimethyl ether complexation (path 1, r(O-O)≥3.9 Å) calculated at the HF/aug-cc-pvDz level within 0.03 kcal/mol! The closer separations, with r(O-O)=3.5 Å (path 1), are described within 0.09 kcal/mol, whereas the average absolute deviation of the force field complex energies from the HF/aug-cc-pvDz energies along the path 2 is 0.09 kcal/mol. The charges from the ϕ-weighted electrostatic potential fit yield a slightly worse but comparable agreement for the complex energy without dispersion, as shown in Figure 3. The excellent agreement between the force field and the HF/aug-cc-pvDz energies along the two paths indicates that the empirical repulsion parameters would be very similar to optimized quantum chemistry-based repulsion parameters and, therefore, could be used in our quantum chemistry-based PEO force field without any further modifications.

Table 7. Repulsion and dispersion parameters for the many-body polarizable and two-body PEO force fields. Reproduced with permission from Ref. 49.

Atom pair	A (kcal/mol)	B ($Å^{-1}$)	C^a (kcal $Å^{-6}$ /mol)
C-C	14976.0	3.090	595.94 (637.6)
C-O	33702.4	3.577	470.18 (503.0)
C-H	4320.0	3.415	128.56 (137.6)
O-O	75844.8	4.063	370.96 (396.9)
O-H	14176.0	3.9015	97.16 (104.0)
H-H	2649.6	3.740	25.44 (27.22)

[a] Dispersion parameters in parentheses are for the two-body PEO force field.

Accurate determination of the dispersion parameters from ab initio quantum chemistry calculations requires the use of large basis sets, generally beyond triple zeta, and the inclusion of higher-order correlations beyond MP2, making it a daunting task. Nevertheless, it is interesting to compare the correlation (dispersion) energy of the empirical force field to that from the ab initio quantum chemistry calculations. The empirical dispersion parameters are taken from the POM force field[67] and are scaled by 7 % to yield the correct density of diglyme liquid at 293 K obtained from MD simulations using the many-body polarizable force field. These modified empirical dispersion parameters are given in Table 7. The correlation energy from the quantum chemistry calculations for the dimethyl ether dimer was obtained by performing a double extrapolation (to the complete basis set limit and to the improved treatment of electron correlations) from the BSSE-corrected MP2/aug-cc-pvDz energies. First, the BSSE-corrected correlation energy at the MP2 level (i.e. MP2 minus HF) was extrapolated to the complete basis set limit using the previously found X^{-3} scaling,[16] based on the MP2/aug-cc-pvXz calculations with X=D,T, yielding the complete basis set limit correlation energy (MP2-HF) of -2.30 kcal/mol for the MP2/aug-cc-pvDz geometry. The MP2-HF, X=D,T,Q,etc correlation energy, using the complete basis set extrapolation, was 14.2 % larger than that for the aug-cc-pvDz value. The second extrapolation takes into account the deficiency of the MP2 level in the treatment of electron correlations and involves scaling by the factor of (MP4/aug-cc-pvDz)/(MP2/aug-cc-pvDz)=1.07. Thus, the double extrapolation for the correlation energy using these two scaling

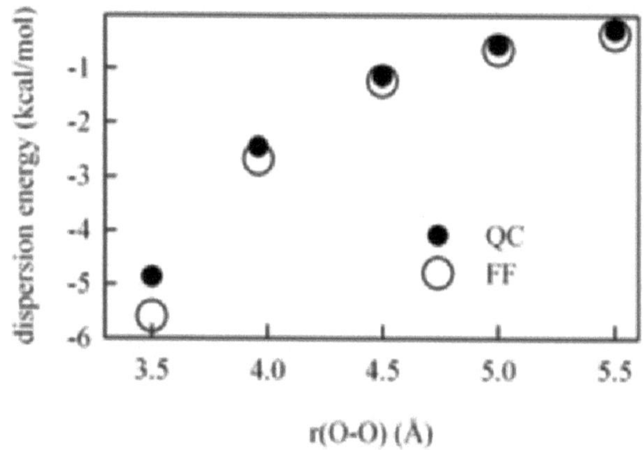

Figure 4. Dispersion (correlation) energy of the dimethyl ether dimer as a function of the distance between ether oxygen atoms r(O-O) from the quantum chemistry calculations and from the force field. Reproduced with permission from Ref. 49.

factors (MP4/aug-cc-pvDz)/(MP2/aug-cc-pvDz) and (MP2/complete basis set limit)/(MP2/aug-cc-pvDz) indicates that the correlation energy at the MP2/aug-cc-pvDz level should be multiplied by a factor of 1.22.

The correlation energy of a dimethyl ether dimer along path 1 for r(O-O)≥3.5 Å obtained by double extrapolation is shown in Figure 4 together with the dispersion energy from the force field. The quantum chemistry predictions are 8 % to 23 % (13% on average) lower than the dispersion contribution from the empirical force field. These results agree well with the conclusions of benchmark calculations on He_2, Ne_2, and Ar_2 that reported that the BSSE-corrected complete basis set extrapolation of the MP4 complex energy was ~10 % below the empirical values,[69] which indicated that the dispersion parameters could be obtained with an accuracy on the order of 10 % from ab initio quantum chemistry calculation, using previously described double extrapolation scheme. Moreover, scaling of the dispersion parameters obtained this way by a factor of 1.1 would yield dispersion parameters that are accurate to within a couple percent.

3.3.4 Dihedral, Bending, and Bonding Parameters

Although bond and bend force constants can be obtained by fitting vibrational frequencies of model PEO compounds, in this work, we simply took them from the previously developed nonpolarizable PEO force field,[27] whereas the equilibrium bond lengths and bending angles were fit to yield the best fit of equilibrium geometries from molecular mechanics to those from the B3LYP/aug-cc-pvDz geometry optimizations for the *tgt*, *ttt* and *tgg* conformers. The valence force field parameters are summarized in Table 8.

The dihedral parameters for the H-C-O-C angle were obtained by fitting quantum chemistry energies for the CH_3-group rotation of dimethyl ether at the MP2/aug-cc-pvDz//B3LYP/aug-cc-pvDz level (not shown). The O-C-C-H and H-C-C-H torsional parameters were taken from the previous nonpolarizable PEO force field.[27]

Table 8. Valence force field parameters for PEO. Reproduced with permission from Ref. 49.

Bond type		r_0 (Å)
C-O	constrained	1.4115
C-C	constrained	1.5075
C-H	constrained	1.1041
Bend type	k^{BEND}(kcal mol^{-1} rad^{-2})	θ_0 (deg)
C-C-H	85.8	110.10
H-C-H	77.0	109.47
O-C-H	112.0	109.48
C-O-C	149.0	108.05
O-C-C	172.0	108.54

We originally fitted the O-C-C-O and C-O-C-C torsional parameters shown in Table 9 to reproduce the conformational energies of the most-important DME conformers and barriers between them, using two sets of the previously obtained charges, together with the polarizabilities and other nonbonded parameters denoted as force fields FF-1 and FF-2. The resulting force fields are able to accurately describe the relative conformational energies of DME shown in Table 1 with an average deviation of 0.15 kcal/mol and 0.13 kcal/mol from the MP2/aug-cc-pvTz//MP2/aug-cc-pvDz quantum chemistry energies, respectively. Geometries of the most-important DME conformers from the force fields, shown in Tables 2, agree with those from the MP2/aug-cc-pvDz optimization with an average error of 3.5° and 3.7° for FF-1 and FF-2, respectively. These average deviations of the torsional angles from those at MP2/aug-cc-pvDz geometries are similar to the average deviations of the HF/aug-cc-pvDz geometries (4.4°) and the B3LYP/aug-cc-pvDz geometries (3.6°) from the MP2/aug-cc-pvDz geometries, which suggests that the force fields do a good job in reproducing the geometries of the most important DME conformers.

Table 9. Torsional parameters (kcal/mol) for PEO in order FF-1, FF-2 and FF-3. If one parameter is given, it is the same for all force fields. Reproduced with permission, Ref. 49.

Type	$k^t(1)$	$k^t(2)$	$k^t(3)$	$k^t(4)$	$k^t(5)$
O-C-C-H	0.0	0.0	0.28[a]	0.0	0.0
H-C-C-H	0.0	0.0	0.28[a]	0.0	0.0
C-O-C-H	0.0	0.0	-0.73	0.0	0.0
O-C-C-O	0.25,0.41, 0.47	-1.87,-2.10, -2.43	-0.43, -0.60, -0.36	-0.69, -0.82, -0.95	0.0, 0.0, -0.45
C-O-C-C	1.76, 1.76, 1.87	0.67, 0.67, 1.17	0.04, 0.04, 0.46	0.0, 0.0, -0.37	0.0, 0.0, 0.0

[a] taken from ref. [67]

The description of the t-ϕ-t and t-g-ϕ rotational isomerization paths is shown in Figures 5. The lowest energy barriers tgt/ttt, tg^+g^-/tgt and ttt/ttg (not shown) were described within 0.2 kcal/mol from the MP2/aug-cc-pvDz//B3LYP/aug-cc-pvDz energies, whereas the energies of the higher barriers tgt/tgg, ttg/tg^+g^- were underestimated by the force fields by 0.3-0.5 kcal/mol. We generally consider the FF-1 and FF-2 force field fits to the quantum chemistry conformational energies to be good.

Note that the partial charges in the FF-1 force field noticeably underestimate electrostatic potential at the region of the most probable Li^+ cation binding ($r_{O\text{-}Li}\approx$1.9-2.0 Å) indicating that the FF-1 force field markedly underestimates PEO/Li^+ binding energy and therefore overestimates the extend of ion aggregation. MD simulations of PEO with the FF-2 force field, which has improved description of the electrostatic potential in the vicinity of EO, indicate that PEO dynamics using this force field are somewhat slow. In order to combine accurate PEO dynamics and electrostatic in the same force field, we refitted the torsional parameters of the FF-2 force field to the barriers for conformational transitions that we reduced by 0.3-0.9 kcal/mol while keeping the partial charges and other force field parameters the same as those in the FF-2 force field. This force field is called the FF-3 force field. This barrier reduction is higher than the expected accuracy of the quantum chemistry calculations and, therefore, is considered to be an empirical adjustment. The FF-3 force field described the relative conformational energies and geometries with an accuracy that was similar to that for the FF-1 and FF-2 force fields (see Tables 1 and 2). The conformational isomerization paths for the FF-3 force field are shown in Figure 5 demonstrating the effect of lowering the barriers for conformational transitions in the FF-3 force field, compared to the FF-1 and FF-2 force fields.

The ability of the developed force fields to predict the relative conformational energies of a diglyme (a three-repeat-unit oligomers of ethylene oxide, CH_3-terminated, $C_6O_3H_{14}$) is shown in Table 10. The average deviation of the force fields conformational energies from the quantum chemistry values is 0.11-0.19 kcal/mol with the maximum deviation of 0.41 kcal/mol. This accuracy is comparable to the accuracy of the quantum chemistry calculations for the diglyme molecule to be expected ~0.3-0.4 kcal/mol.

Table 10. Relative conformational energies of diglyme in kcal/mol from ab initio quantum chemistry and the force fields. (Reproduced with permission from Ref. 49).

Conformer	MP2/D95+ (2df,p)// MP2/D95*	FF-1 MB	FF-1 TB	FF-2 MB	FF-3 MB	FF-3 TB
ttt ttt	0.0	0.0	0.0	0.0	0.0	0.0
$tg^{+}t\ ttt$	0.03	-0.02	-0.02	-0.02	0.0	-0.01
$tg^{+}g^{-}\ ttt$	0.01	0.27	0.28	0.21	0.15	0.13
$tg^{+}g^{+}\ ttt$	0.83	1.04	0.98	1.07	0.87	0.80
$ttg^{+}\ ttt$	1.15	1.14	1.03	1.11	1.31	1.13
$tg^{-}t\ g^{-}tt$	0.95	1.37	1.28	1.25	1.36	1.28
$tg^{-}g^{+}\ g^{+}tt$	1.10	1.35	1.18	1.36	1.46	1.22

3.3.5 *Two-Body Force Field Approximation*

Analysis of the many-body polarizable contribution to the relative conformational energies of DME, using molecular mechanics, showed that the polarizability energy is small and contributes less than ±0.15 kcal/mol to the conformational energies relative to the *ttt* conformer. The induced dipole moment of the most-important DME conformers contributes 10%-20 % to the total dipole moment, which also indicates a relatively small effect of atomic polarizabilities on the electrostatic interactions and suggesting that it is possible to turn off atomic polarizability and slightly increase the dispersion parameters to compensate for the induced dipole-permanent partial charge attraction that is not considered by the two-body force fields. Therefore, it reasonable to suggest that a nonpolarizable (or two-body force field, called TB) can be created from the corresponding many-body polarizable force field (called MB) by zeroing the polarizabilities and increasing the dispersion parameters by 7% to compensate in a mean-field sense for a decreased attraction and to keep the density of PEO oligomers the same for many-body polarizable and nonpolarizable force fields.

We investigated the effect of increasing the dispersion and turning off the polarizability on equilibrium dihedral angles and conformational energies for the force field FF-1, as presented in Tables 2 and 1, respectively. The equilibrium dihedrals angles are different by <1° between the MB and TB force fields. Conformational energies of DME and diglyme are also very similar between the TB and MB force fields with a maximum deviation of 0.15 kcal/mol for DME and 0.17 kcal/mol for diglyme for the FF-1 force field and 0.24 kcal/mol for diglyme for the FF-3 force field, whereas the

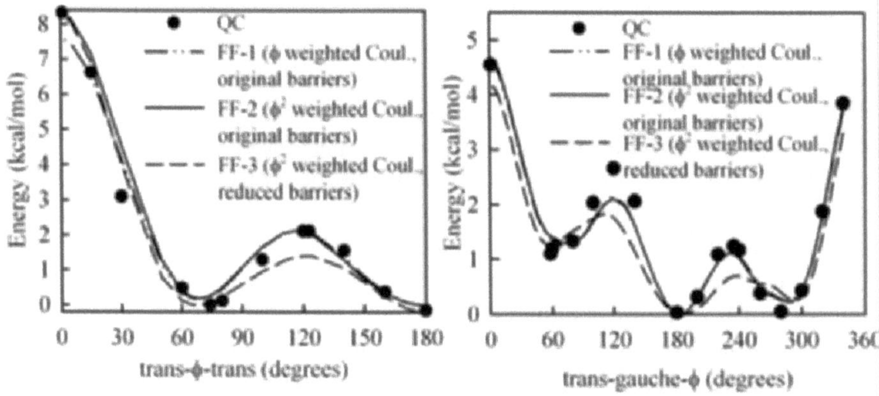

Figure 5. The relative conformational energy along the low energy rotational isomerization paths for DME. Quantum chemistry points correspond to the MP2/aug-cc-pvDz//B3LYP/aug-cc-pvDz level energies.

average deviation of conformational energies of the MB force field from the TB force field were 0.05-0.1 kcal/mol for DME and diglyme. Therefore, we conclude that so far, there is little difference between the MB and TB force fields seen so far. In the section, we investigate, in detail, the effect of many-body polarizable interactions on static, dynamic and thermodynamic properties of PEO and its oligomers in a liquid (melt) phase.

3.4 Molecular Dynamics Simulations Of Poly(Ethylene Oxide) And Its Oligomers

3.4.1 MD Simulations Methodology

MD simulations were performed on CH_3-capped oligomers of ethylene oxide (EO) H-(CH_2-O-CH_2)$_n$-H) for n=1,2,3,54 and a polydisperse system with M_n=398 and M_w=475 (referenced here as "PEO-500"). The linear dimension of the simulation cubic box varied from 23.5 to 34 Å depending on the system. The MD simulations code Lucretius was used.[71] All systems were created in the gas phase. The box was shrunk in MD simulations, using a Brownian dynamics algorithm[72] over a period of 1-3 ns to yield the densities estimated from experimental data, with subsequent equilibration in the NPT ensemble for 0.5-5 ns using the velocity Verlet algorithm with a 1 fs timestep.[73] A 1 ns constant pressure run was performed at each temperature, in order to determine the equilibrium density. Subsequent sampling runs in an NVT ensemble were 1-45 ns. Table 11 summarizes the run lengths for n=1,2,3 and PEO-"500", whereas the NVT runs for n=54 were 25 ns at 393 K (FF-3) and 15 ns for FF-1 and FF-2 and 15 ns at 343 K for all force fields. A Nose-Hoover thermostat[74] and a barostat[75] were used to control the temperature and pressure, whereas bond lengths were constrained using the Shake algorithm.[76] The Ewald summation method[77] was used for treatment of long-range electrostatic forces between partial charges with partial charges and partial charges with induced dipoles for the many-body polarizable potential, and the particle-mesh Ewald (PME) technique[78] was used for the simulations using the two-body nonpolarizable potential. A tapering function[79] was used to drive induced dipole-induced dipole interactions to zero at the cutoff of 10 Å. A multiple-time-step reversible reference system propagator algorithm was employed,[80] with a time step of 0.75 fs for bonding, bending, and torsional motions, a time step of 1.5 fs for nonbonded interactions within a 6.5 Å sphere, and a time step of 3.0 fs for nonbonded interactions within 6.5 Å and 10.0 Å and the reciprocal space part of the Ewald and PME summation.

3.4.2 Thermodynamic Properties

Densities of the EO oligomers from the many-body polarizable force field (MB) and two-body (TB) nonpolarizable force field are compared with available experimental data[81-83] in Table 11.

Table 11. Density and heat of vaporization of PEO oligomers from experiment and MD using FF-3 including many-body polarization (MB) and two-body (TB) nonpolarizable force field FF-3 unless indicated otherwise. Reproduced with permission from Ref. 38.

Com-pound	FF	T (K)	P atm	run length (ns)	ρ^{MD} (kg/m^3)	ρ^{exp} (kg/m^3)	ΔH^{MD} (kcal mol^{-1})	ΔH^{exp} (kcal mol^{-1})
dimethyl ether	MB	254	1.35	0.9	679.3	715.6	4.93	5.05
	TB	254	1.35	0.9	688.1	715.6	4.89	5.05
	TB	293	5.24	0.9	631.2	661.0	4.50	4.61
DME	TB	298	1	5,5[a],5[b]	850, 843[a],848[b]	861.3		
	MB	298	1	8[a]	830[a]	861.3		
	MB	318	1	1.2	824	840.8		
	TB	318	1	8, 4[a],4[b]	826,818[a], 822[b]	840.8		
	TB	358	1	1.5	784		8.42	8.65
diglyme	MB	293	1	1.5	938	943.4		
	TB	293	1	1.5	942	943.4		
PEO-500	TB	300	1	15	1064	1069		
	TB	318	1	42,9[a],9[b]	1048,1038[a], 1049[b]	1054		
	TB	328	1	32,16[a],18[b]	1039, 1030[a],1040[b]	1045		
	TB	363	1	35,6[a],6[b]	1012,1002[a], 1013[b]			

[a] for the FF-1 force field
[b] for the FF-2 force field

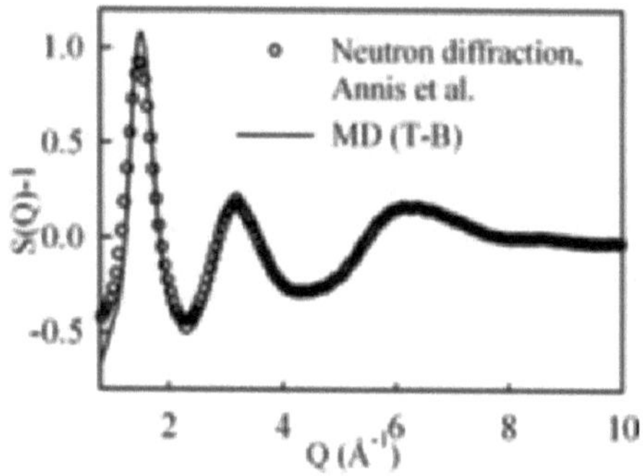

Figure 6 Normalized static coherent structure factor from neutron diffraction experiment and MD simulations with TB FF-3.

As the dispersion parameters in the force field were adjusted to reproduce the density of diglyme for the force fields, the simulation density of diglyme is in good agreement with the experimental density. The density of the "PEO-500" is within 0.5-1.5 % of the experimental data, whereas predicted DME densities are within 1-2.8 % of the experimental values. The force fields (FF-2 and FF-3) predict the densities of dimethyl ether on the saturation curve with the worst accuracy of 3.8-5 %.

Also shown in Table 11 is a comparison for the enthalpies of vaporization. The MD simulations predict enthalpies of vaporization within 3.3% of the experimental values. This agreement is considered excellent, because the experimental heats of vaporization were not used during the parametrization of the force field. MD simulations with the MB polarizable force fields yield densities and heats of vaporization that are within 1% of the data from the simulations with the TB nonpolarizable force fields, which indicates that inclusion of the many-body polarizability is not important for prediction of the density and the heat of vaporization.

3.4.3 PEO Structure Factor

Neutron diffraction experiments, which were performed on high molecular weight PEO at 363 K, are described in detail in ref. [35]. The structure factor from our previous MD simulations of the nonpolarizable force field with the different charges and valence potential was found in excellent agreement with the neutron diffraction data of Annis et al.[36] We calculated the structure factor for the current nonpolarizable and many-body polarizable force fields using eq. a2 from Appendix A. The coherent static structure factor S(Q) from MD simulations using all force fields and neutron diffraction experiment are shown in Figure 6.

The structure factor from MD simulations using FF-1, FF-2, FF-3 many-body polarizable (MB) and two-body (TB) nonpolarizable force fields are almost indistinguishable from the each other, therefore only one data set is shown in Figure 6 for clarity. The agreement between the experimental S(Q) and that from the MD simulations is very good, further validating the developed PEO force field.

3.4.4. Conformations of PEO and Its Oligomers

We proceed with investigation of the influence of many-body polarizable interactions on DME in the gas and liquid phases. Conformational populations of DME in the gas phase from MD simulations using the FF-1 TB and MB force fields are shown in Table 12 They are essentially the same, within the errors shown, indicating that the inclusion of the many-body polarization results in insignificant changes of the gas phase populations and therefore relative conformational free energies. This is in agreement with the small (<0.2 kcal/mol) contribution of polarization to the relative conformational energies found in molecular mechanics studies reported in the Section 3.3.5.

Table 12 Populations of DME liquid and gas phase from MD simulations and liquid DME from IR experiments at 298 K. Reproduced with permission from Ref. 38.

conf.	FF-1 TB liquid, (FF-1 TB gas phase)	FF-1 MB liquid, (FF-1 MB gas phase)	FF-2 TB liquid	FF-3 TB liquid	IR exp.
ttt	16.1±1.4, (24.2 ±2.8)	10.1±1.3, (24.1±2.9)	10.5±1.2	15.0±1.6	12
tgt	36.5±1.9, (24.3±2.2)	46.2±2.1, (26.2±1.9)	42.1±2.3	49.0±2.3	49
tg^+g^-	17.5±1.0, (26.7±2.7)	17.3±1.1, (27.0±2.8)	18.3±1.2	14.5±1.2	33
tgg	12.2±1.0, (5.7±1.0)	14.9±1.3, (6.0±1.2)	15.6±1.3	11.2±1.1	11
ttg	11.7±1.1, (13.5±1.9)	6.7±1.0, (11.9±1.5)	8.0±1.0	7.4±0.9	7
σ(MD -IR)[b]	7.0	2.6	4.2	1.5	

[a] Standard deviations for 100 ps block averages are given as error bars.

[b] Mean-square deviations σ(MD-IR) of MD populations from IR populations for the ttt, tgt, tgg and ttg conformers are also given (in %).

Conformational populations of DME liquid from MD simulations with the TB force field, however, are different from those with the MB force field as shown in Table 12. The inclusion of the polarization terms results in a more significant stabilization of the tgg and, especially, tgt conformers in DME liquid compared to the gas phase, and destabilization of the ttt, tg^+g^- and ttg conformers. Interestingly, a slight increase of the DME dipole moment and the magnitude of the electrostatic potential around DME in the FF-2 force field, compared to the FF-1 force field, also stabilizes tgt and tgg conformers in the liquid phase in comparison to the gas phase, suggesting that the increased Coulomb interaction can be used to capture the effect of the many-body polarization on conformational populations in the condensed phase in a mean-field sense.

Conformational populations for the most important DME conformers from

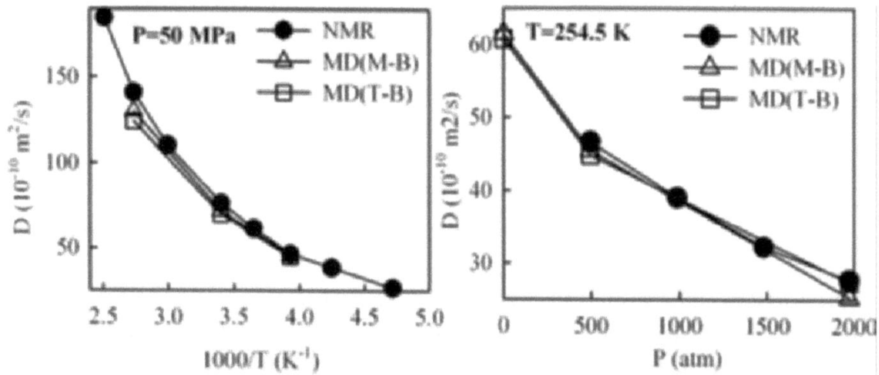

Figure 7. Self-diffusion coefficient of dimethyl ether as a function of temperature and pressure from MD simulation and NMR experiments.[86] (Reproduced with permission from Ref. 38).

MD simulations of DME liquid at 298 K are compared to those estimated from the analysis of IR spectra at 300 K.[85] Excellent agreement between the conformers predicted from MD simulations and from IR spectra is seen for the ttt, tgt, tgg and ttg conformers. The FF-1 MB and FF-3 force fields yield the best agreement, with mean-squared deviations of MD values from IR estimates of 2.6% and 1.5 %. The FF-1 TB force field yields somewhat worse agreement, because turning off the polarization resulted in underestimating tgt stabilization in the condensed phase compared to the gas phase. The only conformer, where a significantly different population from IR experiments is predicted by the MD simulations, is the tg^+g^- conformer. The difference between the conformational populations from MD simulations and the IR spectra for the tg^+g was attributed[85] to overestimation in the MD simulations of the intermolecular CH…O interaction, which is responsible for destabilization of the tg^+g^- in the liquid phase relative to the bulk phase. We believe that overestimation of the intermolecular CH…O interaction is rather unlikely, because the force field binding energy of the dimethyl ether dimer is rather sensitive to the intermolecular CH…O interaction, that is in excellent agreement with the high level quantum chemistry calculations as shown in Figures 3 and 4. However, it is possible that the difference of the tg^+g^- populations from MD and IR may be due to a large change in IR intensity for the tg^+g^- conformer upon transferring DME from the gas phase to the liquid phase.

The characteristic ratio of PEO serves as a measure of global polymer dimensions. At 393 K the characteristic ratio of PEO(M_w=2380) is found 5.9±0.4 for the FF-3 TB force field. It is in good agreement with the experimental value of 5.3-5.9 at 363 K from recent SANS experiments[35] and a value of 5.6 from other SANS measurements[34] at 390 K, indicating that the global PEO conformations are realistically represented by the present force field. The trajectories from MD simulations with the MB force fields and FF-1 and FF-2 are less than a half of the PEO Rouse time and, therefore, unreliable for a determination of the global PEO dimensions.

3.4.5 *Dynamic Properties of PEO Oligomers*

The self-diffusion coefficient of dimethyl ether has been measured as a function of temperature and pressure by pulse-field gradient (PFG) spin-echo NMR.[86] The dimethyl ether self-diffusion coefficient as a function of pressure and temperature is calculated from MD simulations using the Einstein relationship eq. a10 from Appendix A. The self-diffusion coefficients from the MD simulations with the MB and TB force fields are compared with the NMR data in Figure 7. MD simulations with both force fields yield self-diffusion coefficients that are in excellent agreement with

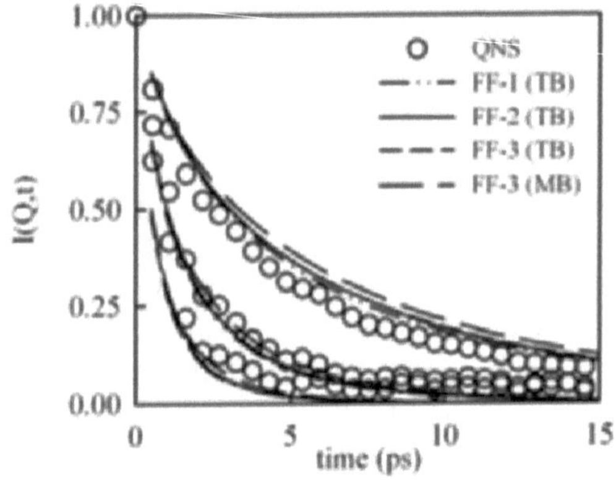

Figure 8. Incoherent intermediate dynamic structure factor I(Q,t) for DME at 318 K from MD and QNS experiments for Q=0.575, 0.909, 1.227 Å^{-1}. Reproduced with permission from Ref. 38.

the experimental data. The maximum deviation of the self-diffusion coefficient from the MB force field from the experimental data was 7.8 %, whereas the TB force field showed deviations less than 12 %. The difference between the MD simulations predictions using the MB and TB force fields is small.

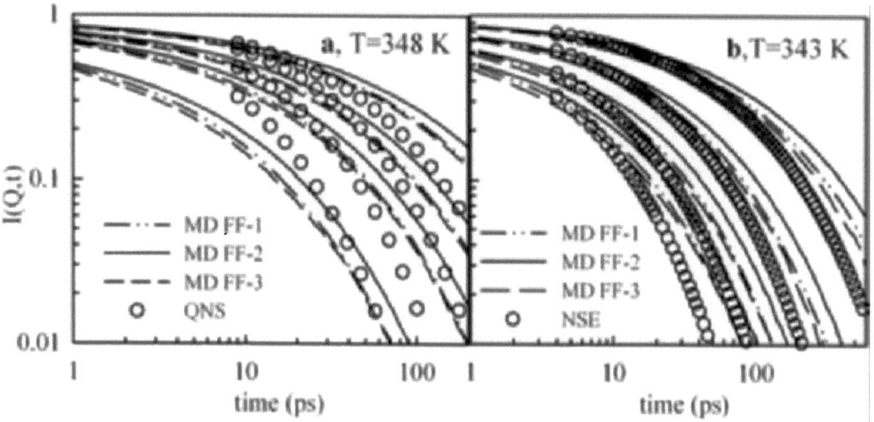

Figure 9. Incoherent intermediate dynamic structure factor I(Q,t) for PEO (M_w=2380) from MD simulations compared with the results for high molecular weight PEO from the QNS experiments for Q=0.72, 0.91, 1.10, 1.55 $Å^{-1}$ at 348 K, NSE experiments for Q=0.7, 1.0, 1.3, 1.55 $Å^{-1}$ at 343 K^{-1}.

Incoherent quasielastic neutron scattering (QNS) experiments of hydrogenated polymers probe hydrogen motion. The most direct comparison between experiment and simulations is achieved by comparing the intermediate scattering functions (ISF) I(Q,t) obtained from the QNS experiment with those calculated from the MD simulations. As scattering from hydrogen is the dominant contribution, and it is essentially entirely incoherent scattering, only the incoherent ISF is relevant. The ISF's from MD simulations were calculated using eq. a3 from Appendix A, whereas the time Fourier transform of the experimentally measured dynamic structure factor S(Q,ω) yields the ISF's from QNS measurements. Unlike the QNS experiments, neutron spin-echo experiments measure ISF I(Q,t) directly.

A comparison of the incoherent ISF for DME at 318 K obtained from MD simulations and QNS experiments[38] performed on the disk chopper spectrometer is shown in Figure 8. Good agreement between the MD simulations for all force fields and QNS data is observed, with MD simulations predicting decay of ISF up to 20 % slower than the QNS experiments for the lowest Q investigated. Simulations with the MB force field yield predict a slightly slower (within 8 %) decay of ISF compared to the corresponding TB force field, indicating that inclusion of the many-body polarizable interactions results in some but insignificant slowing down of the motion of short ethylene oxide oligomers. The DME ISF is also similar from MD simulations using the two-body force fields with the quantum chemistry barriers (FF-1 and FF-2) and reduced barriers (FF-3), indicating that the conformational barrier heights have little influence on DME dynamics.

The effect of the many-body polarizability interactions, description of the electrostatic potential and differing barriers on PEO dynamics is further

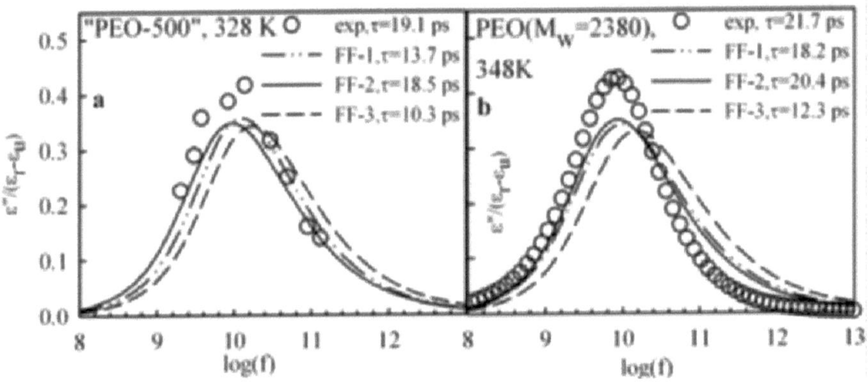

Figure 10. Normalized dielectric loss from the MD simulations for "PEO-500"(a) and PEO(M_w=2380) (b) and from the dielectric spectroscopy experiments for "PEO-500" and high molecular weight PEO ref. [88]

explored by comparing ISF's for high molecular weight PEO (M_w=2380) from MD simulations with all developed force fields at 393 K. Rather high temperature is chosen in order to ensure proper system equilibration. Comparison of the time integrals of ISF, i. e. relaxation times, for Q=1.0 $Å^{-1}$ revealed the following relation between the relaxation times τ(FF-1 TB):τ(FF-2 TB):τ(FF-3 TB): τ(FF-3 MB): =9.2 ps : 10.8 ps : 8.6 ps: 9.99 ps=1.07 : 1.26 : 1.0 : 1.16. These relaxation times indicate that inclusion of polarizability (MB vs. TB) and increase of the DME electrostatic potential (Coulomb interactions) (FF-2 vs. FF-1) both slow down polymer dynamics by about 16 %, suggesting the effect of inclusion of polarizability on PEO dynamics can be reasonably modeled by increasing DME Coulomb interactions as was found for DME conformational populations. Because the PEO dynamics with the MB and TB force fields are rather similar, we use only the TB force fields from here on for investigation of PEO dynamics at lower temperatures, because MD simulations of PEO with the TB force fields are approximately three times faster to compute than those with the MB force fields.

QNS experiments[87] on high molecular weight PEO (M_w=40000) and neutron spin-echo (NSE) experiments[40] on PEO(M_w=8000) have also been previously performed, probing hydrogen dynamics on the timescale of 0.01-1.7 ns. The ISF from these measurements are compared with the results of MD simulations in Figure 9(a-b). Both sets of experimental data predict a slower decay than the MD simulations at short times and high Q-values, whereas faster decay at long times and low Q-values is observed in NSE and QNS experiments compared to MD simulations predictions. MD simulations using the FF-1 and FF-3 force fields yield ISF's in the best agreement with

QNS experiments at long times. The NSE data shown in Figure 9b agree with the ISF's from the MD simulations slightly better than the QNS data shown in Figure 9a. The dynamic correlation times were calculated as time integrals of the stretched exponential fits to ISF's from NSE experiments and MD simulations. Dynamic correlation times were about 20-30 % higher for ISF's from simulations using the FF-1 and FF-3 force field and 65% for the FF-3 force field as compared to the NSE experiments at Q=1 Å^{-1}, indicating good agreement between the NSE experiment and MD simulations with the FF-1 and FF-3 force fields, and fair agreement for the FF-2 force field.

Dielectric spectroscopy of polymer molecules also allows one to probe polymer dynamics through frequency-dependent dielectric loss, which is a Fourier transform of dipole moment autocorrelation function (see Appendix A). The dielectric loss from MD simulations for "PEO-500" is compared in Figure 10a with the results of the dielectric spectroscopy measurements on "PEO-500" with a similar molecular weight distribution. Our MD simulations with the FF-2 force fields yield the best agreement with experiments while the FF-1 and FF-3 force fields predict position of the maximum of dielectric loss shifted by 0.1 and 0.25 on a log10 scale towards higher frequencies relative to FF-2 force field results. Figure 10b compares dielectric loss of PEO(M_w=2380) from MD simulations with that from the Cole-Cole fit to the dielectric spectroscopy measurements[88] performed for frequencies lower then the frequency of the maximum of the loss of high molecular weight PEO (Carbowax 20M). As seen for "PEO-500", the dielectric loss predictions from MD simulations using the FF-1 and FF-2 agree with experiments the best, while the FF-3 force field predicts the maximum frequency position shifted by 0.26 towards higher frequencies indicating that the dielectric spectroscopy experiments are consistent with each other.[88] All force fields predict slightly broader relaxation spectrum than dielectric experiments consistent with the more stretched ISF functions observed during comparison of MD simulations results with NSE experiments.

Spin-lattice relaxation times measured by ^{13}C NMR experiments also serve as a measure of polymer conformational motion through the decay of C-H vector ACF and spectral density functions as described Appendix A and ref.[38]. The T_1 spin-lattice relaxation times for interior carbons of "PEO-500" and PEO(M_w=2380) from the MD simulations are compared with the 500 MHz NMR results for "PEO-500" and PEO(M_w=1854) in Figure 11. This figure indicates that the FF-1 and FF-3 force field predict T_1 spin-lattice relaxation times in best agreement with experiments, slightly underestimating experimental values by 15-25 % for "PEO-500" and by about 30% for PEO(M_w=2380). The FF-2 force field predicts slightly lower T_1 relaxation times (6-10%) and higher C-H vector relaxation times (15-20%) than the FF-1 and FF-3 force fields consistent with the ISF for the corresponding force fields as shown in Figures 9, but slightly different from the predicted tendencies of the dielectric behavior, which are shown in Figure 10. Figure 11 also shows results from the previous 300 MHz NMR study of PEO(M_w=531)[37] that yields T_1 spin-lattice relaxation times approximately two times lower than the present 500 MHz NMR data for the "PEO-500". Our estimates suggest that the molecular weight difference between "PEO-500" and PEO(M_w=531) could account for up to 10 % of the discrepancy in T_1, whereas the difference in frequency (300 MHz vs. 500 MHz) accounts for a few percent at 363 K and less then 25 % at 328 K, indicating that the previous 300 MHz NMR T_1 data are inconsistent with the

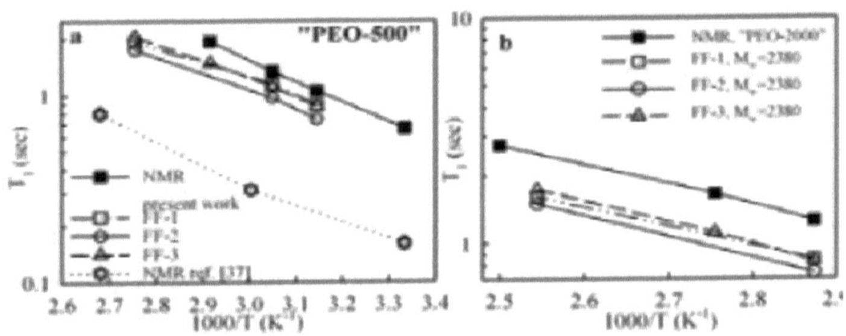

Figure 11. The T1 spin-lattice relaxation times for carbon atoms that are further than three carbons away from the chain ends. The dotted line shows the T1 spin-lattice relaxation times from the previous ^{13}C experiments.

present NMR data. As the present NMR data are in good agreement with the MD simulations predictions, which are also in good agreement with the QNS, NSE, dielectric spectroscopy data, we conclude that the previous NMR data are inconsistent with all experimental data on PEO dynamics presented here.

3.4.6 Transport Properties of PEO Oligomers

The viscosity η of DME and "PEO-500" has been measured experimentally[38] and is accessible from MD simulations (see Appendix A). The viscosities of 0.41, 0.43 and 0.40 mPa·s at 298 K are calculated in MD simulations for DME using the TB FF-1, FF-2 and FF-3 force fields. These values are in good agreement with the experimental value of 0.424 mPa·s. This is consistent with the accurate prediction of DME dynamics by MD simulations, as demonstrated by the agreement between MD and the QNS experiments. The independence of the DME viscosity from the barrier heights (FF-3 vs. FF-2) indicates that, unlike for PEO-500 and PEO(M_w=2380), torsional barriers play a minor role in DME transport and diffusion. MD simulations also accurately (within 8-10 %) predicted the viscosity of "PEO-500" as a function of temperature.[38]

4. MOLECULAR MODELING OF POLY(ETHYLENE OXIDE)/$LiBF_4$ POLYMER ELECTROLYTES

4.1 Previous Studies of Polymer Electrolytes

Solid polymer electrolytes (SPEs) are ionically conducting, solvent-free materials that are usually composed of alkali salts dissolved in a polymer matrix. They combine an ease of fabrication, good mechanical and electrochemical stability, and low flammability and toxicity with the ability to form good interfacial contact with electrodes.[11,89] These factors position SPEs among the promising candidates for use in secondary lithium batteries for automotive, aerospace, and portable electronics applications.

Molecular dynamics simulations are well suited for studying structure on the nanometer length scale and dynamics on the scale from subpicoseconds to tens of nanoseconds, positioning them as ideal candidates for the study of cation environment and transport in SPEs. The first comprehensive MD simulations study of PEO/LiI[31,32] indicated that PEO acted as a polydentate ligand, wrapping around the Li^+ cation and allowing Li^+ transport along the PEO segments, hopping from one segment to another, and associating-dissociating with anions. The influence of the addition of lithium-salt on the PEO conformations and dynamics was also thoroughly discussed.[31,32] Despite the insight provided by the previous studies,[31,32] the dominant mechanism for cation transport in high-molecular-weight polymer electrolytes is not completely resolved.

Little attention was also given to exploration of the ability of MD simulations with the "off-the-shelf" force field to predict structural and dynamic properties. Moreover, when the "off the shelf" potentials were used for the PEO/LiI simulations, the ions did not move on the nanosecond time scale, forcing authors to made ad hoc modifications of the potential.[31,32] The MD simulations of NaI[90] with the force field parameters tuned to reproduce PEO and PEO/NaI crystal structures yielded qualitatively similar information about the SPEs structure and dynamics to those found for the PEO/LiI SPEs, [31,32] A modified all-atom potential for PEO/NaI was used in simulations by the de Leeuw group,[91] leading to a phase separation and forcing the authors to reduce ion charges by one-half to facilitate ion motion. Halley's group[92] investigated the temperature dependence of ion aggregation in SPEs, employing the two-body quantum chemistry-based force field,[93] whereas the Wheeler group performed comprehensive MD simulations, combined with the spectroscopic measurements study of the tetraglyme/$LiCF_3SO_3$ structure.[33] Their simulations with the AMBER 5 two-body force field predicted local anion environments in good agreement with the spectroscopic measurements if the Li^+-triflate coordination cutoff was set to 2 Å; however, no ion transport properties have been investigated.

Our previous MD simulations,[94] with the quantum chemistry-based force field[95] for short PEO chains (M_w=530) that were doped with LiI predicted the Li^+ environment to be similar to that for crown-ether, with 3-6 ether oxygen atoms coordinated around a Li^+ cation, which is in good agreement with neutron diffraction with isotopic substitution (NDIS) experiments, validating the ability of the force field to predict Li^+ complexation accurately.[62] Because of low PEO molecular weight (M_w=530), the dominant mechanism of cation diffusion in that system was the Li^+ cation motion together with PEO oligomers; however, the Li^+ cation motion along the chains with occasional interchain hopping was also observed in agreement with the Li^+ transport mechanism seen for the high-molecular-weight PEO/LiI.[31,32] The structural and dynamic properties of PEO away from the first coordination shell of Li^+ were found to be similar to those in bulk PEO, whereas the PEO segments in the Li^+ first coordination shell exhibited conformational populations and dynamics different from those of the bulk PEO that were only slightly dependent on salt concentration, allowing one to view polymer electrolyte as a composite of pure PEO-like domains and PEO-salt-rich domains. Similar features were observed in MD simulations of PEO(M_w=530)/$LiPF_6$.[2]

On the basis of the previous MD simulations and experimental studies of SPEs, we conclude that the Li^+ cation environment is understood rather well, whereas the dominant cation diffusion mechanism in high-molecular-weight SPEs is not completely resolved and must be studied further. The majority of the existing force fields for PEO-based SPEs are quite often, at best, qualitative[31,32,93] casting doubt on the validity of the results and indicating no predictive capabilities. Moreover, many groups used the two-body nonpolarizable potentials[30-33] for simulations of polymer electrolytes, despite the polarizability being acknowledged to be important for accurate prediction of the structure and dynamics.[96] Attempts to include polarizability in a form of the two-body effective polarizable force field were rather successful for prediction of the cation environment; however, the ion dynamics and conductivity were found to be up to an order of magnitude slower in comparison with experiments.[2,94] A caveat of the previously used methodology for developing the two-body effective polarizable potentials[95] is that the condensed phase effects are estimated on the basis of the gas-phase calculations of small Li^+/ether clusters without any anions present; this is a situation occurring only for "free" Li^+ cation, which constitute only 10-60 % of the total number of cations. Usage of the many-body polarizable force fields, on the other hand, is the most rigorous and elegant approach to the problem of accurate force field development for SPEs, because it naturally considers the many-body nature of the polarizable interactions that account for most of the condensed-phase effects in SPEs.

In this section, we develop the many-body polarizable and two-body force fields for PEO/$LiBF_4$ polymer electrolytes. We begin with derivation of the quantum chemistry-based force field with many-body polarizable and two-body effective polarizability terms for Li^+/BF_4^-/Ethers clusters, then, the MD simulation methodology is presented, followed by the results of MD simulations of PEO/$LiBF_4$ SPEs. At the end of the section, the cation transport mechanism in the PEO/$LiBF_4$ SPE's is analyzed.

4.2 PEO/Li^+ Complexation Energies

We begin development of the Li^+/ether force field by establishing the adequate levels of theory for calculation of the Li^+/ether complexation energy, defined as total energy of the complex minus energy of reactants at optimized geometries. Table 13 shows that total complexation energies of -38.1 and -38.5 kcal/mol for Li^+/ether at the MP2(full) level using aug-cc-pvDz and aug-cc-pvTz basis sets for ether and [5s3p2d][3] basis set for Li agree nicely with the previously reported accurate G2(MP2,SVP) value of -37.9 kcal/mol.[97] B3LYP and HF calculations using the aug-cc-pvDz basis set predict slightly higher (by 1-1.5 kcal/mol) complexation energy for the Li^+/ether complex.

Table 13 Total BSSE corrected complexation energy (E^{tot}), BSSE corrected nonbonded part of complexation energy ($E^{N\text{-}B}=E^{tot}-E^{dist}$) and distortion energy (E^{dist}) for Li^+/ether and Li^+/EO_4 complexes. Reproduced with permission from Ref. 98.

level of theory/basis set for geometry optimization	level of theory/basis set for energy calculation	E^{tot} (kcal/mol)	$E^{N\text{-}B}$ (kcal/mol)	BSSE (kcal/mol)	E^{dist} (kcal/mol)
Li^+/ether					
b3lyp/Dz	b3lyp/Dz	-39.74	-40.79	0.15	1.05
b3lyp/Dz	mp2(full)/Dz	-38.08	-38.91	0.84	0.84
mp2(full)/Dz	mp2(full)/Dz	-38.11	-39.05	0.84	0.95
mp2(full)/Dz	mp2(full)/Tz	-38.47	-39.87	0.51	1.40
b3lyp/Dz	HF/Dz	-39.41	-41.37	0.10	1.96
Li^+/EO_4					
b3lyp/Dz	b3lyp/Dz	-96.7		0.7	
b3lyp/Dz	mp2(full)/Dz	-96.5		2.1	9.4
b3lyp/Dz	HF/Dz	-94.3		0.5	17.8

[a] aug-cc-pvDz basis set is denoted as Dz, aug-cc-pvTz basis set is denoted as Tz, Li basis set [5s2p2d] was is all calculations used

The B3LYP/aug-cc-pvDz level calculations yield better agreement with the MP2/aug-cc-pvTz level calculations than the HF/aug-cc-pvDz level calculations for nonbonded contribution to the complexation energy $U^{N\text{-}B}$, therefore, we will use the B3LYP/aug-cc-pvDz level for further calculations of $U^{N\text{-}B}$.

The repulsion contribution to the Li^+ complexation energies can be determined from quantum chemistry calculations by subtracting the energy of the (charge 1e)/molecule complex from the energies of the Li^+/molecule complex at the same geometry. The repulsion energy from the B3LYP/(aug-cc-pvDz for PEO, and [5s3p2d] for Li^+) calculations along 4 representative paths shown in Figure 1 are presented in Figure 12. The repulsion A and B parameters were determined by fitting the quantum chemistry repulsion energies and are shown in Table 14. The resulting fits, which are shown in Figure 12, are good.

Table 14. Repulsion (A, B) and dispersion (C) parameters for the Li^+/BF_4^-, Li^+/PEO, BF_4^+/PEO, and BF_4^+/BF_4^+ force fields. Reproduced with permission from Ref. 98.

Atom pair	A (kcal mol^{-1})	B (Å^{-1})	C (kcal mol^{-1} Å^{-6})	D (kcal mol^{-1} Å^{-4})
Li-Li	44195[a]	7.277[a]	0.0	0.0
Li-B	1257	2.441	0.0	0.0
Li-F	18636	4.082	0.0	0.0 (50)
Li-O	54403	4.545	0.0	100 (120)
Li-C	3581	2.784	0.0	50 (60)
Li-H	3610.0[c]		0.0	5 (6)
B-B	14976[b]	3.090[b]	1150.0[g]	0
B-F	33702	3.577	487.9	0
B-O	32977	4.323	446.8	0
B-C	41335	2.954	891.5	0
B-H	21690	3.244	177.4	0
F-F	66641[e]	4.301[e]	93.5	0
F-O	19294[d]	3.474	119.0	0
F-C	19294[e]	3.431[e]	237.5	0
F-H	12300[f]	4.143[f]	53.9	0

[a] from ref.[95]

b assumed similar to C-C, the actual value if not significant was B-B close separations are highly improbable

[c] the A/r^{12} form of the potential was used for Li-H interaction

[d] the A(F-O) coefficient was assumed similar to the A(F-C) parameter

[e] from fits to the CF_4 dimer at HF/aug-cc-pvDz level

[f] from the PFDV force field Byutner A.; Smith, G. D.; Jaffe, R.L *Macromolecules* **2000**, *33*, 4264.

[g] ionization potential and polarizability for B^+ were used

The nonbonded contribution to the complexation energy in our force field is approximated by the Coulomb, many-body polarizability, and repulsion terms. Because of an insignificant Li^+ polarizability, the Li^+ dispersion parameters for interactions with PEO carbon, oxygen and hydrogen atoms are set to zero. A negligible Li^+ polarizability also indicates that all the many-body polarization energy comes from Li^+ polarizing the PEO and almost none from PEO polarizing the Li^+. The Coulomb contribution to the DME/Li^+ and DE/Li^+ complexation energy is completely determined by the partial charges obtained in the previous section (i.e. FF-1 and FF-3 force fields) and the Li^+ charge of 1e. The many-body polarizable contributions to the complexation energy are completely determined by the PEO polarizabilities as described in Section 3.

We test the ability of the PEO force fields FF-1 and FF-3 developed in the previous section and the Li^+/PEO repulsion parameters developed here to predict the quantum chemistry complexation energy for the Li^+/(triglyme= EO_4) complex reported in Table 13. The FF-1 force field predicts a total complexation energy of Li^+/EO_4 of -84.6 kcal/mol, whereas the FF-3 force field yields a higher complexation energy of -91.5 kcal/mol. The FF-1 force field significantly (by 11.9 kcal/mol) underestimates quantum chemistry complexation energy of the Li^+/EO_4 complex (-96.5 kcal/mol at the MP2/aug-cc-pvDz level), whereas the FF-3 force field does a much better job in predicting the Li^+/EO_4 complex energy underestimating it by 5 kcal/mol. These results are consistent with the more significant underestimation of the electrostatic potential in the proximity of ether oxygen by the FF-1 force field than by the FF-3 force field shown in Figure 2b.

The Li^+/BF_4^- and BF_4^-/PEO force fields were derived similarly to the PEO/Li^+ force field, their development is described in details in ref.[98].

We performed the simulations of PEO/$LiBF_4$, EO:Li=15:1 at 393 K using FF-1 and FF-3 PEO force fields to compare the effect of underestimating PEO/Li^+ binding energy by PEO FF-1 compared to the FF-3 PEO force field. We found that the fraction of "free", e.g. not complexed by BF_4^-, Li^+ cation is below 20 % for the FF-1 PEO force field, whereas the fraction of "free" Li^+ cations for the FF-3 PEO force field is in the range of 50% to 80%. Because the FF-3 force field yields a more accurate description of the PEO/Li^+ binding energies, it is used in the investigation of the PEO/$LiBF_4$ solutions from here on. However, we would like to note that the reduction of the PEO/Li^+ repulsion energies below the quantum chemistry values shown in Figure 12, would allow one to increase the PEO/Li^+ binding energies for the FF-1 PEO force field, thus making PEO(FF-1)/$LiBF_4$ force field more suitable for simulations of polymer electrolytes.

4.3. Two-Body Force Field for PEO/$LiBF_4$

As MD simulations with the many-body polarizable terms are approximately 2-4 times more expensive compared to the MD simulations of identical systems with force field containing only two-body interactions, it is desirable to develop a two-body force field for PEO/$LiBF_4$ interactions that takes into account the many-body polarizable effects in a "mean-field" sense. Considering only the lowest-order term in r_{ij} in eq. 3.3 – 3.4, the polarization energy between two charged atomic centers with dipolar polarizability is given by eq. 4.1

$$U^{POL}(r_{ij}) = -332.07\,[q_i^2\,\alpha_j + q_j^2\,\alpha_i]/2r_{ij}^4 = -D_{ij}/r_{ij}^4\,, \tag{4.1}$$

where α_i are α_j the atomic dipole polarizabilities. This expression for U^{POL} neglects many-body interactions.

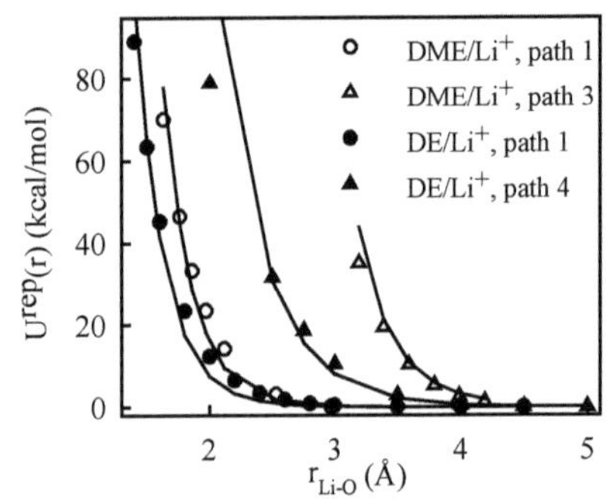

Figure 12. The repulsion contribution to the total binding energy of dimethyl ether(DE) with Li^+, and 1,2-dimethoxyethane (DME) with Li^+ along the paths from Figure 1 using B3LYP/aug-cc-pvDz quantum chemistry calculations and from the force field.

In our previous work on PEO/LiI[95] and PEO/$LiPF_6$,[2] we included the effective two-body polarization terms in the force field using 4.1. We used D-parameters obtained from the fits to the ab initio quantum chemistry calculations of the Li/ether energies that were uniformly scaled by a factor of 0.71 in order to account for the condensed-phase effects. This factor 0.71 was obtained from a detailed investigation of the incremental Li^+/(ether)$_n$ complex energies, n=1,2,3,4. It was found that the polarization energy of the Li^+/(ether)$_4$ representing the Li^+ coordination in a condensed phase is described well by two-body polarization terms (eq. 4.1) parametrized for the Li^+/single ether complex and then scaled by a factor 0.71 in order to obtain good agreement with quantum chemistry energies for the Li^+/(ether)$_4$ complex. In this work we take a different and more rigorous approach for obtaining D-parameters. First, we perform MD-simulations of PEO/$LiBF_4$ solutions at EO:Li=15:1 concentration and 393 K using the many-body polarizable force field. Then, the induced dipole moments and charges from these simulations will be used to estimate the polarizable energy between the induced dipole moment and the Li^+ cation as a function of separation allowing us to estimate the D-parameters from eq. 4.1. As D_{ij}-parameters were found to be dependent on the distance between the atoms i and j, D_{ij}-parameter at the first peak of the radial distribution function between atoms i and j are taken. Following our previous work,[2] the D_{ij}-parameters are scaled to zero beyond the first coordination shell of Li^+ (3.5 Å) using distance dependent dielectric constant as the PEO and BF_4^- outside the Li^+ coordination shell are likely to be in the first coordination shell of the other Li^+ cations and be strongly polarized by them.

The initial MD simulations of the PEO/$LiBF_4$ solutions, EO:Li=15:1 at 393 K with the two-body potential parametrized as described above indicated that these simulations yield ion dynamics slower and lower fraction of "free" Li^+ cations than those from the MD simulations with the many-body polarizable force field. We subsequently set the $D_{Li\text{-}F}$ parameter to zero and decreased the other D-parameters by about 20 % in order to decrease the strength of interaction between Li^+ with PEO and BF_4^- thus increasing the ion dynamics and improving the agreement for the fractions of "free" Li^+ from MD simulations with the MB and TB potentials within the simulation errors.

4.4 Molecular Dynamics Simulations of PEO/$LiBF_4$ Solutions

4.4.1 Simulated Systems

MD simulations were performed on the PEO(M_w=2380)/$LiBF_4$ solutions for the ratio of ether oxygen to Li of 15:1, (EO:Li=15:1) using the PEO FF-3 many-body polarizable (MB) and the corresponding PEO two-body force field with the effective two-body polarizability (TB). The simulation box consisted of 10 PEO chains and 36 anions and 36 cations. MD simulations methodology used as described in Section 3.4.1 Temperature dependence of polymer electrolyte dynamic properties is investigated by performing simulations at 363 K, 393 K, and 423 K for both two-body and many-body

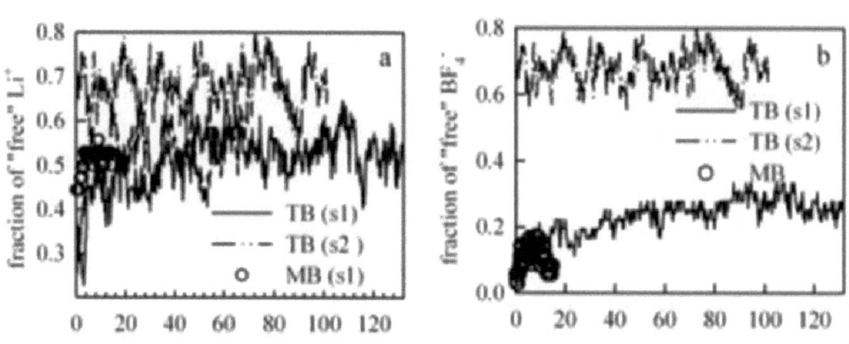

Figure 13. Fraction of "free" Li^+ and "free" BF_4^- during the simulation run for PEO/$LiBF_4$, EO:Li=15:1, 393 K for M-B and T-B force fields.

force fields as described in Ref. 98.

4.4.2 Characteristic Time Scales for Dynamics in Polymer Electrolytes

The dynamics of polymer electrolytes are complex, because of the presence of many characteristic time scales that are dependent on the polymer molecular weight, nature and concentration of salt, temperature and barriers of polymer conformational transitions. For example, in our PEO(M_w=2380)/$LiBF_4$ (EO:Li=15:1, 393 K) the torsional transitions occur on the $\approx 10^{-11}$–10^{-10}sec time scale characterizing local polymer dynamics, while it takes PEO approximately 10^{-7} sec to diffuse its own size, i.e., a radius of gyration (≈14 Å). The characteristic time scale for the Li^+ cation and the BF_4^- anion to diffuse a distance equal to the radius of gyration of a PEO chain is of the order of tens of nanoseconds ($\sim 10^{-8}$-10^{-7} sec). The anion-cation residence time and the cation-EO residence times are on the order of 10^{-8} sec. Analysis of these time scales indicates that the system needs to be equilibrated for a long time, i.e., 10^{-7} sec, in order for PEO to reach equilibrium global conformations, for ions to diffuse a distance of the order of the PEO chain length, and for every cation to associate and dissociate with any anion at least once, on average.

Long equilibration time scales pose challenges for obtaining equilibrium structural and dynamic properties. This issue has not been adequately considered in the previous MD simulations of polymer electrolytes probably due to lack of CPU time required for the investigation of system equilibration. In this contribution we explore the timescales for a polymer electrolyte equilibration by creating two systems of PEO(M_w=2380)/$LiBF_4$ (EO:Li=15:1, 393 K) with their initial configurations corresponding to different ion aggregation states: *system 1* has the majority of ions (>80 %) being a part of ion aggregates or ion pairs; *system 2* has the majority of ions (>80 %) existing as free ions (i.e. ions not having any counterions in its first coordination shell of 4 Å) and little ion aggregates. It is expected that the equilibrium state of ion aggregation will be somewhere in between the initial configurations of systems 1 and 2 allows us to estimate lower and upper boundaries for the structural and dynamic properties.

We begin investigation of equilibration timescales using the TB force fields (FF) as they are computationally 3-4 times less expensive. We performed the 130 ns, and 100 ns NVT runs for systems 1,2 at the box size corresponding to the atmospheric pressure. The fraction of free cation and anions verses time was monitored during these simulations, and is shown in Figure 13, allowing us gauge changes in ion aggregation with time for each system. Figure 13a demonstrates that the fraction of free cations in the first system dramatically increases over the first 10 ns from ≈30% to ≈50% showing only fluctuations around 50-60% for the next 120 ns, whereas the second system has the number of free Li^+ fluctuating in the range of 65%-80% during the 100 ns run.

Monitoring the number of free BF_4^- anions vs. time showed a similar tendency as observed for the behavior of free Li^+. It is interesting that even after 130 ns and 100 ns runs for systems 1 and 2, respectively, the number of free cation and free anion are still different between the systems by 20 % and 40 %, respectively, which indicates that even long 100 ns simulations allow us to obtain fractions of free ions with accuracies of 20% and 40 % for free cations and anions. Use of parallel tempering algorithms is a promising route for acceleration of system equilibration[99] and should be explored for future equilibration of polymer electrolytes.

In order to further explore the evolution of ion aggregation state, the

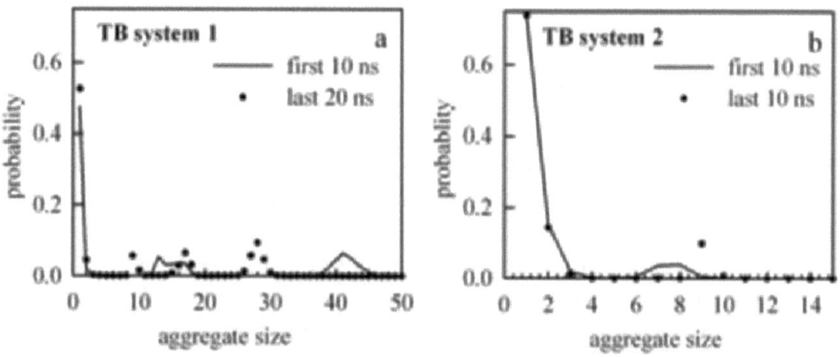

Figure 14. Probability of a ion to belong to the ion aggregate of a particular size in PEO/$LiBF_4$, EO:Li=15:1, T=393 K in the beginning and the end of the simulations for system 1 that has been simulated for 130 ns and system 2 that has been simulated for 100 ns using TB force fields (PEO FF-3 force field).

probability of ions to belong to a cluster of a particular size at the beginning and the end of the simulations is analyzed and presented in Figure 14. The size of largest ion aggregates for the first system noticeably decreased from ≈40 to ≈30 over the 100 ns simulation, whereas a slight increase of the ion aggregate size is observed for the second system over the 80 ns simulation, which was initially set up with ion being the primarily dissociated. A slow decrease of the size of ion aggregates for system 1 and slow increase for system 2 during MD simulations indicates that sub-microsecond simulations might be required to accurately calculate the distribution of sizes of ion aggregates from MD simulations if the systems are set up far from equilibrium conditions and exhibit extensive aggregation. Slow equilibration of PEO/$LiBF_4$ solutions, observed in this work, questions whether the sub-nanosecond simulations of the tetraglyme/Li-triflate[33] solutions were long enough to obtain the equilibrium ion aggregation states.

Fractions of free ions from MD simulations of PEO/$LiBF_4$ with the MB force field are also shown in Figure 13. This system was created from TB system 1 after 5 ns of equilibration. The time dependence of the fraction of

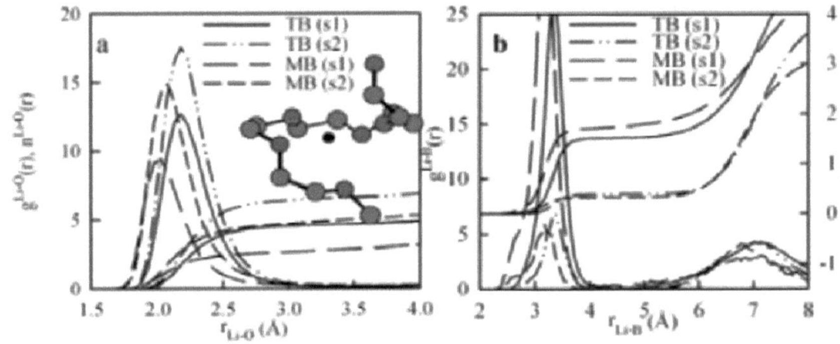

Figure 15. Radial distribution functions and coordination numbers for PEO/$LiBF_4$, EO:Li=15:1 at 393 K. A typical Li^+ coordination by a PEO segment is also shown

free cations and anions for the MB system is similar to that for the TB system 1 indicating similarity of equilibration timescales for the MB and TB force fields.

The differences in ion aggregation shown in Figure 13 and 14 are likely to result in some differences in ion environment and ion transport. Therefore, we perform analysis of the structural and dynamic properties for the both systems 1 and 2 to estimate the effect of initial conditions and ion aggregation on cation environment, ion self-diffusion coefficient and conductivity.

4.4.3 *Structural Properties of PEO/$LiBF_4$*

The structure of the Li^+ environment is of great interest as it is expected to be intimately related to the Li^+ transport mechanism. The last 5 ns of simulations employing the two-body force field and the last 0.5 ns of simulations employing the many-body force field are included in the analysis of the structural properties. We begin analysis of the Li^+ environment by calculating the radial distribution functions (RDF, $g^{Li\text{-}X}(r)$) and distance dependent coordination numbers ($n^{Li\text{-}X}(r)$), shown in Figure 15, for the MD simulations with the MB and TB force fields for systems 1 and 2. Position of the first peak of the Li-O RDF is about 2.02-2.05 Å for the MB force fields and 2.15-2.20 Å for the TB force fields. These most probable distances from MD simulations are similar to the Li-O most probable distance of 2.07-2.1 Å obtained from analysis of the neutron diffraction with isotopic substitution experiments on PEO/LiI and PEO/LiTFSI[100] indicating that both force fields (MB and TB) reasonably reproduce the structure of the Li^+-PEO complexation. The Li-O RDF's for system 2 are systematically higher than those for system 1, because system 2 has a higher fraction of free cation than system 1 allowing more ether oxygen atoms to enter the first coordination shell of Li^+ as observed from $n^{Li\text{-}O}(r)$ also presented in Figure 15. We define ether oxygen atoms being complexed by a Li^+ if they are within 3 Å of a Li^+. Figure 15 demonstrates that the number of ether oxygen atoms complexed by a Li^+ varies significantly (from 2.7 to 6.4) depending on the force field type and system aggregation state. The number of complexed ether oxygen (EO) atoms is higher by about 2 EO for system 1 than for system 2 for both force fields, indicating that despite our long MD simulations we can predict the average number of complexed EO with the accuracy of 2 EO atoms, depending on the initial configuration we choose. The MB force fields predict approximately 2 EO fewer in the first coordination shell of a Li^+ than the MB force fields, because of the shorter Li-O distances for the MB force field than the TB force field resulting in a more compact Li^+ cation first coordination shell for the MB force field.

Analysis of the Li-B RDF and coordination number shown in Figure 15b indicates a strong preference of an anion to be in the first coordination shell of a Li^+ for system 1, the system with the larger fraction of ion aggregates, and a much weaker Li-B affinity for system 2. Coordination numbers of 1.7 (MB system 1), 0.3 (MB system 2) and 1.5 (TB system 1), 0.4 (TB system 2) boron atoms of BF_4^- anions within 4 Å coordination sphere of a Li^+ were found on average. The TB and MB force fields yield a very similar number of BF_4^- anions in the first coordination shell of a Li^+, whereas the greater number of anions is observed for the system that started in the aggregated state (system 1), compared to the system started with the majority of ions dissociated (system 2), again demonstrating property dependence on initial conditions. However, a more detailed analysis of the Li^+ cation coordination revealed a strong preference for the Li^+ cation to be coordinated by 5 EO from the same chain or 4 BF_4^- anion is found regardless of the force field, MB or TB.[98] Thus, a somewhat simplified picture of a mixture of $Li^+(EO)_5$ and $Li^+(BF_4^-)_4$ clusters determining the average configurations shown in Figure 15 is suggested.

The Li^+ complexation is also found to significantly perturb the PEO local and global conformations resulting in an increase of the fraction of the *tgt* and *tgg* conformers of -O-C-C-O- sequences and a decrease in the fraction of *ttt*, tg^+g^- and *ttg* dihedrals. The PEO radius of gyration decreased, as $LiBF_4$ is added Li^+ is typically wrapped around by a PEO segment as shown in Figure 15. These structural changes are in accord with the experimental data and are independent of the choice of the force field, i.e. MB or TB, or initial configuration of the system. Similar findings were reported for our previous PEO/LiI simulations[62] and PEO/LiTriflate simulations.[33]

4.4.4 *Effect of Li^+ on PEO Dynamics*

Quasielastic neutron scattering experiments have shown that the addition of salts slows down polymer dynamics.[91] NMR ^{13}C spin-lattice relaxation experiments also probe polymer dynamics. We have performed the NMR ^{13}C spin-lattice experiments on PEO(Mw=1854) obtained from Sigma-Aldrich with $LiBF_4$ salt EO:Li=15:1 and without salt at 393 K in order to compare with simulation results.[98] The ratio of T_1(pure PEO)/T_1(PEO/$LiBF_4$)=2.9 was found in the NMR experiments performed according to the methodology described elsewhere.[98] The T_1 spin-lattice relaxation times were also calculated from the MD simulations as described in Section 3 and Appendix A. The decay from 1 to 0.005 is fitted to the P_2(C-H) C-H vector autocorrelation function for the interior carbons. The T_1(pure PEO)/T_1(PEO/$LiBF_4$) ratios calculated from MD simulations are: 2.8 (MB system 1) and 4.0 (TB system 2), 3.2 (TB system 1). The T_1(pure PEO)/T_1(PEO/$LiBF_4$) ratio from MD simulations with the MB force field is in the best agreement with experiments, whereas the TB force field predicts the ratio of T_1(pure PEO)/T_1(PEO/$LiBF_4$) that is higher than the experimental one, thus indicating that the TB force field predicts a higher extent of PEO slowing down with addition of salt than the experiments and, therefore, slower PEO dynamics than that observed experimentally. These results are consistent with the greater number of EO coordinated by Li^+ from the TB force field, compared to the MB force field as greater number of EO from the TB force field makes Li^+ motion along the PEO chain slower compared to that for the MB force field.

4.4.5 *Transport Properties of PEO/$LiBF_4$*

Ion transport properties such as cation and anion self-diffusion coefficient and conductivity are of most technological importance. Ion self-diffusion coefficients and conductivity are calculated from MD simulations as described in Appendix A.

Analysis of the fraction of "free" ether oxygen and BF_4^- and is presented in Figure 13 and 14. These figures highlight the dependence of the Li^+ environment on initial configuration of the system. The difference in the state of ion aggregation is expected to translate into the difference in ion transport properties between system 1 and system 2. Table 15 presents the Li^+ and BF_4^- mean-square displacement from MD simulations employing the MB and TB force fields for systems 1 and 2. Ion diffusion coefficients from the MB force field are 6-10 times higher than those for the TB force field indicating that our two-body approximation predicts slower ion dynamics because of the larger number of EO wrapping around a Li^+ and impeding its ability to move along a PEO chain.

Table 15. Ion self-diffusion coefficients (D) and conductivities (λ) for PEO/$LiBF_4$, EO:Li=15:1 at 393 K using MB and TB force fields for systems 1 and 2 (denoted s1 and s2).

System	Trajectory length used for analysis (ns)	D(Li^+) (10^{-11} m^2/sec)	D(BF_4^-) (10^{-11} m^2/sec)	D (10^{-3} S/m)
System 1, T-B	70	0.44	0.98	0.33
System 2, T-B	40	0.71	1.11	0.61
System 1, M-B	8	4.1	8.3	3.1

The Li^+ and BF_4^- self-diffusion coefficients and the conductivity of the systems also exhibited some dependence of the system initial conditions as observed from the comparison of the transport properties for the aggregated system 1 with the system 2, which consisted primarily of free ions. The difference of the ion transport properties between system 1 and 2 is less than 50%, which indicates that ion transport properties can be predicted from MD simulations with errors of less then 50% despite uncertainties in prediction

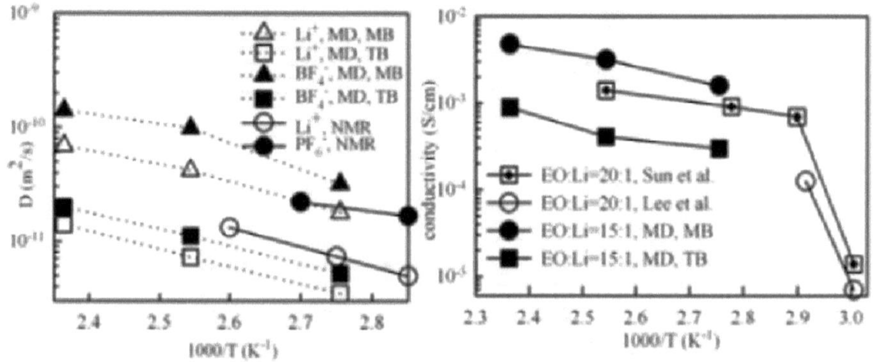

Figure 16. Self-diffusion coefficients of Li^+, BF_4^- from MD simulations and Li^+, and PF_6^- from NMR measurements on the right[101] The scaling factor of 1.3 is based on molecular weight dependence of Li in PEO/$LiCF_3SO_3$, EO:Li=20:1 at 90 °C measured by NMR.[102] Conductivity from MD simulations and experiments are shown on the left.

of the fraction of free ions of 20%-30%. Because of a relatively small difference for ion transport properties between systems 1 and 2, we perform investigation of the temperature dependence of ion transport properties and study mechanism of ion transport only for the system 1 as the longest trajectories have been generated for this system.

The ion self-diffusion coefficients from MD simulations as a function of temperature are compared in Figure 16 with the results of NMR PFG[101] measurements of high molecular weight oxymethylene-linked PEO($M_w=10^5$)/$LiPF_6$ as the ion self-diffusion coefficients are expected to be similar for PEO/$LiBF_4$ and PEO/$LiPF_6$ and no experimental data are available for PEO/$LiBF_4$. Because the cation self-diffusion coefficient is known to be dependent on the PEO molecular weight,[102] we used a factor of

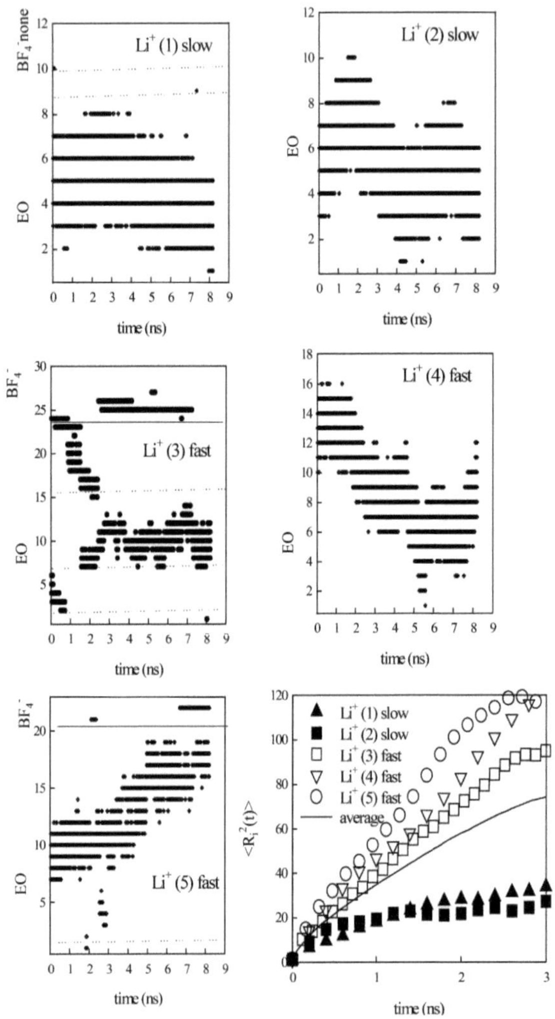

Figure 17 The time dependence of the Li^+ cation environment and cation mean-squared displacements $<R^2_i(t)>$ for PEO/$LiBF_4$ at 393 K with the MB force field. Reproduced with permission from Ref. 98.

1.3 based upon experimental investigation of Li^+ self-diffusion coefficient of PEO molecular weight to account for the difference between the Li^+ diffusion in PEO(Mw=10^5) and PEO(M_W=2380). The scaled Li^+ self-diffusion coefficients from NMR measurements are approximately 2-3 times lower than those from MD simulations with the MB force field and about 3 times higher than that from the MD simulations with the TB force field. A similar quality agreement between the simulation and the experiments[103,104] is seen for the anion. The conductivities from MD simulations with the MB force field are in better agreement with the scaled experimental data[105,106] than those for the TB force field as shown in Figure 16

4.4.6 Li^+ Transport Mechanism

Having validated the ability of the force fields to reasonably represent the structural and dynamic properties of the PEO/$LiBF_4$ polymer electrolytes, we are in a position to tackle the controversial issue of the Li^+ transport mechanism in polymer electrolytes. The possible cation transport mechanism might include: Li^+ motion along PEO chains, Li^+ hopping from one PEO chain to the other, Li^+ diffusion with the PEO chain is PEO molecular weight is below entanglement MW, Li^+ diffusion within ion aggregates. Extensive MD simulations of PEO/LiI[31,32] indicated that all these mechanisms are present. In our MD simulations, we also observe all of these mechanisms of Li^+ diffusion. Naturally the question arises as to what is the dominant and the most important mechanism(s) of Li^+ diffusion?

Following Müller-Plathe,[31,32] we begin our investigation of Li^+ transport mechanism by monitoring changes in the Li^+ environment with time. If a Li^+ has an EO or BF_4^- within 4 Å for 20 ps a continuously filled circle is associated with the EO or BF_4^- index number in Figure 17 for that time. The PEO segments of continuous ether oxygen atoms are separated by dotted lines, whereas the blocks with the BF_4^- anions are separated from the ether oxygen atoms by the solid lines. Figure 17 shows the representative Li^+ motion types: the Li^+(1), Li^+(2), and Li^+(4) cations are complexed primarily by one segment of consecutive ether oxygen atoms. Complexation with one segment is found for more than 90 % of Li^+ cations. The Li^+(1) and Li^+(2) represent Li^+ cations, which do not undergo any significant motion along the PEO chain and, therefore, are moving only with the PEO chains; whereas the Li^+(4) represents the Li^+ cations moving along the PEO chain and not being complexed by the BF_4^- anions during the MD simulations. The Li^+(3) cation represents Li^+ undergoing association-dissociation with different PEO chains and the BF_4^- anions. The Li^+(5) cation represents Li^+ cations moving primarily along the PEO chain, which occasionally associate and dissociate with BF_4^- anions.

Figure 17 also shows individual mean-square displacements for the same five Li^+ (1-5) cations. The Li^+(1) and Li^+(2) cation, which do not move along the chain, have the mean-squared displacements significantly below the average, whereas the Li^+ cations (3-4) undergoing significant displacements along the chain or interchain hopping have mean-squared displacements significantly higher than the average. Similar behavior is seen for the other Li^+ cations not shown in Figure 17 as well as for the Li^+ cations in MD simulations with the TB force field. We have found that the Li^+ self-diffusion coefficient of Li^+ cations moving along the PEO chain is about 5 times (MB) and 2-3 times (TB) faster than the Li^+ self-diffusion coefficient of Li^+ cations undergoing insignificant movement along a PEO chain, allowing us to conclude that Li^+ motion along PEO chains makes an important contribution to the Li^+ transport in PEO/$LiBF_4$.

5. CONCLUSIONS

We have presented a methodology for development of the many-body polarizable and effective mean field type two-body force fields for simulations of polymers and polymer electrolytes using PEO and PEO/$LiBF_4$ as examples. MD simulations of PEO, its oligomers, and PEO/$LiBF_4$ yielded thermodynamic, structural and dynamics properties in good agreement with experiments. Inclusion of the many-body polarizable terms into force field was important for accurate prediction of polymer electrolyte transport and structural properties, whereas the many-body polarization terms are not needed for accurately prediction of structural and dynamic properties of PEO melts.

Slow association-dissociation of ion aggregates in PEO/$LiBF_4$ solutions at 393 K necessitated long equilibration times (~100 ns) in order to approach equilibrium ion aggregation states. MD simulations suggest the following picture of PEO/$LiBF^4$ polymer electrolytes at high temperatures: the Li^+ cations are wrapped around primarily by one PEO segment with ≈5-6 ether oxygen atoms. From 50% to 80 % of the Li^+ cations also have BF_4^- anion in its environment. The main contribution to the "free" Li^+ diffusion coefficient comes from the Li^+ motion along the PEO chains.

ACKNOWLEDGMENT

The authors are indebted to NASA grant NAG3 2624 and a subcontract from LBL subcontract #6515401 for financial support. We also would like to thank our collaborators John Kerr, David Price, Marie-Louise Saboungi, Larry Curtiss, Dmitry Bedrov, Sergio Petriccu, Frans Trouw, Oleksey Byutner, Richard Jaffe, and Michel Armand.

APPENDIX A

1. Enthalpy of Vaporization. Heats of vaporization (ΔH) is calculated using eq. a1

$$\Delta H = U^{liq} - U^{vap} + RT \qquad \text{(a1)}$$

where U^{liq} and U^{vap} are the energies of liquid and gas phases, R is the gas constant, and T is temperature. The gas-phase energy is calculated in MD simulations using Brownian dynamics and no nonbonded interaction between molecules.

2. Static Structure Factor. The static structure factor is calculated using eq. a2

$$S(Q)=1+\frac{1}{<b>^2} n \sum_{\alpha\beta} x_\alpha b_\alpha x_\beta b_\beta \int_0^{r_c} [g_{\alpha\beta}(r) -1] \frac{\sin Q r}{Q r} 4 \pi r^2 dr \qquad \text{(a2)}$$

where $<b>^2 \equiv \sum_\alpha x_\alpha x_\beta b_\alpha b_\beta$, n is a number density, $g_{\alpha\beta}(r)$ is the radial distribution function (RDF) for $\alpha\beta$ atom types, Q is the wave vector, b_α is the coherent scattering length for species α, x_α is a fraction of atom type α, and r_c is the cutoff for integration (15 Å).

3. Intermediate Incoherent Dynamic Structure Factor. For isotropic systems such as liquids, the neutron scattering incoherent intermediate scattering function (ISF), I(Q,t), is given by eq. a3[107]

$$I_{inc}(Q,t) = < \frac{\sin(\Delta r_i(t)\, Q)}{\Delta r_i(t)\, Q} > \qquad \text{(a3)}$$

where $\Delta r_i(t)$ is the displacement of atom i after time t, Q is the magnitude of the momentum transfer vector, and < > denotes an average over all time origins for atoms with a significant incoherent cross-section (i.e. hydrogen atoms).

4. Frequency-Dependent Dielectric Constant. Linear response theory allows us to obtain the complex dielectric permittivity $\varepsilon^*(\omega) = \varepsilon' + i\varepsilon''$ for the PEO melt using the relationship[108]

$$\frac{\varepsilon'+i\,\varepsilon''}{\Delta\varepsilon} = 1- i\,\omega \int_0^\infty \Phi(t)\exp(-i\,\omega t)dt, \qquad \text{(a4)}$$

where the dipole moment autocorrelation function (DACF) is given by

$$\Phi(t)=\frac{<\mathbf{M}(0)\bullet\mathbf{M}(t)>}{<\mathbf{M}(0)\bullet\mathbf{M}(0)>}, \qquad \text{(a5)}$$

where $\Delta\varepsilon$ relaxation strength is equal to $\varepsilon_r - \varepsilon_u$. Here, **M**(t) is the dipole moment of the system at time t, V is the volume of the system, k_B is the Boltzmann constant and T is the temperature, while <> denotes an ensemble average. The unrelaxed dielectric constant ε_u is the dielectric constant that includes all relaxation processes at frequencies higher than the process of interest, i.e., electronic polarization and relaxation due to vibrations and librations, whereas the relaxed dielectric constant ε_r is the value obtained after the relaxation process, i.e., dipole orientational relaxation, is complete.

5. NMR Spin-Lattice Relaxation Times T_1. ^{13}C NMR spin-lattice relaxation times (T_1) probe the rate of decay of the C-H vector which is related to polymer dynamics. The experimentally determined T_1 values are related to the microscopic motion of the C-H vectors through the relationships[109]

$$\frac{1}{nT_1} = K[J(\omega_H-\omega_C)+3J(\omega_C)+ 6J(\omega_H+\omega_C)] \qquad \text{(a6)}$$

where n is the number of attached protons, ω_C and ω_H are the Larmor (angular) frequencies of the ^{13}C and ^1H nuclei, respectively, while γ_C and γ_H are the corresponding gyromagnetic ratios. The constant K is given by[109]

$$K = \frac{\hbar^2 \mu_0^2 \gamma_H^2 \gamma_C^2 <r_{CH}^{-3}>^2}{160\pi^2} \qquad \text{(a7)}$$

where μ_0 is the permittivity of free space and r_{CH} is the carbon-hydrogen bond length. K assumes values of 2.29 X 10^9 s^{-2} for sp^3 hybridized nuclei.[109] The spectral density function $J(\omega)$ is given as[109]

$$J(\omega) = \frac{1}{2} \int_{-\infty}^{\infty} P_2^{CH}(t)e^{i\omega t}dt \qquad \text{(a8)}$$

where

$$P_2^{CH}(i,t) = \frac{1}{2}\{ 3<[\mathbf{e}_{CH}(i,t)\bullet \mathbf{e}_{CH}(i,0)]^2> - 1\} \qquad \text{(a9)}$$

Here, $\mathbf{e}_{CH}$ is a unit vector along a particular C-H bond, and the index i denotes differentiable resonances due to the local environment (methyl, α,β, interior carbons).

6. Self-Diffusion Coefficient. Self-diffusion coefficient D is calculated using Einstein relation,

$$D = \lim_{t \to \infty} \frac{\langle R^2(t)\rangle}{6t} \qquad \text{(a10)}$$

where $R^2(t)$ is mean-square displacement of a molecule center of mass during time t, and < > denotes an ensemble average.

7. Viscosity. The viscosity is calculated using the Einstein relation[110]

$$\eta = \lim_{t\to\infty} \frac{V}{20k_B Tt}\left(\left\langle \sum_{\alpha\beta} \left(L_{\alpha\beta}(t) - L_{\alpha\beta}(0)\right)^2 \right\rangle\right), \qquad \text{(a11)}$$

where $L_{\alpha\beta}(t) = \int_0^t P_{\alpha\beta}(t')dt'$, k_B is the Boltzmann constant, T is temperature, t is time, $P_{\alpha\beta}$ is the symmetrized stress sensor, and V is the volume of the simulation box.

8. Ionic Conductivity. The conductivity (λ) was calculated using Einstein relation

$$\lambda = \lim_{t\to\infty} \frac{e^2}{6\, t\, V\, k_B\, T\, N} \sum_{i\,j}^{N} z_i\, z_j < [\mathbf{R}_i(t)-\mathbf{R}_i(0)][\mathbf{R}_j(t)-\mathbf{R}_j(0)] > \qquad \text{(a12)}$$

where e is the electron charge, V is the volume of the simulation box, k_B is Boltzmann's constant, T is the temperature, t is time, z_i and z_j are the charges over ions i and j in electrons, $\mathbf{R}_i(t)$ is the displacement of the ion i during time t, the summation is performed over all ions, $<>$ denote the ensemble average, N is the total number of ions in the simulation box.

REFERENCES

1. Tuckerman, M. E.; Martyana, G. J. *J. Phys. Chem. B,* **104**, 159 (2000).
2. Borodin, O.; Smith, G. D.; Jaffe, R. L. *J. Comput. Chem.*, **22**, 641 (2001).
3. Smith, G. D.; Jaffe R. L.; Partridge, H. *J. Phys. Chem. A* **101**, 1705 (1997).
4. Cappello, B.; Del Nobile, M.A.; La Rotonda, M.I.; Mensitieri, G., Miro, A.; Nicolais, L. *Il Parmaco* **49**, 809 (1994).
5. Amiji, M.; Park, K. *Biomaterials*, , **13**, 682 (1992).
6. Jeon, S.I.; Lee, J.H.; Andrade, J.D.; de Gennes, P.G. *J. Colloid and Interface Science*, **142,** 129 (1991).
7. Harris, J.M. *Poly(Ethylene Glycol) Chemistry. Biotechnical and Biomedical Applications.* Plenum Press: New York and London, 1992.
8. Harris, J.*M*, Zalipsky, S. *Poly(Ethylene Glycol) : Chemistry and Biological Applications (ACS Symposium Series, No 680)* American Chemical Society, 1998.
9. Morra, M.; Occhiello, E.; Garbassi, F. *Clinical Materials* **14**, 255 (1993).
10. Espadas-Torre, C.; Meyerhoff, M.E. *Anal. Chem.* **67**, 3108 (1995).
11. Gray, F. M. Polymer Electrolytes, The Royal Society of Chemistry, Cambridge, 1997.
12. Warriner, H.E.; Davidson, P.; Slack, N.L.; Schellhorn, M.; Eiselt, P.; Idziak, S.H.J.; Schmidt, H.W. ; Safinya, C.R. *J. Chem. Phys.* **107**, 3707 (1997).
13 Rex., S. ; Zuckermann, M.J.; Lafleur, M. ; Silvius, J.R. *Biophysical Journal* **75**, 2900 (1998).
14. Evans, E.; Rawicz, W. *Phys. Rev. Lett.* **79**, 2379 (1997).
15 Aqueous Biphasic Separations, Rogers, R.D. ; Eiteman, M.A. eds., Plenum Press, 1995, 1-17.
16. Smith, G. D.; Borodin, O.; Bedrov, D. *J. Comput. Chem.* **23**, 1480 (2002).
17. Bedrov, D.; Pekny, M.; Smith, G. D. *J. Phys. Chem. B* **102**, 996 (1998).
18. Trouw, F.; Bedrov, D.; Borodin, O.; Smith, G. D. *Chem. Phys.* **261**, 137 (2000).
19. Bedrov, D.; Borodin, O.; Smith, G. D. *J. Phys. Chem. B* **102**, 5683 (1998).
20. Bedrov, D.; Borodin, O.; Smith, G. D. *J. Phys. Chem. B* **102**, 9565 (1998).
21. Borodin, O.; Bedrov, D.; Smith, G.D. *Macromolecules* **34**, 5687 (**2001**).
22. Bedrov, D.; Smith, G. D. *J. Phys. Chem. B* **103**, 3791 (1999).
23. Bedrov, D.; Borodin, O.; Smith, G. D.; Trouw, F.; Mayne, C. *J. Phys. Chem. B* **104**, 5151 (2000).

24. Smith, G. D.; Bedrov, D.; Borodin, O. *J. Am. Chem. Soc.* **122**, 9548 (2000).
25. Smith, G. D.; Bedrov, D.; Borodin, O. *Phys. Rev. Lett.* **85**, 5583 (2000).
26. Halley, J. W.; Duan, Y.; Nielsen, B.; Redfern, P. C.; Curtiss, L. A. *J. Chem. Phys.* **115**, 3957 (2001).
27. Smith, G. D.; Jaffe, R. L.; Yoon, D. Y. *J. Phys. Chem.* **97**, 12752 (1993).
28. Neyertz, S.; Brown, D.; Thomas, J. O. *J. Chem. Phys.* **101**, 10064 (**1994**).
29. Gejji, S. F.; Tegenfeldt, J.; Liidgren, J. *Chem. Phys. Lett.* **226**, 427 (**1994**).
30. Lin, B.; Halley, J. W.; Boinske, P. T. *J. Chem. Phys.* **105**, 1668 (1996).
31. Müller-Plathe, F; W. F. van Gunsteren *J. Chem. Phys.* **103**, 4745 (1995).
32. Müller-Plathe, F. *Acta Polymer* **45**, 259 (1994).
33. Hyun, J.-K.; Dong, H.; Phodes, C. P.; Frech, R.; Wheeler, R. A. *J. Phys. Chem. B.* **10**, 3329 (2001).
34. Smith, G. D.; Yoon, D. Y.; Jaffe, R. L.; Colby, R. H.; Krishnamoorti, R.; Fetters, L. J. *Macromolecules* **29**, 3462 (1996).
35. Annis, B. K.; Kim, M.-H., Wignall, G. D.; Borodin, O.; Smith, G. D. *Macromolecules* **33**, 7544 (2000).
36. Annis, B. K.; Borodin, O.; Smith, G. D.; Benmore, C. J.; Soper, A. K.; Londono, J. D. *J. Phys. Chem.* **115**, 10998 (2001).
37. Smith, G. D.; Yoon, D. Y.; Wade, C. G.; O'Leary, D.; Chen. A.; Jaffe, R. L. *J. Chem. Phys.* **106**, 3798 (1997).
38. Borodin, O.; Douglas, R.; Smith, G. D.; Trouw, F.; Petrucci, S. *J. Phys. Chem.* B **107** 6813 (2003).
39. Neyertz, S.; Brown, D.; Thomas. J. O. *Computational Polym. Sci.* **5**, 107 (1995).
40. Mos, B.; Verkerk, Pouget, S.; van Zon, A.; Bel, G.-J.; de Leeuw, S. W.; Eisenback, C. D. *J. Chem. Phys.* **113**, 4 (2000).
41. Matsuura, H.; Miyazawa, T.; Machida, K. *Spectrochim. Acta* **29A**, 771 (1973).
42. Yoshida, H.; Kaneko, I.; Matsuura, H.; Ogawa, Y.; Tasumi, M. *Chem. Phys. Lett.* **196**, 601 (1992).
43. Yoshida, H.; Tanaka, T.; Matsuura, H. *Chem. Lett.,* 637 (1996).
44. Inomata, K.; Abe, A. *J. Phys. Chem.* **96**, 7934 (1992).
45. Astrup, E. E. *Acta Chem. Scand.* **A33**, 655 (1979).
46. Yoshida, H.; Matsura, H. *J. Phys. Chem. A* **102**, 2691 (1998).
47. Tsuzuki, S.; Uchimaru, T.; Tanabe, K.; Hirano, T. *J. Phys. Chem.* **97**, 1346 (1993).
48. Jaffe, R. L.; Smith, G. D.; Yoon, D. Y. *J. Phys. Chem.* **97**, 12745 (1993).
49. Borodin O, Smith G. D. J. Phys. Chem. B **107** 6801 (2003).
50. Abe, A.; Furuya, H.; Mitra, M. K.; Hiejima, T. *Comp. and Theor. Polymer Science,* **8**, 253 (1998).
51. Kendall, R. A.; Dunning Jr.,T. H.; Harrison, R. J. *J. Chem. Phys.* **96**, 6796 (1996).
52. Frisch, M. J. et al. GAUSSIAN 98 (Revision A.7), Gaussian, Inc., Pittsburgh, PA, 1998.
53. Becke, A. D. *J. Chem. Phys*, 1993, *98*, 5648. Lee, C.; Yang, W.; Parr, R.G. *Phys. Rev. B*, **37**, 785 (1988).
54. CRC Handbook, 81st Edition, 2000 CRC Press
55. G. Ruocco, M. Sampoli, *Molecular Physics* **82**, 875 (1994).

56. Ma, B.; Lii, J.-H., Allinger, N. L. *J. Comput. Chem.* **21**, 813 (2000).
57. Stout, J. M.; Dykstra, C. E. *J. Am. Chem. Soc.* **117**, 5127 (1995).
58. Applequist, J.; Carl, J. R.; Fung, K.-K. *J. Am. Chem. Soc.* **94**, 2952 (1972).
59. Bode, K. A.; Applequist, J. *J. Phys. Chem.* **100**, 17820 (1996).
60. Thole, B. T. *Chem. Phys.* **59**, 341 (1981).
61. Breneman, C. M.; Wiberg, K. B. *J. Comput. Chem.* **11**, 361 (1990).
62. Borodin, O.; Smith, G. D. *Macromolecules* **31**, 8396 (1998).
63. Cieplak, P.; Coldwell, J.; Kollman, P. J. Comput. Chem. **22**, 1048 (2001).
64. Stone, A. J. "The Theory of Intermolecular Forces", Oxford University Press, Inc., New York, 1996.
65. Ren, P.; Ponders, J. *J. Comput. Chem.* **23**, 1497 (2002).
66. Watkins, E. K.; Jorgensen, W. L. *J. Phys. Chem. A*, **105**, 4118. (2001).
67. Sorensen, R. A.; Liau, W. B.; Kesner, L.; Boyd, R. H. *Macromolecules*, **26**, 5213 (1993).
68. Cui, S. T.; Siepmann, J. I.; Cochran, H.D.; Cummings, P. T. *Fluid Phase Equil.* **146**, 51 (1998).
69. Woon, D. E. *J. Chem. Phys.* **100**, 2838 (1994).
70. van Duijneveldt, F. B.; van Duijneveldt-van de Rijdt, J. G. C. M.; van Lenthe J. H. *Chem. Rev.* **94**, 1873 (1994).
71. Lucretius, www.che.utah.edu/~gdsmith
72. Smith, G.D. ;. Jaffe, R.L ; Yoon, D.Y. *Macromolecules* **26**, 298 (1993).
73. Allen, M. P.; Tildesley, D. J. *Computer Simulation of Liquids*, Oxford University Press, New York, 1987.
74. Nose, S. in: Meyer, M.; Pontikis, V. (Eds.), *Computer Simulation in Materials Science*, Kluwer Academic Pub., Netherlands, 1991, p 21.
75. Martyna, G.J., Tuckerman, M.; Tobias, D.J.; Klein, M. L. *Mol. Phys.* **87**, 1117 (1996).
76. Ryckaert, J.P.; Ciccotti, G.; Berendsen, H.J.C. *J. Comput. Phys.* **23**, 327 (1977).
77. Kittel, C. *Introduction to the Theory of Solids.* 2nd. Ed. Wiley, New York, 1956**,** Appendix A.
78. Deserno, M.; Holm, C. *J. Chem. Phys.* **109**, 7678 (1998).
79. Steinhauser, O. *Molecular Physics*, **45**, 335 (1982).
80. Martyna, G.J., Tuckerman, M.; Tobias, D.J.; Klein, M. L. *Mol. Phys.* **87**, 1117 (1996).
81. Das, B.; Roy, M. N.; Hazra, D.K. *Indian Journal of Chemical Technology*, **1**, 93 (1994).
82. Lide, D.L. *Handbook of Organic Solvents*, CRC Press, London, Tokyo 1995.
83. Flick, E. W. *Industrial Solvent Handbook*, 5th ed. Noyes Data Corporation, New Jersey
84. Densities of "PEO-500" at 298K and 328 K have been measured, while densities at the other temperatures were obtained by linear interpolation of the experimental data at 298 K and 328 K
85. Goutev, N.; Ohno, K.; Matsuura, H. *J. Phys. Chem. A*, **104**, 9226 (2000).
86. Heinrich-Schramm, A.; Price, W. E.; Lüdemann, H.-D.;*Z. Naturforsch*, **50a**, 145 (1995).

87. Saboungi, M.-L.; Price, D. L.; Mao, G.; Fernandez-Perea, R.; Borodin, O.; Smith, G. D.; Armand, M.; Howells, W. S., *Solid State Ionics*, **147**, 225 (2002).
88. Porter, C. H.; Boyd, R. H. Macromolecules **4**, 589 (1971).
89. P. G. Bruce, C. A. Vincent "Polymer Electrolytes" *J. Chem. Soc. Faraday Trans*, **89**, 3187 (1993).
90. Neyertz, S.; Brown, D. *J. Chem. Phys.* **104**, 3797 (1996).
91. Mos, B.; Verkerk, Pouget, S.; van Zon, A.; Bel, G.-J.; de Leeuw, S. W.; Eisenback, C. D. *J. Chem. Phys.* **113**, 4 (2000).
92. Halley, J. W.; Duan, Y.; Curtiss, L. A.; Baboul. A. G. *J. Chem. Phys*. **111**, 3302 (1999).
93. Boinske, P. T.; Curtiss, L.; Halley, J. W.; Lin, B.; Sutjianto, A. *J. of Computer-Aided Mater. Design.* **3**, 385 (1996).
94. Borodin, O.; Smith, G.D. Macromolecules **33**, 2273 (2000).
95. Smith, G.D.; Jaffe, R.L.; Partridge, H. J. Phys. Chem. A. **101**, 1705 (1997).
96. de Jonge, J.J.; van Zon, A.; de Leeuw, S.W. *Solid State Ionics* **147**, 349 (2002).
97. Sutjianto, A.; Curtiss, L. A. *J. Phys. Chem. A.* **102**, 968 (1998).
98. Borodin, O.; Smith, G. D.; Douglas, R. *J. Phys. Chem. B.* **107**, 6824 (2003).
99. Bedrov, D.; Smith, G. D. *J. Chem. Phys.* 115, 1121 (2001).
100. Mao, G.; Saboungi, M.-L.; Price, D.L.; Armand, M.; Howells, W. S. *Phys. Rev. Lett.* **84**, 5536 (2000).
101. Arumugam, S.; Shi, J.; Tunstall, D. P.; Vincent, C. A. *J. Phys.: Condens. Matter*, **5**, 153 (1993).
102. Shi, J.; Vincent, C. A. *Solid State Ionics*, **60**, 11 (1993).
103. Sun, H. Y.; Takeda, Y.; Imanishi, N.; Yamamoto, O.; Sohn, H.-J. *J. Electrochem. Soc.* **147**, 2462 (2000).
104. Lee, J. Y.; Liu, Y.; Hong, L. IMLB 11, Montery, USA, June 23-28, 2002, abstract 179
105. Sun, H. Y.; Takeda, Y.; Imanishi, N.; Yamamoto, O.; Sohn, H.-J. *J. Electrochem. Soc.* **147**, 2462 (2000).
106. Lee, J. Y.; Liu, Y.; Hong, L. IMLB 11, Montery, USA, June 23-28, 2002, abstract 179
107. Higgins, J. S.; Benoît, H. C. *Polymers and Neutron Scattering,* Clerendon Press, Oxford, 1996.
108. Edwards, D.M.F.; Madden, P.A.; McDonald, I.R. *Molec. Phys.* **51**, 1141 (1984).
109. Gisser, D.J.; Glowinkowski, S.; Ediger, M.D. *Macromolecules* **24**, 4270 (**1991**).
110. Haile, J. M.; *Molecular Dynamics Simulations*, Wiley, New York, 1992.

Chapter 3

NANOSTRUCTURE FORMATION AND RELAXATION IN METAL(100) HOMOEPITAXIAL THIN FILMS:ATOMISTIC AND CONTINUUM MODELING

K.J. Caspersen,[a] Da-Jiang Liu,[b] M.C. Bartelt,[c] C.R. Stoldt,[d] A.R. Layson,[a] P.A. Thiel,[a] and J.W. Evans[e]
a Department of Chemistry and Ames Laboratory, Iowa State University, Ames, IA 50011
b Ames Laboratory – U.S. Department of Energy, Iowa State University, Ames, IA 50011
c Department of Chemistry and Materials Science, Lawrence Livermore National Laboratory, Livermore, CA 94550
d Department of Mechanical Engineering, University of Colorado, Boulder, CO 80309
e Department of Mathematics and Ames Laboratory, Iowa State University, Ames, IA 50011

1. INTRODUCTION

Much of nanotechnology and microelectronics is based upon the feature that collections of atoms or molecules deposited onto surfaces can cooperatively self-organize into desired far-from-equilibrium nanostructures.[1] A first-principles modeling strategy for these important processes might utilize Molecular Dynamics (MD) to follow nanostructure evolution. However, it is typically not viable (even with recent MD-acceleration techniques)[2] to access the appropriate time and length scales. Furthermore, MD is to some extent a "black box" which incorporates much irrelevant fast vibrational dynamics, and which may not readily provide insight into what controls behavior. A particularly natural alternative, especially for epitaxial systems, is to use atomistic lattice-gas models with appropriate stochastic dynamics. These models are analyzed via Kinetic Monte-Carlo (KMC) simulation.[3] However, sometimes even these cannot provide the type of insight which comes from "coarse-graining", and analyzing corresponding continuum models for evolution. (The latter are often applicable on surprisingly short length scales of a few nanometers.) It should also be noted that for both MD and KMC modeling, one would ideally use as input precise *ab-*

L.A. Curtiss and M.S. Gordon (eds.),
Computational Materials Chemistry: Methods and Applications, 91–124.

initio electronic structure values for relevant interaction energies and activation barriers. This is often not possible, but this problem may be avoided at least for KMC by systematically extracting these parameters from comparison with experiment.[4]

A generic goal is to develop models with quantitative predictive capabilities often connecting-the-length-scales between atomistic and continuum pictures. If successful, these models will facilitate development of strategies to control nanostructure evolution. We have had considerable success in this endeavor for the "simple" systems, which we now describe.

Metal(100) homoepitaxial films are perhaps the simplest of all thin film systems. Nonetheless, they continue to present a rich variety of novel new phenomena.[1,4] In these systems, the atoms reside at a regular array of crystalline sites, which form square array within each layer. Deposited atoms are typically supported by four atoms in the layer beneath at so-called four-fold hollow (4fh) adsorption sites. See Fig.1. Adatoms make "rapid" transitions between neighboring sites due to thermally activated diffusive hopping, as described in detail below. As suggested above, atomistic lattice-gas models with appropriate stochastic prescription of the deposition process and of hopping can effectively describe film growth and relaxation on the correct time and length scales. It is appropriate to note that we do sometimes take advantage of input from very specific MD studies for deposition dynamics, and for probing possible concerted many-atom moves.[5]

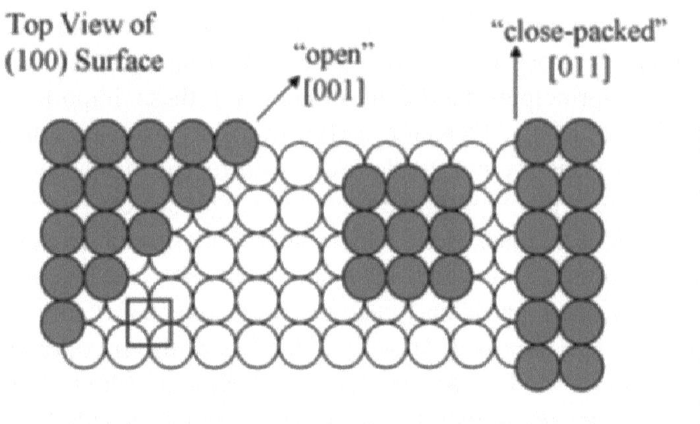

Figure 1. A bird's eye view of a fcc(100) metal surface. The schematic shows: a four-fold hollow (4fh) adsorption site, indicated by a square box; a "typical" square-shaped 3×3 island; and two different types of step edges which separate consecutive layers. On the left is an open or kinked [001] step edge. On the right is a close-packed [011] step edge.

Another feature of these metal(100) homoepitaxial systems is that the equilibrium states are also particularly simple. For submonolayer films (on rigid substrate), the equilibrium state is (loosely speaking) a single large 2D island. This follows since typical substrate temperatures, T, are well below the critical temperature, T_C, for phase separation into a 2D condensed island phase and a very dilute 2D "gas" phase. For these systems, a reasonable first approximation is that adatoms interact with only nearest-neighbor (NN) interactions of strength $\phi=\phi_{NN}$. Then, one has that $k_B T_C \approx 0.57\phi$, borrowing from Onsager's famous solution of the 2D Ising model. Thus, one has that T_c=1650K for typical ϕ=0.25eV (whereas experimental T are usually below 500K).[6] For multilayer films, the equilibrium state is a smooth film. This follows since again experimental T are well below the critical temperature, T_R, for the thermal roughening transition (where $T_R \approx T_C$).[7] We now sketch the basic characteristics of film growth and relaxation processes, which it should be emphasized involve non-equilibrium film configurations.

The picture for epitaxial film growth[1] is simplest for "lower" T where lateral bond-breaking between adatoms is effectively absent. It involves: "random" deposition; "rapid" diffusion of adatoms across the surface nucleating new islands when two adatoms meet, and growing of existing 2D islands when adatoms reach with their boundaries (either by diffusing within the same layer, or by hopping down from a higher layer); nucleation and growth of islands on-top of existing islands also occurs for continued deposition. Ultimately 3D multilayer stacks of 2D islands (commonly referred to as "mounds") can form. These grow taller leading to kinetic roughening of the film, and also coarsen laterally in some as yet incompletely understood fashion. Also important for growth are the features that shapes of individual islands are often controlled by "periphery diffusion" (PD) of adatoms at island edges, and that the details of interlayer transport control multilayer morphologies. Extending this picture to higher T, bond-breaking becomes important, specifically dissociation of small clusters, and detachment from island edges.

Next, we provide a picture of post-deposition relaxation[4] of the overall film morphologies as well as of individual nanostructures. Deposition drives the system out-of-equilibrium creating "many small" islands in submonolayer regime versus one large one. Rough films are created in the multilayer regime rather than smooth ones. Thus, after deposition ceases, the system tries to relax back to its simpler equilibrium state.

Consider first the submonolayer regime. Even in the absence of bond-breaking, atoms can still hop around the periphery of islands, producing diffusion of the entire island. As a result, islands can diffuse, collide, and coalesce (i.e., sinter), reducing the total number of islands, thus coarsening the film. This is termed Smoluchowski Ripening (SR),[4] by analogy with Smoluchowski's classic description of coagulation. Of specific interest here is the individual coalescence or sintering events wherein 2D islands can collide either corner-to-corner, creating dumbbell-shaped clusters, or side-to-side, creating rectangular shaped clusters. These will relax back to a near-square equilibrium shape. Even if bond-breaking is not significant during deposition, it may be crucial on the longer time-scale of relaxation. Indeed, the common (but here often incorrect) expectation is that the number of islands is reduced primarily by Ostwald Ripening (OR):[8] dissolution of smaller islands, and incorporation of their adatoms into larger islands (rather than by SR). Consider next the multilayer regime. Here bond-breaking is necessary for mass to be transferred from islands in higher layers down to lower layers, thus leading to film smoothening. The nature of the smoothening dynamics for the entire film, as well as the decay of individual 3D mounds, are specific issues of interest.

In Sec.2, we describe the general ingredients of atomistic lattice-gas modeling of epitaxial thin film systems. The details of our more tailored approach to such modeling for the Ag/Ag(100) system are presented in Sec.3. The key features and issues for film growth are described in Sec.4 for both the submonolayer and multilayer regimes. Specifically, we compare the results of both atomistic and continuum modeling against experimental observations. An analogous discussion of post-deposition relaxation is provided in Sec.5. Finally, in Sec.6, we describe how the above growth and relaxation processes (and the associated nanostructure evolution) are significantly affected by the presence of a chemical additive (oxygen), even though the perfect Ag(100) surface is relatively unreactive towards oxygen.

2. GENERAL FEATURES OF ATOMISTIC LATTICE-GAS MODELS FOR METAL(100) HOMOEPITAXY

In growth models for metal(100) homoepitaxy, atoms impinge on the surface at a rate or flux of F per site per second (or ML/s) at randomly chosen lateral locations. If impinging directly at a 4fh adsorption site, they adsorb there. If impinging upon step edges, or upon larger micro-protrusions, then they typically "funnel down" to lower 4fh sites.[5] This downward funneling (DF) model was proposed to explain smooth growth at low T, and has been confirmed by MD simulations. Recently, significant deviations from DF for growth of films at very low T below 130K (specifically, capture of depositing atoms on the sides of microprotrusions) were shown to lead to overhangs and internal voids in growing homoepitaxial films.[9,10] However, these features will not be relevant for film growth above 150K, which is considered exclusively in this work.

The greater challenge for both growth and post-deposition relaxation is the treatment of the numerous possible types of surface diffusion events, the rates (h) for which depend on the local environment. Usually, one assumes an Arrhenius form, $h=\nu\exp[-E_{act}/(k_BT)]$, based on transition state theory, and further that rates have common prefactor $\nu=10^{12}$-10^{13}/s. Then, it suffices to determine the many activation barriers, E_{act}. We should note the *detailed-balance constraint*:[3] the difference in activation barriers for forward and reverse hoping processes is given by energy difference of initial and final configurations. As indicated above, often configuration energies (and thus these energy differences) can be reasonably estimated by assuming NN interactions, thus providing important estimates of differences between activation barriers. For realistic modeling of non-equilibrium phenomena, it is necessary to avoid a common shortcoming in generic statistical physics treatments of kinetic processes which assume that the E_{act} are also simply determined in terms of $\phi=\phi_{NN}$, e.g., by counting the number of bonds in the initial state. This can be quite inaccurate. Activation barriers must be assessed independently, and their detailed form often controls behavior.

Some examples of key hopping processes are illustrated in Fig.2, including terrace diffusion, PD, detachment from step edges, and interlayer diffusion. Samples of complete Potential Energy Surfaces (PES) – from which detailed-balance is self-evident – are also shown. Perhaps the most important of all parameters is the activation barrier for terrace diffusion of $E_d \approx 0.40$eV for Ag on Ag(100),[11,12] as determined in Sec.4, which controls the lateral length scale of nanostructures for irreversible island formation. See Sec.4.1. We now discuss two basic approaches to input appropriate rates into atomistic models and KMC simulations.

The generic approach was pioneered by Voter,[13] and recently adopted by several groups.[14,15] Suppose that the rates or barriers for lateral hopping depend on the occupancy of the 10 sites neighboring the initial and final site. Then, there are a total of $2^{10} = 1028$ configurations, and thus barriers to be determined (ignoring reduction due to symmetry). Of course, there are additional distinct rates for interlayer diffusion processes. In fact, slight differences (reduction) in rates for hopping down between layers (relative to hopping in layer) are critical for the roughening of growing films. In any case, the barriers for all these configurations can be determined with semi-empirical potentials such as the Embedded-Atom-Method (EAM), and increasingly some are supplied by Local Density Approximation (LDA) or Generalized Gradient Approximation (GGA) calculations.[11,16] All these rates then provide input to the atomistic model or KMC simulations.

The tailored approach was adopted and developed in our work.[12,17-19] The idea here is that the actual values of many barriers are irrelevant: either they are so high, that the process does not occur on the relevant time scale, or they are so low that the hop rate is effectively infinite. Thus, one can simplify the simulation model by either forbidding these processes, or making them infinitely fast (i.e., implementing them instantaneously). Ideally, one is just left with a few key barriers, which could be evaluated by GGA, or (often better) determined one at a time by comparison with data from an appropriate sequence of experiments. The latter has been our strategy. We emphasize that the simplifications which are adopted depend very much on the process of interest (e.g., growth versus relaxation).

Standard procedures for kinetic Monte Carlo simulation of these types of models are discussed briefly in the Appendix.

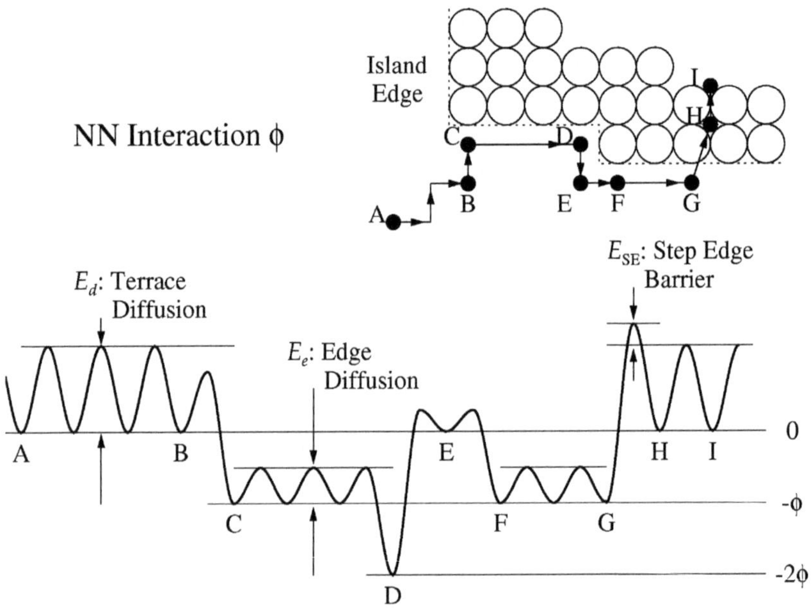

Figure 2. The upper schematic gives a bird's eye view of the edge of an island on an Ag(100) surface showing an additional single adatom undergoing terrace diffusion, step edge attachment-detachment, various periphery diffusion (PD) processes, and interlayer transport. The lower schematic shows the associated PES constructed assuming that configuration energies are determined by NN interactions ϕ.

3. DETAILS OF TAILORED ATOMISTIC LATTICE-GAS MODELING FOR AG/AG(100)

In most of our modeling of *film growth* in the Ag/Ag(100) system, we forbid bond-breaking processes (which include dissociation of dimers, and detachment of adatoms from island edges).[9,12,17,18] Thus, island nucleation and growth are irreversible. This assumption of irreversibility is justified if bond-breaking is slow on the time scale of aggregation of atoms with other atoms or with islands. This is reasonable for Ag/Ag(100) below about 320K,[12] as discussed in Sec.4. (It is often incorrectly assumed that it suffices for bond-breaking to be slow on the longer time scale of deposition.) We also assume that islands of two or more adatoms are immobile. In our simplest "canonical" *square-island model* for metal(100) homoepitaxy,[20] we assume that individual growing 2D islands within each layer maintain near-square shapes due to efficient PD: the growth sequence for each island forms an Archimedian spiral. New aggregating atoms are immediately added to the next growth site in the sequence, typically the single kink site on a near-square island, except when starting a new edge.

A more refined, but still tailored treatment of PD is possible for these systems. In fact, such refinement is necessary to realistically describe island coalescence during growth, or to describe nanostructure relaxation. For simplicity, the following discussion of PD assumes that configuration energies can be determined by counting NN bonds with strength ϕ. The four key elementary PD processes shown in Fig.3 are:[19,21,22] diffusion along close-packed edges with "low" barrier $E_e \approx 0.25$eV; escape from kinks along step edges with barrier $E_k \approx E_e+\phi$; diffusion around kinks with barrier $E_r = E_e+\delta$; "core breakup" with barrier $E_c \approx E_e+\phi+\delta$. The third and fourth proc-

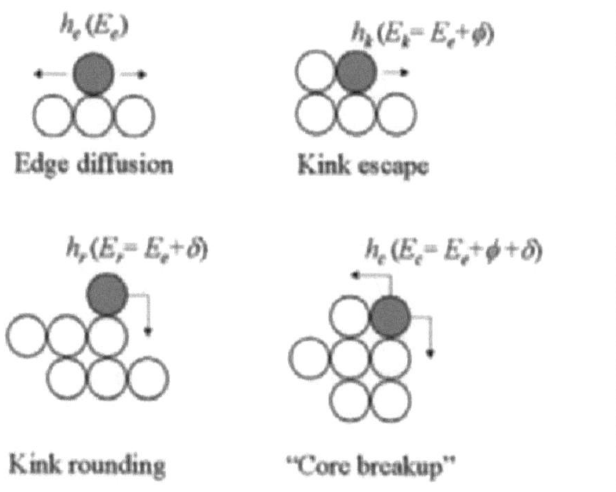

Figure 3. A bird's eye view of key periphery diffusion (PD) processes and the associated barriers assuming NN interactions ϕ: straight edge diffusion (barrier E_e); kink escape (barrier $E_e+\phi$); kink rounding (barrier $E_e+\delta$); core breakup (barrier $E_e+\phi+\delta$).

esses actually involve two hops, the first (slow) step to a diagonal NN site to the cluster, and the second "fast" step back to a NN site, but we combine these into a single effective hop. It is also conceivable that one should distinguish rounding of single-atom-high kinks with barrier E_r from rounding of multiple-height kinks or "corners" with barrier $E_r' = E_e+\delta'$. The former could more easily occur by a concerted exchange process. Note that $E_e = 0.25$eV is far below $E_d = 0.40$eV (the opposite of the simple bond-counting prescription for E_{act}), a feature which will be important below.

One important aside to the above discussion relates to the barrier for dimer dissociation and diffusion. See Fig.4. Dissociation in the direction of the bond has an activation barrier of $E_{diss} \approx E_d+\phi$. It has been proposed that "easy" dissociation is possible by first shearing with barrier roughly equal to E_r, and then separating. This suggests that the barrier for dissociation may be comparable to E_d for small δ. This would obviously create problems for our picture of irreversible island formation. However, the argument is incorrect. The overall or effective barrier for this "shearing" pathway is again E_{diss}, as determined by detailed-balance arguments.[12] It is however the case that dimer diffusion can occur via this shearing pathway (followed by easy hops back to configurations with NN adatoms). Then, for small δ, one might expect that dimer diffusion plays an important role in the island nucleation and growth process.[23] For the Ag/Ag(100) system, experimental evidence suggests otherwise[12] (perhaps the above approximation for the activation barrier is inadequate), although the situation is not completely resolved. In any case, we do not include dimer diffusion in our simulations of growth.

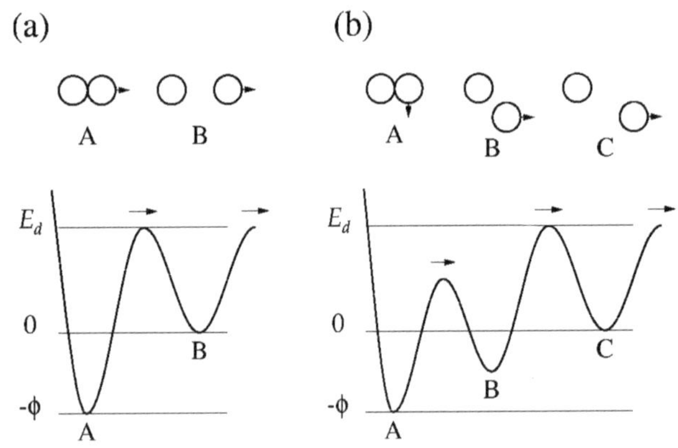

Figure 4. Dimer dissociation pathways: (a) direct dissociation in the direction of the bond, with barrier $E_{diss}=E_d+\phi$; (b) dissociation via a shearing pathway in two steps, each with a lower barrier than E_{diss}. However, the effective barrier for this overall dissociation process is E_{diss}. If the first step in (b) is followed by return to a NN configuration, this corresponds to dimer diffusion, with "low" barrier given by the first step in (b).

A related issue pertains to the barriers for detachment from step edges. The barrier for detachment of a single atom at a close packed step edge (in the orthogonal direction) is $E_{diss} = E_d + \phi$. However, the effective barrier for detachment from islands is given by the barrier for detachment from kink sites of $E_{evap} = E_d + 2\phi$. The latter follows from noting that the effective detachment rate can be determined from the population of edge atoms with a single bond of $\sim\exp[-\phi/(k_BT)]$ *times* the detachment rate for these atoms with barrier E_{diss}.[19]

Next, we discuss a more realistic treatment of PD during film growth (compared to the canonical square-island model mentioned above). In its simplest form, we immediately move atoms reaching island edges to a kink site neighboring the attachment point, and forbid escape from these doubly-bonded kink sites. More precisely, in moving the atom to doubly-bonded sites, it is allowed to round kinks, but not corners (multiple-height kinks). If it cannot reach a kink site (because that would require rounding corners), or if no kink sites exist (as on a perfectly rectangular island), it remains at the point of impact. This has been described as the "Efficient Kink Rounding (EKR)" model.[17,18] A number of observations can be made regarding this treatment:

(i) It assumes that straight edge diffusion (barrier E_e) and kink rounding (barrier $E_k = E_e+\delta$) are facile, as is terrace diffusion (barrier E_d). It neglects corner rounding (barrier $E_e+\delta'$) and kink escape (barrier $E_k = E_e+\phi$), as well as processes with higher barriers such as dimer dissociation, step-edge detachment, and core breakup. This requires that[24] $\delta<\phi$, and $\delta<\delta'$.

(ii) It is critical in our modeling that kink rounding is facile in order to obtain compact islands observed during growth above 220K, as opposed to irregular islands.[17,18,25,26] In our studies of growth at lower T, we do refine the model to inhibit kink rounding in order to obtain irregular islands.

(iii) Individual islands have near-rectangular shapes. Deviations from near-square shapes occur due to difference in fluxes of diffusing atoms to the different sides of the island. See Sec.4.1.

(iv) When islands collide during growth, the model realistically describes the formation of a neck between them during continued growth (i.e., growth coalescence). This is important for mound formation during multilayer growth, as mounds are built upon bases of coalesced 2D islands.[17,18]

In contrast to the above, our treatment of post-deposition relaxation via PD necessarily includes both kink escape and core breakup.[19,21,22] These are needed for relaxation to near-square shapes. It typically still neglects other PD processes such as escape of triply-bonded atoms from straight edges with barrier $E_e+2\phi+\delta$, as well as detachment. As noted above, the effective barrier for detachment of $E_{EC}=E_d+2\phi$ exceeds the largest barrier for the relevant PD processes of $E_{PD}=E_e+\phi+\delta$ (core breakup), since it is clear that $E_e<E_d$ and $\delta\leq\phi$. Thus, PD has an energetic advantage over detachment pathways for relaxation.[19,21]

Finally, we comment on the treatment of *downward interlayer hopping*, which is critical for multilayer growth (and relaxation). It has long been recognized that there often exist additional Ehrlich-Schwoebel or step-edge barriers inhibiting downward transport of atoms at step edges (relative to intra-layer terrace diffusion).[27] Only more recently was it appreciated that these lead to "unstable growth" manifested by mound formation.[28] In our modeling, we distinguish between close-packed step edges and open (or kinked) step edges. Specifically, we assume that the former have an ES barrier of E_{ES} (and thus a barrier for downward hopping of E_d+E_{ES}), but that the latter have no additional barrier.[17,18] This is motivated by EAM results and by our own detailed modeling for Ag/Ag(100) growth.

4. ATOMISTIC AND CONTINUUM MODELING OF AG/AG(100) FILM GROWTH

Throughout the following, we measure lateral distances in units of the surface lattice constant, a ≈ 0.29 nm for Ag(100), so the terrace diffusion rate, $D=a^2h$, corresponds to h. Correspondingly, island and adatom densities are measured per adsorption site (rather than per unit area).

4.1 SUBMONOLAYER GROWTH

In Fig.5, we present simulation results for submonolayer growth using our Efficient Kink Rounding (EKR) model. By adjusting the terrace diffusion barrier to $E_d \approx 0.40$eV (for $\nu=10^{13}$/s), we match the experimentally observed island densities, N_{isl}, for deposition of Ag/Ag(100) at 300K in the regime of lower coverages of $\theta \leq 0.25$ML.[4,12] (Here we also used the experimental deposition flux, F.) From Fig.1, for higher coverages, one can see the growth coalescence morphologies predicted by the model.[17] One must reach a coverage of about 0.75-0.8ML for percolation of clusters of submonolayer islands.[12]

For lower T around 200K, one might expect that kink rounding becomes inhibited leading to irregular (rather than compact) individual islands. To quantify this transition, it suffices to compare the rate of aggregation of diffusing atoms with an island edge, $R_{agg} \sim F/N_{isl}$, with the rate of kink rounding $R_{kr} \sim \nu \exp[-E_k/(k_BT)]/L$. The factor of 1/L in R_{kr} gives the likelyhood of finding a rapidly diffusing adatom on an outer edge of length L at a site adjacent a kink where it can attempt to round. The transition to irregular islands occurs when atoms arriving on an outer terrace do not have enough time to round a kink to reach a doubly-bonded kink site before getting "caught" by another aggregating atom and "nucleating" a new outer terrace. Thus, the transition occurs when $R_{agg} \approx R_{kr}$, yielding $\delta \approx 0.17$eV assuming that the transition T≈200K (and using $N_{av} \approx 6\times10^{-3}$, L≈6-7, $\nu=10^{12}$/s, and F=0.02ML/s). It is possible to refine our growth model to include inhibited kink rounding without doing a full multiple-hopping-atom simulation of PD by just incorporating information about the probability of kink rounding, as calculated from $P_{kr}=R_{kr}/(R_{agg}+R_{kr})$. See Refs. 18 and 29 for more details

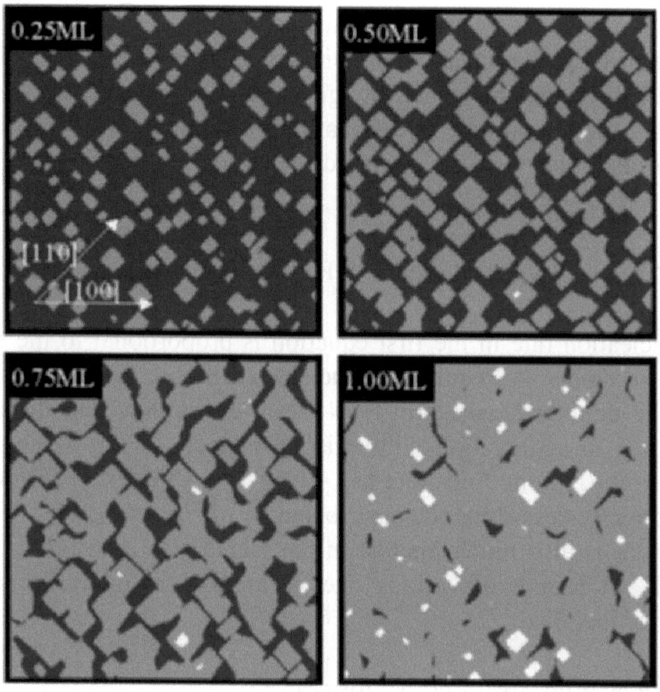

Figure 5. Results from our atomistic EKR simulation model for submonolayer growth of Ag/Ag(100) at 300K with F=0.055 ML/s. Images are 102×102 nm^2. Dark gray denotes the substrate, light gray denotes the first deposited layer, and white denotes second layer islands. Directions of [110] and [100] steps are indicated in the first frame. Reprinted with permission from Ref. 17. Copyright 2001 American Institute of Physics.

For higher T above 300K, bond-breaking becomes operative and island formation becomes reversible. To quantify this transition, it suffices to compare R_{agg} with the rate of dimer dissociation, $R_{diss}=h_{diss}=\nu \exp[-(E_d+\phi)/(k_BT)]$. Reversibility sets in when dissociation occurs before additional aggregating atoms have a chance to reach the dimer causing it to grow into a larger more stable island. Thus, the transition is determined by setting $R_{agg} \approx R_{diss}$, which yield $\phi\approx0.3$eV assuming that the transition occurs at T≈320K (and using $N_{av}\approx6\times10^{-4}$, $\nu=10^{13}$/s, and F=0.06ML/s). This simple analysis neglects the feature that dissociating dimers are quite likely to recombine based on the statistical properties of random walks (Polya's theorem).[12,30] This effect would lead to downward revision of the above estimate of ϕ.

Next, we will briefly comment on some fundamental issues in nucleation theory, and on continuum formulations of adatom capture and island growth. Below, we let σ_s denote the mean "capture number" for islands of size s, which gives the mean propensity for islands of that specific size to capture diffusing adatoms. Then σ_{av} gives the average over all sizes. First, we note that the classic rate equations for irreversible island formation describing mean island density, N_{isl}, and mean adatom density, N_1, have the form[1,31]

$$d/dt\, N_{isl} \approx \sigma_1 h(N_1)^2 \text{ and } d/dt\, N_1 \approx F(1-\theta) - \sigma_{av} h N_1 N_{av}. \tag{1}$$

The nucleation rate in the first equation is proportional to the square of N_1 since two atoms are required to nucleate an island. The second equation includes the dominant gain-term to N_1 from deposition, and the dominant loss-term due to aggregation with existing islands. In the steady-state approximation for N_1, one finds that $N_{isl} \sim (h/F)^{-1/3} \sim (F/\nu)^{1/3} \exp[1/3 E_d/(k_B T)]$, for fixed θ, correctly describing the basic behavior of N_{isl}.

There is also much interest in the island size distribution, i.e., the densities N_s of islands of various sizes s (atoms). The corresponding rate equations for low θ are[1,31]

$$d/dt\, N_s \approx h\sigma_{s-1}N_{s-1} - h\sigma_s N_s, \text{ for } s>1. \tag{2}$$

The shape of the size distribution (N_s versus s) depends critically on the variation of σ_s with s.[32] This s-dependence is traditionally calculated from the diffusion equation for deposited atoms aggregating with an island, subject to the *mean-field assumption* that the environment of an island is *independent* of its size.[31] A surprising discovery of our recent work is that this assumption is *invalid*.[32] Larger islands have larger "capture zones" surrounding them, where one imagines that the area of the capture zone (CZ) is proportional to the capture number (or island growth rate). This assertion can be tested by KMC simulation. However, an effective and more instructive alternative (which also yields a precise definition of CZ's) is to solve continuum diffusion equations describing adatom deposition and capture in the complex and disordered geometry of growing islands,[33,34] i.e.,

$$\partial/\partial t\, N_1 = F + h\,\nabla^2 N_1 \approx 0, \text{ with } N_1 \big|_{\text{island edges}} = 0. \tag{3}$$

Capture numbers or growth rates are calculated from integrating the normal component of the flux around island boundaries. Equivalently, they are determined from the areas of capture zones constructed so that following the lines of diffusive flux from all points within the CZ leads to the appropriate island. Fig.6 shows an example of such CZ's constructed from experimental island distributions created by deposition on Ag(100) at 300K.

Experimental and simulation results for σ_s versus s are also shown in Fig.7 (increasing far more quickly than the mean-field $\sigma_s \sim s^{1/2}$, for large s).[34]

We now provide some closing general comments. First, our continuum formulation of adatom diffusion and island growth has provided a foundation for (potentially) more efficient computational modeling of film growth (compared to KMC). Indeed, this approach has been implemented in combination with a level set formulation to handle the evolution of island boundaries.[35] The other fundamental open question is whether a sound theory can be provided for the non-trivial dependence of σ_s on s. Some attempts have been made to provide such a theory,[36,37] but we are convinced that a correct theory must properly describe spatial aspects of the nucleation which preferentially occurs along the boundaries of existing CZ's.[38,39]

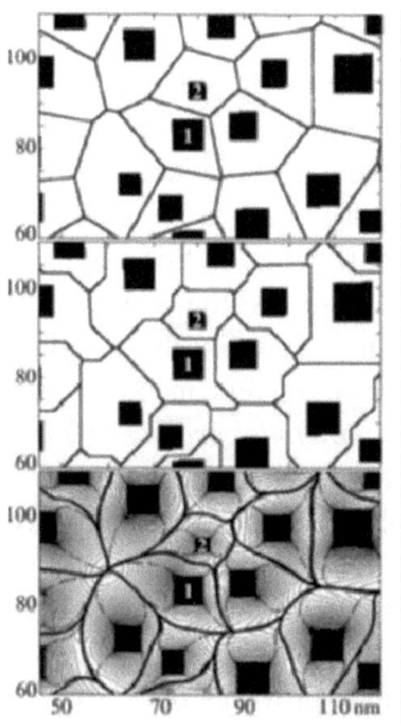

Figure 6. Capture zones (CZ's) for an experimental distribution of Ag islands deposited on Ag(100) at 300K: a crude Voronoi cell approximation to CZ's, based on distance to island centers (top); a reasonable "edge cell" approximation to CZ's, based on distance to island edges (middle); exact CZ's, based on steady-state solution of the diffusion equation for deposited adatoms (bottom). The thin lines are contours of constant adatom density; the thick lines are CZ boundaries; and the dashed lines indicate boundaries for sub-CZ's for individual island edges. Reprinted with permission from Ref. 34. Copyright 1999 American Institute of Physics.

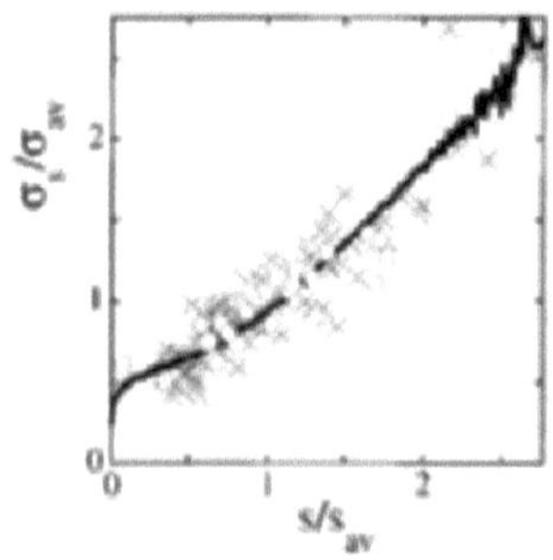

Figure 7. Capture numbers (symbols) versus island size for an experimental distribution of Ag islands deposited on Ag(100) at 300K. Also shown are results for these quantities from simulations run under the same conditions as in the experiment (solid curve). Reprinted with permission from Ref. 34. Copyright 1999 American Institute of Physics.

4.2 MULTILAYER GROWTH

To model multilayer growth, we must first determine the key parameter E_{ES} (the step-edge barrier) for close-packed step edges. One approach is to match experimental observations for the second layer population after deposition of 1ML at 300K, say, as shown in Fig.5.[12] Sec.5.2 describes some subtleties of these experiments. In this way, one determines that E_{ES} is below 0.1eV,[17] and we shall use a value of E_{ES}=0.070eV (based on further comparison with experimental data for the roughness of 25ML films for T>230K).[17,18] Now, all the parameters in our model are determined. This raises a basic question. *Can the model predict quantitatively behavior during multilayer growth in the regime of 100's and 1000's of layers?* We claim that our model is remarkably successful in this respect, in fact providing the first such successful example for a specific thin film system. It explains previous "anomalies" in experimental data; it reveals behavior completely opposite to previous expectations for kinetic roughening of Ag/Ag(100) at 300K versus 230K, say; and it provides basic new insight into dynamics of 3D mounds during growth.[18]

To clearly illustrate the capabilities of the model, Fig.8 shows its predictions for film roughness, W, versus θ up to 2000ML for growth of Ag/Ag(100) at 230K, 260K, and 300K. Note that W corresponds to the RMS width of the film height distribution, and is measured here in units of the interlayer spacing. Simulations match experimental observations for W up to about 100ML revealing that growth is smoother at 300K than at 200K. In particular, it should be noted that the model does an excellent job of matching the rather complicated variation of W with θ at 300K. Fig.9 shows simulated film morphologies which are compared with STM images. One can see that our model also correctly recovers the lateral size of the mounds, together with the slow increase in this size with θ. Together, these successes give us confidence to trust the predictive capabilities of the models up to the regime of 1000's of layers.[18]

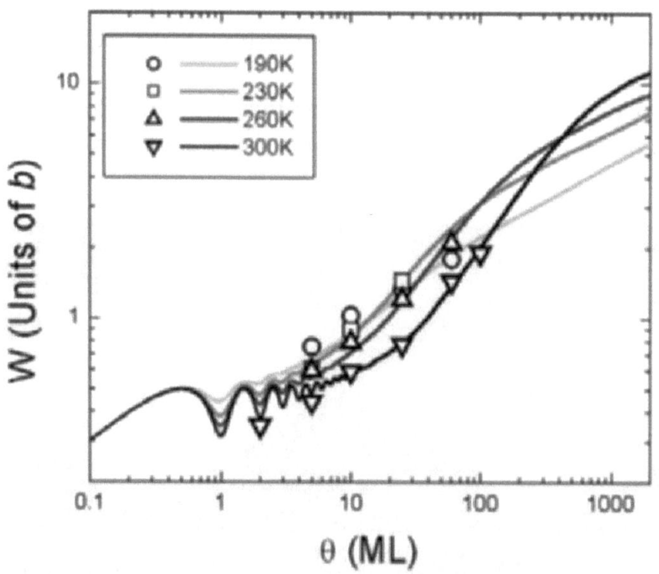

Figure 8. Roughness, W (in units of the interlayer spacing b=0.20 nm) versus coverage, θ (in ML) for Ag films deposited on Ag(100) at 190K, 230K, 260K, 300K with F≈0.02ML/s. Symbols show experimental data. Curves show simulation results (bottom to top – on right – corresponds to increasing T). At the three higher temperatures, kink rounding is efficient, and behavior can be accurately predicted by the EKR model. At 190K, kink rounding is significantly inhibited, and modeling must include an additional kink rounding barrier of δ≈0.16eV. Reprinted with permission from Ref. 18. Copyright 2002 American Institute of Physics.

Remarkably, in this regime, our model predicts that roughening ultimately occurs faster at 300K than at lower T, totally opposite to traditional expectations. The only previous clue as to this behavior was a single data point from surface X-ray scattering studies suggesting that thick Ag/Ag(100) films grown at 300K were very rough.[40] This observation was not explained, and plausibly attributed to some sort of "anomalous" growth behavior. Our modeling shows that it is a direct consequence of conventional growth with mounding. After the discussion below of continuum models for growth, we shall explain in more detail the unusual roughening behavior, and also describe our observations on mound coarsening dynamics.

Finally, we note that for lower T below 230K, it is essential to incorporate inhibited kink rounding. The above "efficient kink rounding" model predicts films that are too rough. The explanation is clear. EKR produces individual islands which are always compact (near-rectangular). In reality, at lower T, islands are irregular with a larger proportion of open or kinked step edges have no step-edge barrier. This explains why growth is smoother than predicted, and in fact is very sensitive to the kink rounding barrier δ. We find that δ≈0.16eV gives the best fit to experimental film roughness values. See Ref.s 18 and 29.

Next, we describe coarse-grained approaches to modeling multilayer growth. The natural extension to the submonolayer approach is to retain dis-

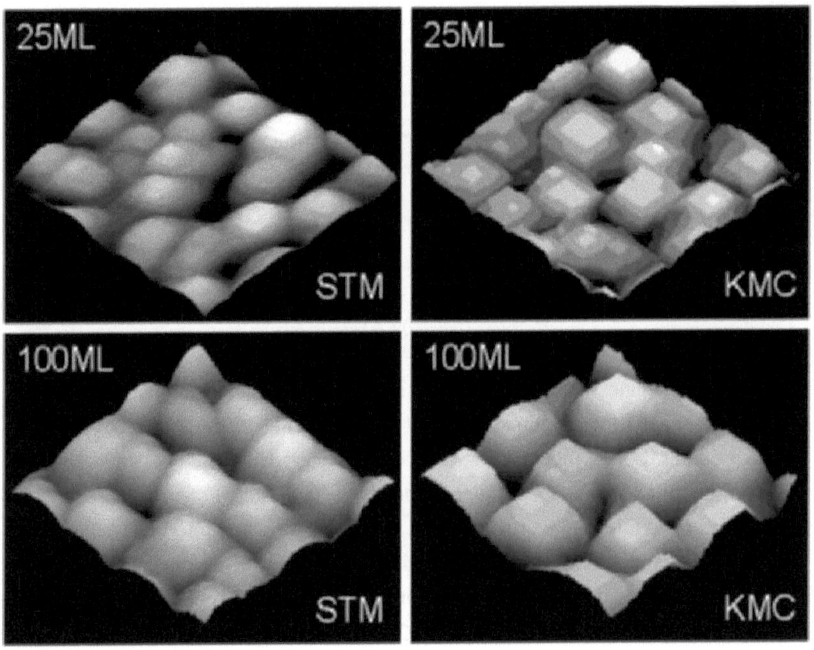

Figure 9. Film morphologies (50×50 nm^2) for Ag deposited on Ag(100) at 230K with F≈0.02ML/s: experimental STM images (on left) are compared with predictions of atomistic modeling (on right). Reprinted with permission from Ref. 18. Copyright 2002 American Institute of Physics.

crete vertical levels in describing film structure, but to treat step edge locations with a continuous variable. One would also solve a continuous diffusion equation in each layer (with appropriate boundary conditions) for the density of diffusing adatoms. A level set version of this approach has been implemented.[35] However, what is more useful for our analysis is a fully continuum treatment where the film height, h($\underline{x}$,t), is regarded as a continuous function (measured in units of interlayer spacing) of a continuous lateral variable. One then writes down a stochastic evolution equation of the form[41]

$$\partial/\partial t\, h = F - \nabla\cdot\underline{J} + \eta, \qquad (4)$$

where $\underline{J}$ denotes the lateral mass current of deposited atoms, and η is a mean-zero random variable describing noise in deposition, etc. (and is delta-function correlated in time and space). The key is to determine an appropriate form for $\underline{J} = \underline{J}_{up} + \underline{J}_{down} + \underline{J}_{relax}$ +... Here, $\underline{J}_{up}$ is a destabilizing uphill-current resulting from the step-edge barriers[42,43] as proposed by Villain;[28] $\underline{J}_{down} \approx -\nu_{DF}\, F\, \nabla h$ is a downhill current due to downward funneling deposition dynamics as proposed by our group;[44] $\underline{J}_{down}$ is proportional to the density of step edges ($\propto\nabla h$); $\underline{J}_{relax}$ denote relaxation terms which have the form of the Mullins-surface-diffusion-type term ($\nu\nabla\cdot\nabla^2 h$) and generalizations.[43] The origin and nature of these terms is a somewhat open question. For example, our model does not include conventional surface diffusion with bond-breaking, so the Mullins-type term derives from other sources.

The formulation (4) provides a framework with which to understand the basic features of multilayer growth observed in our system.[39] First consider growth at 300K. In the initial stage of growth up to ~30ML, mounds are formed from 2D islands as growth evolves from a quasi-layer-by-layer mode. In the next extended stage up to ~1500ML, the sides of these mounds steepen as $\underline{J}_{up}$ dominates $\underline{J}_{down}$. Thus, the film roughens quickly while the mounds coarsen slowly (in the lateral direction). Finally, above ~1500ML, there is a regime of "slope selection" for the slope of the sides of the mounds where $\underline{J}_{down}$ counterbalances $\underline{J}_{up}$. Here, film roughening (due to increased mound height), and mound coarsening, necessarily occur at the same rate. At lower T, the initial regime and the steepening regime are contracted, which ultimately leads to smoother growth. Why the contraction in these regimes? At lower T, the individual submonolayer islands, and thus the lateral dimension of the mounds, is much smaller. Consequently, mounds reach their selected slope more quickly (even though this selected slope increases somewhat at lower T).[17,44]

There is much interest in the mound coarsening dynamics in this asymptotic slope-selection regime. Sophisticated theoretical analyses of the continuum equations (4) incorporating up-down symmetry of the film surface suggested that mound dynamics depends sensitively on the relaxation terms, and the substrate symmetry.[45,46] In these studies, alternating arrays of mounds and inverted mounds form, the evolution of which is controlled by the evolution of "rooftop defects". Our simulations reveal a very different dynamics.[18] Ordered arrays of mounds form (with no inverted mounds) reflecting strong up-down symmetry breaking, so valley floors between mounds are preferred over rooftops. Coarsening occurs by a highly cooperative, fluctuation-dominated mechanism, an example of which is illustrated in Fig.10.[18]

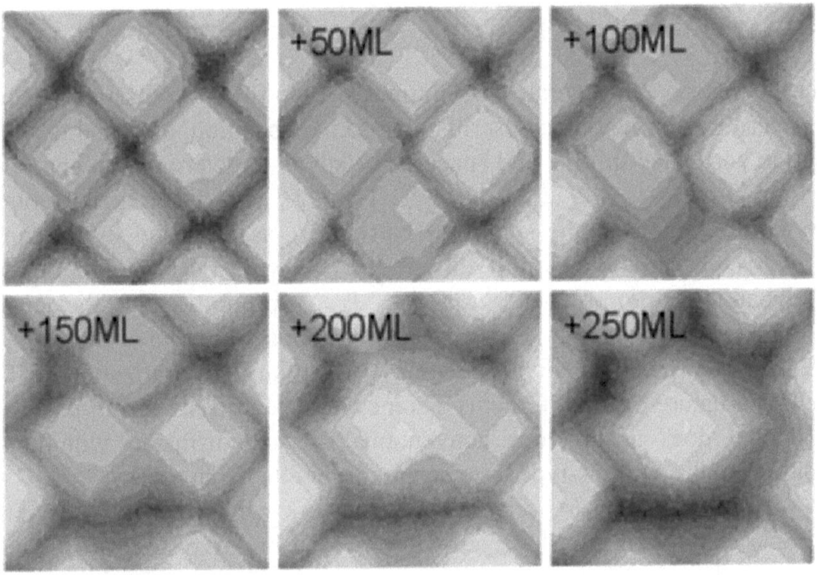

Figure 10. Bird's eye view of complex mound coarsening dynamics within ordered 1×1 arrays of square mounds (aligned with [110] edges). These form during prolonged deposition at 230K with F≈0.02ML/s in our atomistic model. Dynamics involves fluctuation-mediated disappearance of single mounds, triggering corner-to-corner coalescence, and simultaneous disappearance of nearby mounds. Images are 35×35 nm^2. Coverage increments are 50ML. Reprinted with permission from Ref. 18. Copyright 2002 American Institute of Physics.

5. ATOMISTIC AND CONTINUUM MODELING OF POST-DEPOSITION RELAXATION FOR AG/AG(100)

5.1 SUBMONOLAYER RELAXATION

For submonolayer homoepitaxial systems, where the substrate is "completely wet" by the 2D islands, the natural and traditional expectation was that post-deposition coarsening of the adlayer would occur via Ostwald Ripening (OR).[47] The remarkable discovery first made for the Ag/Ag(100) system at 300K was that large 2D clusters or islands of 100's to 1000's of adatoms have significant diffusive mobility.[48] Furthermore, cluster diffusion and subsequent coalescence, i.e., Smoluchowski Ripening (SR) rather than OR actually dominates the coarsening process at 300K (which occurs on the time-scale of hours).[49] The diffusion coefficient of a cluster with linear size ~L scales like $D(s) \sim L^{-\alpha}$, with $\alpha \approx 2.2$ for L=10-20 at T=300K, and the mechanism was proposed to be periphery diffusion (PD).[50] In fact, these and related studies have prompted much theoretical analysis of cluster diffusion (and, specifically, of the size scaling) via both atomistic[51-53] and continuum[54] models. Simplistic analysis of cluster diffusion via PD notes that the center-of-mass of the cluster moves by $\delta d=L^{-2}$ for each time one of the ~L periphery atoms makes a hop. Thus, one has $D\sim L(\delta d)^2\sim L^{-\alpha}$, with $\alpha=3$. The deviation of the experimental α from 3 might suggest a mechanism other than PD,[55] however the arguments given above indicate that PD has an energetic advantage and thus should dominate (at least for "smaller" sizes).[19] Instead, the deviation more likely derives from the feature that cluster diffusion is limited by nucleation of a new edge[52,56] (a process which will lower exponent to 1), and also the scaling is modified by the presence of a kink rounding barrier[22] (which also reduces exponent). We do caution, however, that in a different regime (larger L or higher T), a different mass-transport mechanism could dominate cluster diffusion.[57]

Rather than discuss further cluster diffusion, here we focus on the interesting PD-mediated sintering or coalescence process, which occurs after collision of two diffusing clusters. This process is a 2D analogue of conventional 3D sintering. Specifically, we consider the relaxation of the irregular cluster thus formed back to its near-square equilibrium shape. If its typical linear size is L, then the characteristic relaxation time, τ, scales like $\tau\sim L^n$, where it is often proposed that $n\approx\alpha+1$. Thus, the "expected" value for PD-mediated relaxation is $n\approx4$. However, it is now recognized that actual behavior can deviate from this prediction, depending on cluster size and shape,[22,56] and on the presence of a kink rounding barrier.[22]

The only experimental data available for size-scaling involves the "typical" case of corner-to-corner collision of two islands for Ag/Ag(100) at 300K forming dumbbell-shaped clusters. See Fig.11. The experiments suggest that $\tau\sim L^n$, with $n\approx3$ deviating from the expected value.[19] Detailed simulation studies of the atomistic model for PD described in Sec.3 reveal that the value of the exponent for dumbbell relaxation equals the classic value of n=4 for in the *absence* of a kink rounding barrier ($\delta=0$). However, n decreases toward 3 upon increasing δ, crossover occurring when the "kink Ehrlich-Schwoebel length" $L_r=\exp[\delta/(k_BT)]$ increases above L.[22] Our original modeling of the experimental data choose $\delta=\phi$ and found that $E_{PD}=E_e+\phi+\delta\approx0.75$eV matched experimental relaxation rates (so that $\phi\approx\delta\approx0.25$eV choosing $E_e=0.25$eV).[19] Our subsequent studies suggested lower values for E_{PD} and δ.[21,22]

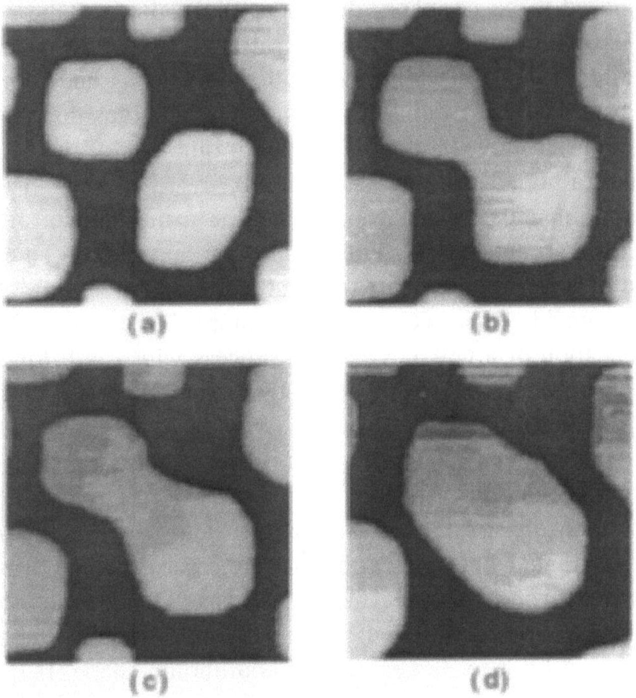

Figure 11. Experimental STM images (50×50 nm^2) of dumbbell formation and relaxation after corner-to-corner collision of Ag islands on Ag(100) at 300K. Image times are: (a) 0 min.; (b) 46 min. (<10 min. after collision); (c) 91 min.; and (d) 271 min.

To gain a simplistic understand the size-scaling[22] of τ in dumbbell relaxation, consider the removal of a complete "outer layer" of ~L atoms towards the neck by transferring them around kink sites a distance ~L away. As indicated in Fig.12, this overall process is essentially a "random walk" between configurations with different numbers of transferred atoms. However, the energy decreases upon transfer of the last atom biasing the walk, so this configuration is an (imperfect) adsorbing state for the "walk". Thus, based on Einstein's relation for random walks, the time to transfer ~L atoms scales like the square of the number of transferred atoms, i.e., $\tau_{layer} \sim L^2 \tau_0$, where τ_0 is the typical time to transfer a single atom. Detailed analysis shows that $\tau_0 \sim (L_r+L)\tau_k$, where $\tau_k = 1/h_k$. The overall relaxation process requires transfer of ~L layers so that[22]

$$\tau \sim L\tau_{layer} \sim L^3\tau_0 \sim L^3(L_r+L)\tau_k. \qquad (5)$$

This expression reveals the behavior $\tau \sim L^4$ for $L >> L_r$, and $\tau \sim L^3$ for $L << L_r$.

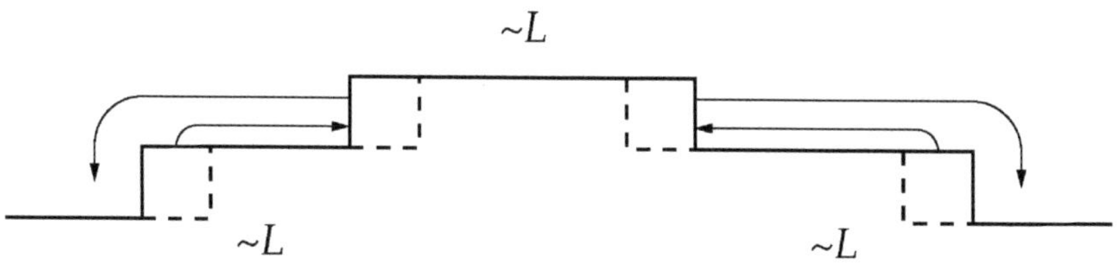

Figure 12. Schematic of adatom mass transfer from an outer terrace of a non-equilibrium nanostructure during shape relaxation via periphery diffusion. Typical terrace lengths are ~L sites. Atoms are transferred reversibly between this outer terrace and the kinks on the lower terrace, as the system undergoes a "random walk' in configuration space. However, complete removal of the outer terrace lowers the energy, thus biasing evolution in this direction.

We briefly mention that relaxation of rectangular clusters formed by side-to-side collision of square islands is distinct from the above case, being mediated by nucleation of new edges. Deviation from classical behavior occurs for L below the characteristic length $L_c=\exp[\frac{1}{2}\phi/(k_BT)]$, which gives the separation of kinks on a close-packed step edge.[22,56] The same processes control cluster diffusion. For $L<<L_c$, one finds that $n\approx2$ for $\delta=0$, with lower values for $\delta>0$.[22,56] The experimental value of $\alpha\approx2.2$ (or $n\approx3.2$) presumably reflects crossover from the classic value of n=4 to this lower value.

Finally, we comment on the Mullins-type continuum theory for relaxation via PD. Here, the cluster perimeter is described by a continuous curve. Let V_n denote its normal velocity, $\underline{J}_{PD}$ the mass current around the perimeter, μ the chemical potential of step edge atoms, σ_{PD} the mobility of step edge atoms, $\tilde{\beta}$ the step edge stiffness, and κ the step edge curvature. Then, one has

$$V_n \propto -\nabla\cdot\underline{J}_{PD}\text{, with } \underline{J}_{PD} \propto -\sigma_{PD}\nabla\mu \text{ and } \mu \propto \tilde{\beta}\,\kappa\,, \qquad (6)$$

the first equation imposing mass conservation, and ∇ denoting the derivative along the perimeter. Since σ_{PD} and $\tilde{\beta}$ depend only on step-edge orientation, it is straightforward to show that this continuum formulation always produces exactly the scaling relation $\tau \propto L^4$. In fact, the continuum formulation reproduces remarkably well results of atomistic simulations for relaxation of dumbbells when $\delta=0$, even for small L (see Fig.13), significant deviation only occurring when $\delta>0$. We have also successfully used the continuum formalism to model observed "pinch-off" phenomena observed for large "worm-like" nanovacancies formed within an adlayer created by deposition of about 0.8ML.[21] This pinch-off phenomenon is actually shown to be a signature of relaxation via PD.[21]

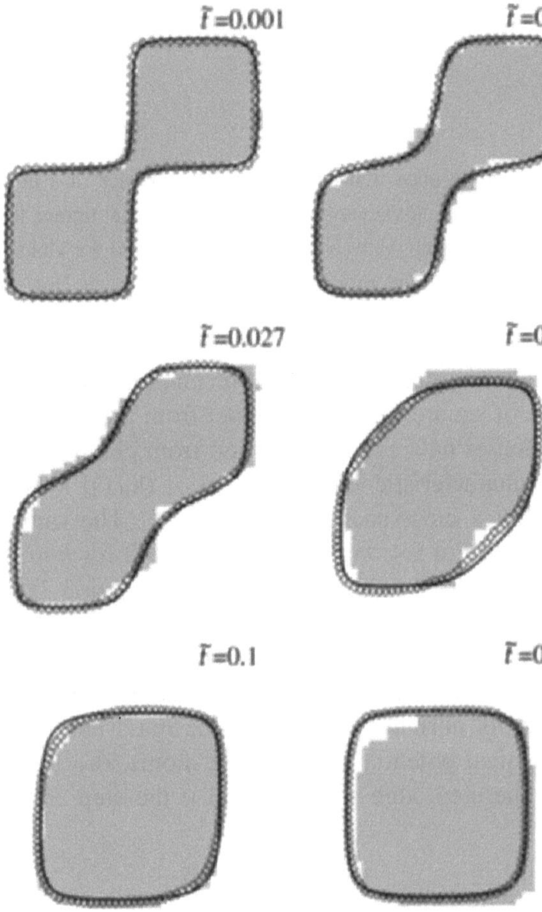

Figure 13. Comparison of predictions of atomistic and continuum modeling for relaxation of dumbbell-shaped clusters when δ=0. Parameters are T=300K, ϕ=0.235eV (and δ=0). In the atomistic simulations, two 20×20 atom clusters initially touch corner-to-corner. Symbols show simulation results for the shape averaged over 100 KMC runs, and the gray area shows the configuration for a single run. Results of continuum modeling with constant σ_{PD}, but with the correct anisotropic step edge stiffness, are shown as solid curves. Scaled times shown are in units of τ=2.1 min, using E_e=0.26 eV.

5.2 MULTILAYER RELAXATION

In Sec.4.2, we indicated that the second layer coverage, θ_2, after deposition of ~1ML at, say, 300K, was quite sensitive to the magnitude of the step-edge barrier. Thus, E_{ES} in our model could be determined by comparing model predictions for with experimental observations. One complication, however, is that one must determine θ_2 immediately after deposition, and it takes at least several minutes to obtain the first STM image. During this time, the film smoothens, mass being transferred from the second layer to vacancies in the first layer via dissolution of the second layer islands (a two-layer OR-type process). The difference in behavior from submonolayer relaxation, which is dominated by SR, presumably reflects the feature that the edges of first layer vacancies provide stronger sinks for atoms than the edges of same-layer islands.

Thus, if δt denotes the time since deposition, we monitor the decrease in $\theta_2(\delta t)$ with δt and extrapolate back to $\delta t=0$. Using this procedure for the data in Fig.14 yields a best fit of $E_{ES} \approx 0.085$eV for the close-packed step edge (but values between 0.07eV and 0.10eV are plausible given the uncertainty in $\theta_2(0)$ from extrapolation and statistical factors).

One can quantify the rates of decay of individual second layer islands by solving a steady-state (continuum) diffusion equation for N_1 with the appropriate boundary conditions. This approach has been used previously.[58] One assumes that the density of adatoms at the edge of second layer islands of "radius" R is determined by the Gibbs-Thompson equation,

$$N_1 \approx N_{eq+}(R) \approx N_1(eq)\cdot\exp[A\beta^{[110]}/(Rk_BT)]. \qquad (7)$$

Here, $N_1(eq)$ is the equilibrium adatom density on an island free terrace, $\beta^{[110]} \approx 0.13$ eV/a is the free energy of [110] step edges, and A is a geometrical factor of order unity. The boundary condition at the edges of vacancy islands is potentially more complicated because of the presence of a step-edge barrier. The effect of this barrier is measured by the "Schwoebel length", $L_{ES}=\exp[E_{ES}/(k_BT)]$ ($\approx$15 at 300K). If L_{ES} is much smaller than the characteristic separation between islands and vacancies, then the step-edge barrier will not significantly hinder interlayer transport. Then, one has $N_1 \approx N_{eq-}(R)$, at the edge of vacancy islands of "radius" R, where N_{eq-} and N_{eq+} differ only in the sign of the exponent.[4,58] Adopting this approximation, we have solved the steady-state diffusion equation for N_1 for an experimental film configuration, results for the mass flow being shown in Fig.15.

Finally, we comment on the interesting problem of characterizing the smoothening of more complex multilayer mounded film morphologies. There has been long-standing interest in the continuum formulation of this problem, which is particularly challenging due to faceting below the thermal roughening transition.[59,60] Atomistic simulations of the process also exist.[61] The general conclusion is that the film roughness often tends to decrease (roughly) linearly with time.

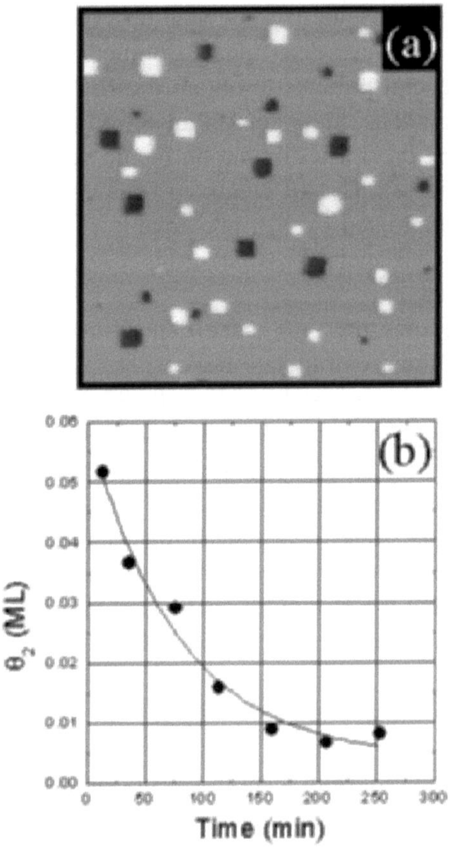

Figure 14. Smoothening of a 1.02 ML Ag film deposited on Ag(100) at 300K with F=0.02 ML/s: (a) STM image (100×100 nm^2) taken 13 min. after deposition; (b) Decrease of the second layer coverage, $\theta_2(\delta t)$, with time since deposition, δt. Extrapolation back to $\delta t=0$ yields $\theta_2(0)\approx0.059$ML.

Analogous to our discussion of submonolayer relaxation, rather than considering global aspects of the smoothening of the entire film, one can focus on the decay of individual 3D nanostructures (i.e., mounds). Most studies of this problem either develop an appropriate atomistic model to be analyzed by KMC simulation,[62] or adopt a formulation where lateral positions of step edges within each layer are treated as continuous curves (but discrete vertical layers are retained).[63] Here, we just elaborate on one atomistic study which is specific to the decay of mounds in metal(100) homoepitaxial systems. An atomistic model was developed which input appropriate attachment-detachment kinetics for atoms in each layer. It was used to explore the difference in decay behavior for uniform versus non-

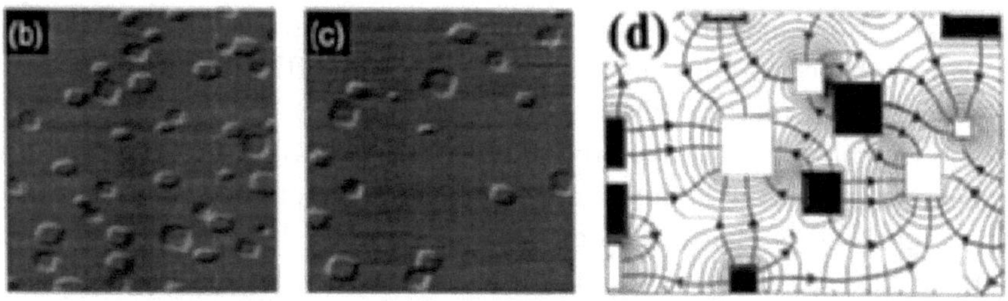

Figure 15. Smoothening of a 1ML Ag film deposited on Ag(100) at 300K: STM images (100×100 nm^2) taken 45min. (b) and 139 min. (c) after deposition; solution of the continuum steady-state (Laplacian) diffusion equation for interlayer mass transport (d), as described in Sec.5.2. The film configuration corresponds to the lower right corner of (b). Dark square are second layer islands; open squares are first layer vacancies; the thin lines are contours of constant adatom density; the curves with arrows indicate lines of diffusive flux from second layer islands to first layer vacancies. Reprinted with permission from Ref. 4. Copyright 2000 American Chemical Society.

uniform step-edge barriers. (In the latter case, kinked or open steps were assigned the lower barrier, entirely consistent with our multilayer growth studies.) The main conclusion is that decay preserves the slope of the mounds for a uniform step-edge barrier, but that the slope steepens during decay for preferential downward transport at kinks.[62] The latter feature seems to apply to the decay of Cu mounds on Cu(100) at 300K.[62]

6. EFFECT OF A CHEMICAL ADDITIVE (OXYGEN) ON AG/AG(100) FILM EVOLUTION

Chemical additives have long been used to manipulate overall film growth behavior. Typically, a “surfactant” is used to convert rough growth to smooth growth.[64,65] To understand this effect, attention is paid to the influence of the surfactant on the formation of individual nanostructures (e.g., islands) during growth. Much less attention has been paid to the influence of chemical additives on post-deposition relaxation. One notable exception is a study of multilayer Au mounds on Au(111) which were found to be quite stable in ultra-high vacuum conditions, but which quickly decayed when exposed to air.[66]

We are thus motivated to explore how chemical additives can influence nanostructure evolution in the Ag/Ag(100) system. We choose oxygen for the additive since the flat Ag(100) surface is relatively unreactive to this species,[67] thereby facilitating our control of its influence relative to the pristine system (without oxygen). Specifically, we expose the surface to molecular oxygen. It has been shown that a chemisorbed molecular state is stable on the perfect Ag(100) surface below ~140K, but that dissociative adsorption of O_2 is thermally activated only at or above that temperature. Thus population of the Ag(100) surface with oxygen is difficult except for carefully chosen T or alternatively for very high exposures. On the other hand, it was shown that dissociative adorption of O_2 is enhanced at step edges, likely at kink sites, providing an alternative pathway for population of the surface with oxygen.[68] It is plausible step-edge sites provide the energetically most favorable sites for atomic oxygen, but that O can detach, diffuse across the terraces and reattach to these sites.

Our first experiments[69] found no significant influence of oxygen on submonolayer island formation if the surface was simultaneously exposed to depositing Ag and to O_2. However, this probably does not allow for sufficient buildup of oxygen. In contrast, in subsequent studies[70] where the surface is pre-exposed to up to 30L of O_2, and then Ag is deposited at 250K show a decrease in the island density by a factor of ~20. See Fig.16. (There is evidence that the extent of this effect depends on the global Ag sample morphology. Oxygen may attach to the Ag sample in faceted regions far from the perfect terrace examined by STM, and then migrate to this region.) Additional studies show that the exponent describing the variation of N_{isl} with F increases from 1/3 with no oxygen [see the expression following equation (1)] to ~0.9 for 30L exposure. The latter increase is evidence for a transition from irreversible to strongly reversible island formation.

The first experiments noted above did however use surface-sensitive diffraction techniques to show that exposure to oxygen did activate coarsening of (submonolayer) adlayers at "low" temperatures around 250K.[69] For the oxygen-free system SR dominates at 300K, and would continue to dominate upon lowering T. However, by 250K, coarsening would be slower than at 300K by a factor of ~300, and thus be effectively inactive. Subsequent STM experiments[71] revealed that not only was submonolayer coarsening activated by oxygen at 250K, but that the dominant pathway is changed from SR to OR. Analogous studies of the (potential) decay of 3D mounds reveal that this process is inactive at 250K for oxygen-free systems, but is activated by exposure to O_2.[71] For submonolayer coarsening, the mechanism should be reflected in the time-dependence of the decreasing island density. Analysis of the appropriate rate equations reveals that for long times t after deposition one has that $N_{isl} \sim t^{-2/3}$ for OR, but that $N_{isl} \sim t^{-2/(2+\alpha)}$ for SR, where α is the exponent describing the size-dependence of cluster diffusion.[72,73] In practice, these times are difficult to achieve experimentally,[73] so instead we have examined the width of the island size distribution,[71] and shown that it evolves to values consistent with OR.[8]

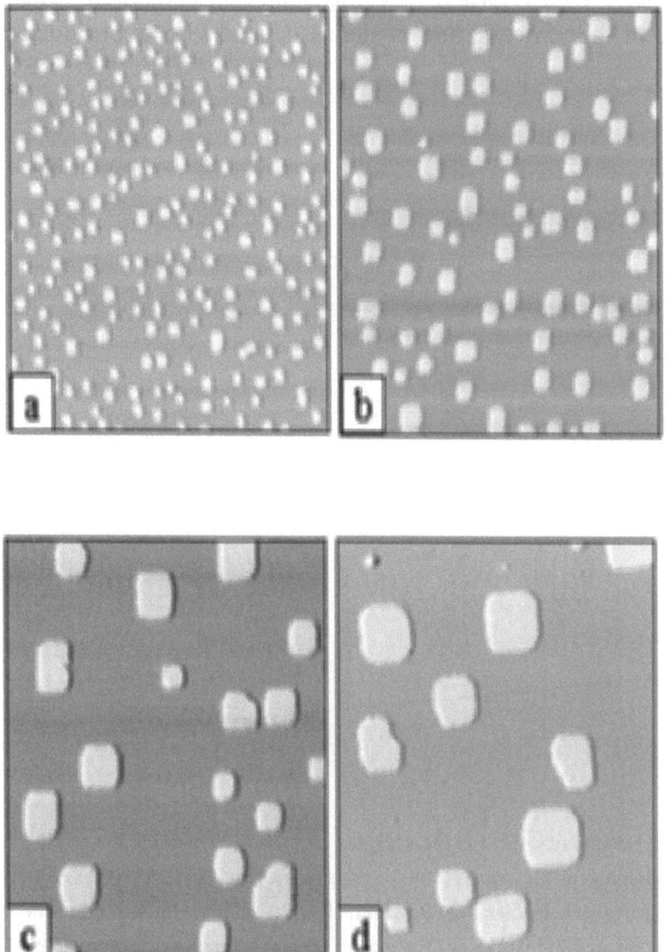

Figure16. Islands formed on an Ag(100) surface pre-exposed to various amounts of oxygen prior to Ag deposition at 250K with F=0.002ML/s. All images span 125×125 nm^2 and correspond to O_2 exposures in Langmiurs (L) of: (a) 0L (clean); (b) 5L; (c) 15L; (d) 30L.

Next, we discuss the likely mechanism underlying the influence of oxygen on post-deposition coarsening and smoothening, and then note that this picture is consistent with observations of its effect on submonolayer island formation. The central idea in our explanation for enhanced coarsening is that the presence of oxygen in the system allows for the generation of "volatile complexes" which can facilitate the transport of Ag across the surface. Specifically, we propose that a species of the form Ag_nO, with n=1 or 2, can "readily" detach from step edges, diffuse across terraces, and reattach at other step edges (where "readily" means compared with Ag).[71] A schematic of this process, and of the associated PES is provided in Fig. 17.

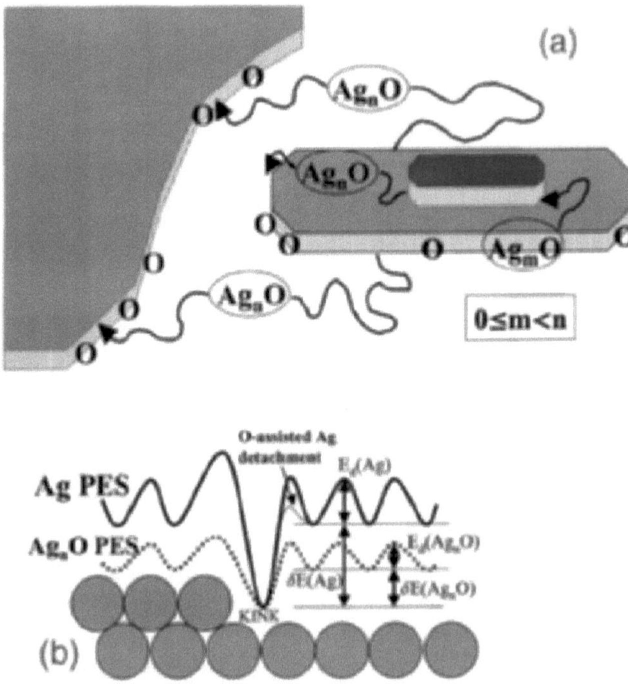

Figure 17. Schematics for: (a) the complex mass-flow dynamics underlying coarsening and smoothening in the Ag/Ag(100) + oxygen system; Ag is transported between step edges by an Ag-rich species, Ag_nO, with n=1 or 2; islands with step edges depleted of O can be replenished by diffusion of a second species, Ag_mO, with m<n (likely O, or perhaps AgO if n=2); (b) the PES for adspecies detaching from a kink site at a step edge: for Ag in an O-free-system (solid curve); for Ag with detachment assisted by O at the step edge, which might lower the barrier for the first hop, but not the overall detachment barrier (thin dotted curve); and for an Ag_nO species exhibiting easy detachment and/or terrace diffusion - relative to Ag (thick dotted curve). Reprinted with permission from Ref. 71. Copyright 2002 American Institute of Physics.

Since the effective barrier for mass transport via this mechanism is the sum of the terrace diffusion barrier, and the binding energy of the diffusing species at kink sites, it follows that one or both of these quantities must be reduced for Ag_nO compared to Ag.[71] One other caveat is that the oxygen supply at the edge of "small" islands could be depleted by detachment of Ag_nO

potentially freezing out further evolution. Since this is not observed, we propose that O (and/or AgO if n=2) can also detach from step edges repopulating any depleted islands.

This same model could explain the influence on island formation of pre-exposure to oxygen. If O can detach from step edges, it can attach to depositing Ag modifying their diffusion characteristics, and more significantly destabilize small islands of Ag as they form (the fragments likely including Ag_nO). The challenge for modeling these complex mass-flow dynamics are that there are several models consistent with experimental observations. Key input allowing unique determination of the correct picture would be *ab-initio* values for key binding energies and diffusion barriers for the different possible participating complexes.

Finally, we note that the concept of "volatile complexes" participating in and enhancing mass transport between metal nanostructures is familiar in the field of catalysis,[74] and elsewhere.[75] Exposure of supported Pt catalysts to oxygen leads to enhanced sintering of Pt-nanoclusters, i.e., catalyst degradation, presumably due to the formation of a Pt_2O complex.[74] Indeed much of the recent interest in oxygen + Ag surface systems derives from the use of Ag as an industrial catalyst in epoxidation of ethylene.[67,68]

7. CONCLUSIONS

We have shown that rather "simple" metal(100) homoepitaxial thin film systems can display a rich and unexpected variety of nanostructure formation and evolution phenomena. Furthermore, we are often able to accurately model this behavior on the correct time and length scales with suitable atomistic lattice-gas models exploiting KMC simulation. Indeed, either by using tailored atomistic models, or by developing appropriate "coarse-grained" continuum models, we have been particularly successful in connecting-the-length-scales in describing behavior of 2D nanostructures spanning 10's of nanometers, or 3D mounded film morpholigies in the regime of 1000's of layers (thicknesses of 100's of nanometers). With the addition of chemical additives, a much more diverse variety of complex-mediated mass-flow dynamics is possible. More extensive input from *ab-initio* calculations of the relevant energetics is required to develop appropriate atomistic models of nanostructure evolution in these systems.

ACKNOWLEDGEMENTS

KJC, ARL, PAT, and JWE were supported for this work by NSF Grant CHE-0078596. DJL and JWE were also supported by NSF Grant EEC-0085604. This work was performed at Ames Laboratory, which is operated for the U.S. Department of Energy by Iowa State University under Contract

No. W-7405-Eng-82. Work by MCB was performed under the auspices of the U.S. Department of Energy by the University of California, Lawrence Livermore National Laboratory under contract No. W-7405-Eng-48.

KJC current address: Department of Chemistry and Biochemistry, University of California - Los Angeles, Los Angeles, CA 90095, and California Institute of Technology, Pasadena, CA 91125

ARL current address: Department of Chemistry and Biochemistry, Denison University, Granville, OH 43023.

APPENDIX: KINETIC MONTE CARLO (KMC) SIMULATION PROCEDURES

The basic task of KMC is to mimic the actual film growth process, tracking in the computer the evolution of the configuration of the film by simulating all the relevant deposition and diffusion events as random or stochastic processes with the *correct relative rates*. Consider a simple example where deposition and terrace diffusion are the only processes with finite rates. The naïve way to do this is to randomly pick a site, and then attempt one of these processes with probabilities proportional to their rates, i.e., attempt to hop with probability $4h/(4h+F)$, and to deposit with probability $F/(4h+F)$. Here, h is the hop rate per direction and there are 4 directions (one of which is chosen randomly).[3] This is very ineffective since most sites are empty, but $h>>F$. Thus, typically one picks a site, attempts to hop and fails since the site is empty.

The alternative Bortz-type approach[76] is to keep lists of all N_{hop} hopping atoms, and of all N_{dep} sites available for adsorption. Then, the total rate for hopping is $R_{hop}=4hN_{hop}$ and for deposition is $R_{dep}=FN_{dep}$, so $R_{tot}=R_{hop}+R_{dep}$. In this approach, at each step one chooses to hop with probability $P_{hop}=R_{hop}/R_{tot}$ and to deposit with probability $P_{dep}=R_{dep}/R_{tot}$. If the former, one randomly chooses a hopper from the list, and a direction, and implements the hop. The lists are updated and the time incremented by $\delta t=1/R_{tot}$. In reality, since most sites are empty, it is not worthwhile to maintain a list of these sites in treating adsorption. One simply replaces the above total deposition rate by $R_{dep}=FL^2$, and attempts to adsorb at randomly chosen sites, accepting some small fraction of failures.

It should be clear to the reader how this Bortz-approach generalizes to the typical cases where there are several types of processes with distinct rates, and lists are kept of atoms in each of these categories.[76] In this way one can readily calculate the total rate for each type of process, and if required implement the process by randomly choosing a member from the list. The complicated part of the algorithm is the "book-keeping" involved with updating the lists.

REFERENCES

1. *Morphological Organization in Epitaxial Growth and Removal* Eds. Zhang Z. and Lagally M.G. (World Scientific, Singapore, 1998).
2. Voter A.F. Phys. Rev. Lett. (1997) **78**: 3908; J. Chem. Phys. (1997). **106**: 4665.
3. Binder K. and Landau D.P. *a Guide to Monte Carlo Simulations in Statistical Physics* (Cambridge U.P., Cambridge, 2000).
4. Thiel P.A. and Evans J.W. (2000) J. Phys. Chem. B **104**: 1663
5. Evans J. W., Sanders D. E., Thiel P. A. and DePristo A. E. (1990) Phys. Rev. B **41**: 5410.
6. Baxter R.J. *Exactly Solved Models in Statistical Mechanics* (Academic, London, 1982).
7. Mueller-Krumbhaar H. in *Monte-Carlo Methods in Statistical Physics*, Topics in Current Physics, Vol. 7 Ed. Binder, K. (Springer, Berlin, 1986).
8. Ardell A.J., Phys. Rev. B (1990) **41**:2554.
9. Stoldt C.R., Caspersen K.J., Bartelt M.C. Jenks C.J., Evans J.W., and Thiel P.A., Phys. Rev. Lett. (2000) **85**: 800.
10. Botez C.E., Elliott W.C., Miceli P.F., and Stevens P.W., Mat. Res. Soc. Symp. Proc. (2001) **672**: O2.7.1.
11. Yu B. D. and Scheffler M. (1997) Phys. Rev. Lett. **77**: 1095.
12. Zhang C.-M., Bartelt M. C., Wen J.-M., Jenks C. J., Evans J. W. and Thiel P. A. (1998) Surf. Sci. **406**: 178.
13. Voter A. F. (1987) SPIE Modeling of Optical Thin Films **821**: 214; Phys. Rev. B (1986) **34**: 6819.
14. Barkema G.T., Biham O. Breeman M., Boerma D.O, and Vidali G. (1994) Surf. Sci. Lett. **306**: L569.
15. Mehl H., Biham O., Furman I., and Karimi M., Phys. Rev. B (1999) **60**:2106.
16. Bogecivic A., Stromquist J., and Lundqvist B.I., Phys. Rev. Lett. (1998) **81**: 637.
17. Caspersen K.J., Stoldt C.R., Layson A.R, Bartelt M.C., Thiel P.A., and Evans J.W., Phys. Rev. B (2001) **63**: 085401.
18. Caspersen K.J., Layson A.R., Stoldt C.R. Fournee V., Thiel P.A., and Evans J.W., Phys. Rev. B (2002) **65**: 193407
19. Stoldt C. R., Cadilhe A. M., Jenks C. J., Wen J.-M., Evans J. W. and Thiel P. A. (1998) Phys. Rev. Lett. **81**: 2950; Cadilhe A. M., Stoldt C. R., Jenks C. J., Thiel P. A., and Evans J. W. (1998) Phys. Rev. B (2000) **61**: 4910.
20. Bartelt M. C. and Evans J. W. (1993) Surf. Sci. **298**: 421.
21. Pai W.W., Wendelken J.F., Stoldt C.R., Thiel P.A., Evans J.W. and Liu D.-J., Phys. Rev. Lett. (2001) **86**: 3088.
22. Liu D.-J., and Evans J.W., Phys. Rev. B (2002) **66**: 165407.

23. Furman I., Biham O., Zuo J.-K., Swan A.K., and Wendelken J.F., Phys. Rev. B (2000) **62**: R10649.
24. If island structure does not change much if one switches on kink escape (but still forbids core breakup), then our simple treatment of PD applies as long as core breakup is much slower than other processes.
25. Bartelt M.C. and Evans J.W., Surf. Sci. Lett. (1994) **314**: L829.
26. Zhong J., Zhang T., Zhang Z., and Lagally M.G., Phys. Rev. B (2001) **63**:113403.
27. Ehrlich G. and Hudda H. (1966) J. Chem. Phys. **44**: 1039; Schwoebel R. L. and Shipsey E. J. (1966) J. Appl. Phys. **37**: 3682.
28. Villain J. (1991) J. Physique I (France) **1**: 19.
29. Caspersen K.J. and Evans J.W., in *Atomistic Aspects of Epitaxial Growth*, Proceedings of NATO ARW in Corfu, Greece, June, 2001 (Kluwer, Dordrecht, 2002).
30. Bales G.S. and Zangwill A., Phys. Rev. B (1997) **55**: 1973.
31. Venables J. A. (1973) Phil. Mag. **27**: 697
32. Bartelt M. C. and Evans J. W. (1996) Phys. Rev. B **54**: R17359.
33. Bartelt M. C., Schmid A. K., Evans J. W. and Hwang R. Q. (1998) Phys. Rev. Lett. **81**: 1901.
34. Bartelt M. C., Stoldt C. R., Jenks C. J., Thiel P. A. and Evans J. W. (1999) Phys. Rev. B **59**: 3125.
35. Gibou F.G., Ratsch C., Gyure M.F., Chen S., and Caflisch R.E., Phys. Rev. B (2001) **63**: 11505; Ratsch C., Gyure M.F., Chen S., Kang M., and Vvedensky D.D., *ibid* (2000) **61**: R10598; Ratsch C., Gyure M. F., Caflisch R. E., Gibou F., Petersen M., Kang M., Garcia J., and Vvedensky D. D., *ibid* (2002) **65**: 195403.
36. Mulheran P.A., and Robbie D.A., Europhys. Lett. (2000) **49**: 617.
37. Amar J.G., Popescu M.N., and Family F. Phys. Rev. Lett. (2001) **14**: 3092.
38. Evans J.W., and Bartelt M.C., Phys. Rev. B (2001) **63**: 235408.
39. Evans J.W., and Bartelt M.C., Phys. Rev. B (2002) **66**: 235410; Li M., Bartelt, M. C., and Evans J. W., Phys. Rev. B (2003) **68**: 121401 (R).
40. Elliot W. C., Miceli P. F., Tse T. and Stephens P. W. (1996) Phys. Rev. B **54**: 17938.
41. Barabasi A.-L. and Stanley H. G. *Fractal Concepts in Surface Growth.* (Cambridge University Press, Cambridge, 1995).
42. Johnson M.D., Orme C., Hunt A.W., Graff D., Sudijono J., Sander L.M., Orr B.G., Phys. Rev. Lett. (1994) **72**: 116.
43. Politi P., and Villain J. Phys. Rev. B (1996) **54**: 5114.
44. Bartelt M. C., and Evans J. W. (1995) Phys. Rev. Lett. **75**: 4250; (1999) Surf. Sci. **423**: 189.
45. Siegert M. Phys. Rev. Lett. (1998) **81**: 5481.
46. Moldovan D. and Golubovic L., Phys. Rev. B (2000) **61**: 6190.
47. Zinke-Allmang M., Feldman L.C., and Grabow M.H., Surf. Sci. Rep. (1992) **16**: 377.

48. Wen J.-M., Burnett J. W., Chang S.-L., Evans J. W. and Thiel P. A. (1994) Phys. Rev. Lett. **73**: 2591.
49. Wen J.-M., Evans J. W., Bartelt M. C., Burnett J. and Thiel P. A. (1996) Phys. Rev. Lett. **76**: 652.
50. Pai W. W., Swan A. K., Zhang Z. and Wendelken,J. F. (1997) Phys. Rev. Lett. **79**: 3210.
51. Sholl D.S., and Skodje R.T., Phys. Rev. Lett. (1995) **75**: 3158.
52. Mills G., Mattson T.R., Mollnitz L., and Metiu H., J. Chem. Phys. (1999) **111**, 8639.
53. Pal S., and Fichthorn K.A., Phys. Rev. B (1999) **60**: 7804.
54. Khare S.V., Bartelt N.C., and Einstein T.L, Phys. Rev. Lett. (1995) **75**: 2148.
55. Heinonen J., Koponen I., Merikoski J., and Ala-Nissila T., Phys. Rev. Lett. (1999) **82**: 2733.
56. Combe N., and Larralde H., Phys. Rev. B (2000) **62**: 16074.
57. Hannon J.B., Klunker C., Giesen M., Ibach H., Bartelt N.C., and Hamilton J.C., Phys. Rev. Lett. (1997) **79**: 2506.
58. Icking-Konert G.S, Giesen M., and Ibach H., Surf. Sci. (1998) **398**: 37.
59. Rettori A. and Villain J., J. Physique (Paris) (1988) **49**: 257.
60. Israeli N., and Kandel D., Phys. Rev. Lett. (2002) **88**: 116103.
61. Ramana Murty M.V., Phys. Rev. B (2000) **62**: 17004.
62. Li M., Wendelken J.F., Liu B.-G., Wang E.G., and Zhang Z., Phys. Rev. Lett. (2001) **86**: 2345.
63. Thürmer K., Reutt-Robey J.E., Williams E.D., Uwaha M., Emundts A., and Bonzel H.P., Phys. Rev. Lett. (2001) **87**: 186102.
64. Egelhoff W.F., Chen P.J., Powell C.J., Stiles M.D., and Mc Michael R.D., J. Appl. Phys. (1996) **79**: 2491.
65. Fiorentini V., Oppo S., and Scheffler M., Appl. Phys. A (1995) **60**: 399.
66. Cooper B.H., Peale D.R., McLean J.G., Phillips, R., and Chason E., Mat. Res. Soc. Symp. Proc. (1993) **280**: 37.
67. Buatier de Mongeot F., Cupolillo C., Valbusa U., and Rocca M., Chem. Phys. Lett. 270, 345 (1997); Buatier de Mongeot F., Rocca M., Cupolillo C., Valbusa U., Kreuzer H.J., and Payne S.H., J. Chem. Phys. (1997) **106**: 711; Schinkte S., Messerli S., Morgenstern K., Nieminen J, Schneider W.-D., J. Chem. Phys. (2001) **114**: 4206.
68. Costantini G., Buatier de Mongeot F., Rusponi S., Boragno C., Valbusa U., Vattuone L., Burghaus U., Savio L., and Rocca M., J. Chem. Phys. (2000) **112**: 6840; Savio L., Vattuone L., and Rocca M., Phys. Rev. Lett. (2001) **87**: 276101.
69. Layson A.R., and Thiel P.A., Surf. Sci. Lett. (2001) **472**: L151.
70. Layson A.R., Evans, J.W., Fournee V., and Thiel, P.A., J. Chem Phys. (2002) 118: 6467.
71. Layson A.R., Evans, J.W., and Thiel, P.A., Phys. Rev. B (2002) **65**: 193409.
72. Sholl D.S., and Skodje R.T., Physica A (1996) **231**: 631.

73. Stoldt C.R., Jenks C.J., Thiel P.A., Cadilhe A.M., and Evans J.W., J. Chem. Phys. (1999) **111**: 5157.
74. Harris P.J.F., Int. Materials Rev. (1995) **40**: 97.
75. K. Pohl, J. de la Figuera, M.C. Bartelt, N.C. Bartelt, P.J. Feibelman, and R.Q. Hwang, Bull. APS (1999) **44**: 1716.
76. Bortz A.B., Kalos M.H., and Lebowitz J.L., J. Comput. Phys. (1975) **17**: 10.

Chapter 4

THEORETICAL STUDIES OF SILICON SURFACE REACTIONS WITH MAIN GROUP ABSORBATES

C. H. Choi[a] and M. S. Gordon[b]

[a] *Department of Chemistry, College of Natural Sciences, Kyungpook National University, Taegu 702-701, South Korea*

[b] *Ames Laboratory, Department of Chemistry, Iowa State University, Ames Iowa 51001*

1. INTRODUCTION

Surface chemistry on semiconductor surfaces has gained enormous popularity recently and the interest is still growing. This may be partially due to the tremendous potential of the new functionalities of synthetically modified surfaces. Furthermore, the well-ordered silicon surface, especially Si(100)-2x1, provides a unique environment in which a great deal of existing chemical knowledge can be tested.

It is conceivable that, as the size of electronic devices based on crystalline silicon wafers shrinks, the surface characteristics of semiconductor materials become crucial[1,2] for the proper functioning of devices. Thus, the understanding and tailoring of interfacial chemistry of silicon surfaces in particular is increasing in importance. In addition to the traditional modifications of the silicon surface such as etching, doping and film deposition, much attention is being directed to synthetically modified surfaces[3] in the pursuit not only of enhanced properties for microelectronics, but other applications including sensors and biologically active surfaces. With the help of traditional organic and organometallic chemistry, a wide variety of new chemically modified silicon surfaces can be synthesized to provide fine tailoring of surface characteristics for a broad range of applications. By studying to what degree our knowledge of organic chemistry may be applied to silicon surfaces, insight into the fundamental characteristic features of surface reactions can be gained.

L.A. Curtiss and M.S. Gordon (eds.),
Computational Materials Chemistry: Methods and Applications, 125–190.

In addition, to gain the control needed to fabricate an organic function into existing semiconductor technologies and ultimately to make new molecule-scale devices, a detailed understanding of the adsorbate surface as well as interfacial chemical reactions and their products at the atomic/molecular level is critical. To accomplish this, both novel experiments and theoretical investigations need to play a significant role in the advance of this emerging field.

This chapter begins with a summary of the theoretical methodologies adapted for surface studies and then proceeds to a consideration of the unique features of clean silicon surfaces. Then, the main focus is directed to the structures and reaction mechanisms of organic molecules on silicon surfaces. Surface chemical reactions involving the Si-X interactions (X: various main group elements) are presented focusing on the mechanistic aspects.

2. THEORETICAL METHODS AND SURFACE MODELS

2.1 Slab Model

Traditional band structure programs that allow one to perform electronic structure calculations of solids, have been used to model surfaces.[4,5,6,7] In this approach, one models a real system using the basic repeating unit cell (or slab) with two-dimensional translational symmetry for the surface and then employs periodic boundary conditions. Such a model introduces the reciprocal or *k* vectors, leading to the *k*-dependent energy levels of the system. The calculations involve the evaluations of interactions over a certain range of cells (lattice sum). Non-self consistent field (non-SCF) methods such as Hückel theory[4], and the tight binding method (or Extended Hückel Theory)[5] were used early on to provide the potential, and this method is still frequently employed. With the advance of methodologies, more sophisticated density functional theory methods have been used in combination with plane-wave basis sets[6] or with Gaussian basis sets[7] to expand the wavefunctions. There are advantages to both approaches. Plane waves appear to be most useful for incorporating periodic boundary conditions, whereas gaussian expansions are more appropriate for describing mechanistic and energetic details.

The conjugate gradient[8] method has most frequently been used to find stable geometries. Recent analytic gradient development[9] based on the periodic fast multipole method (FMM)[10] may provide more promising approaches.

The main advantage of periodic boundary conditions is the elimination of "edge effects" or terminal atom problems that occur in the finite cluster approach. The Hamiltonian in slab calculations is generally limited to density functional theory[11], an approach that is not always an appropriate

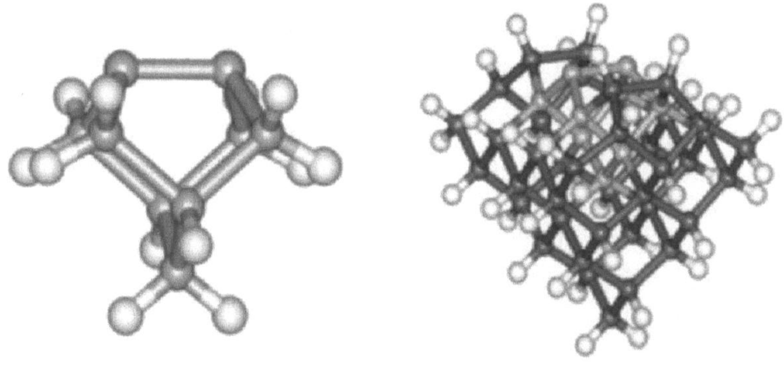

(a) (b)

Figure 1. Model clusters for Si(100)-2X1 surface. (a) 9-Si surface model. (b) 66-Si model.

choice.

2.2 Cluster Approach

The cluster approach suffers from the finite edge effects noted above. On the other hand, this approach can take advantage of well-developed high-level quantum chemical methods, which in some cases is essential for even a qualitative accounting of surface phenomena. In the cluster approach, the choice of a cluster that is representative of a real system is crucial, and not always feasible. Computational considerations limit the size of the cluster model, but the model needs to retain the essential features of the system of interest. Dangling bonds are usually capped with hydrogens[12] or simple capping atoms.[13] The most common cluster model that has been used for the Si(100)-2x1 surface is comprised of nine silicon atoms capped with hydrogens, Si_9H_{12} (see Figure 1a).[12] Since this cluster contains only one dimer, it will clearly be inadequate for investigating phenomena for which dimer-dimer interactions are important or in which two or more dimers are involved in a surface reaction mechanism. Shoemaker, et al.[14] systematically studied clusters of varying size up to $Si_{66}H_{52}$ with the GVB-PP1[15] (Generalized Valence Bond) level of theory, the simplest correct model for a singlet diradical, to investigate the utility of such clusters as prototypes for the study of silicon surfaces. They revealed that surface rearrangement due to dimer bond formation is "felt" several layers into the bulk, so small clusters, such as Si_9H_{12}, cannot necessarily adequately represent bulk behavior.

2.3 QM/MM approach

Hybrid quantum mechanics/molecular mechanics (QM/MM) techniques have gained popularity for modeling large molecular systems. In this approach, one assumes that a large molecular system can be partitioned into a small, chemically active part where a reaction will occur, and a larger, chemically inactive piece. Such methods have gained considerable popularity for the study of solvent effects[16] in which the solute and solvent parts of the system are well separated. More recently, QM/MM methods have been developed across covalent bonds, in order to represent bulky substituents with molecular mechanics.[17] A particularly promising approach is the IMOMM (integrated molecular orbital/molecular mechanics) method. In an attempt to minimize time-consuming electronic structure computations while maintaining the effect of the "bulk" in surface chemistry calculations, Shoemaker et al. developed the IMOMM method into an embedded cluster approach called SIMOMM[18] (Surface Integrated Molecular Orbital/Molecular Mechanics). In this method, the "action" region where the actual chemical reaction occurs is treated quantum mechanically, while the spectator region that primarily provides the effect of bulk is treated using molecular mechanics. It has been shown that SIMOMM is especially effective for systems, such as semiconductors, in which the surface reaction is localized.[19] An important consideration in such embedded cluster or QM/MM methods is the way in which the transition is made from the QM to the MM region. This is not an issue for solvation, where solute and solvent are well separated, but it is important when the transition is made across a covalent bond. As noted above, the common approaches are to use hydrogens or model atoms as "place holders" in the link region, and this is the approach used in both IMOMM and SIMOMM. Recently, several groups have developed a more promising approach, based on the use of frozen localized orbitals to represent the link region[20] for calculations on large biological species. This approach may also be a viable alternative for embedded cluster surface methods, and is being explored in this laboratory.

2.4 Linear Scaling methods

It is, of course, desirable, to develop a fully quantum mechanical method for the study of surface science, since such a method is certain to be more generally applicable to a wide range of surface problems, including metallic surfaces. A very promising method is the linear scaling quantum technique[21], which has the potential to overcome the traditional high-order scaling barrier. This method has been applied to all levels of ab initio (Hartree-Fock, MP2, Coupled Cluster) and density functional theories.[21]

Combined with high-level quantum mechanics methods, this new technique may present a very appealing alternative to QM/MM approaches. This technique has the potential to extend our ability to study more realistic surface models.

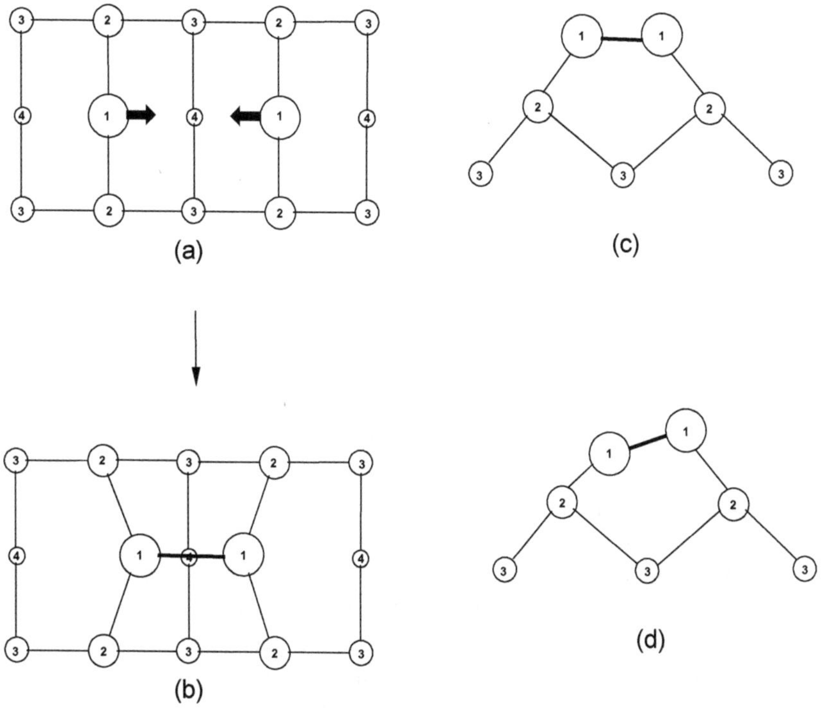

Figure 2. 2X1 surface relaxation. (a) Si(100) unreconstructed surface where the topmost layer (layer 1) atoms have two dangling bonds. (b) 2X1 reconstructed surface where the topmost atoms make chemical bonds with each other. (c) Side view (d) Side view of (buckled) 2X1 reconstruction.

3. 2X1 RECONSTRUCTION OF A CLEAN SILICON SURFACE

The most common crystal of silicon has the diamond structure, such that all atoms make four bonds with nearest neighbors in tetrahedral coordination. One of the most well studied and technologically most important silicon surfaces is Si(100). Figure 2(a) illustrates the Si(100) surface in which the topmost layer of the unreconstructed surface possesses silicon atoms (layer **1**) bonded only to two silicon atoms (layer **2**) yielding two singly occupied orbitals at each surface Si. The latter are called surface dangling bonds. Since this is an unstable arrangement, the entire surface undergoes reconstructions in order to stabilize the dangling bonds. The commonly accepted reconstruction is "dimerization" of adjacent surface Si atoms. Schlier and Farnsworth proposed the first model of this kind on the basis of their observation of the (2x1) LEED (Low Energy Electron Diffraction) pattern.[22] Figure 2(b) shows this (2x1) reconstruction of the surface Si along the (100) direction yielding dimerized silicon atoms.

The question regarding whether the dimer is symmetric (Figure 2c) or buckled (2d) has been controversial since Levine[23] and Chadi[24] proposed that the dimers could be asymmetric. Many experimental[25] and theoretical[26] papers predict the Si(100) surface to be buckled, while others[14,27,28,29] predict symmetric dimers. Data from room temperature STM suggested that no buckling occurs on terraces, but it occurs on steps and defects.[27a,b,e] However, more recent STM data showed that large regions of the Si(100) surface consist of buckled dimers far from defect sites.[25e] The observation of predominantly symmetric dimers at room temperature may be due to rapid flipping of buckled dimers between different buckling directions in defect-free regions. Weakiem et al[30] have indeed argued that a static picture of the Si(100) surface is insufficient and that the dynamics as a function of temperature is critical. However, the most recent low-temperature experiments show definitively that the surface is symmetric (not buckled).[31]

Several calculations based on density functional theory, using both the extended slab and cluster approaches, and employing both the local density (LDA) and generalized gradient (GGA) approximations, as well as plane wave basis sets,[26] predict the surface dimer to be buckled. However, it is important to recognize that because of the distance between the two silicon atoms in a dimer and the weakness of Si-Si π bonding[32], the dimer pair has significant diradical character. To the extent that this is true, a dimer requires a multi-reference wave function for a proper description.

Redondo and Goddard[28] first reported that the lowest energy configuration for a dimer in a small silicon cluster model, Si_9H_{12} shown in Figure 1, is the symmetric geometry. They showed that the dimerized bond should be considered as a singlet diradical. Therefore, a qualitatively correct description of the dimer requires at least a generalized valence bond, GVB-PP(1)[33], or two configuration self-consistent field (TCSCF)[34] wavefunction. More recently, Paulus[29] performed a more exhaustive multi-reference analysis of silicon clusters and reconfirmed this conclusion.

Shoemaker et al[14] used TCSCF calculations and clusters of increasing size to show that the surface dimer is symmetric with a small buckling frequency on the order of ~ 190cm^{-1}, in agreement with the earlier results of Redondo and Goddard and Paulus. More recently, these same authors showed that the addition of dynamic correlation (via second order perturbation theory) to either single reference or multi-reference wave functions does not alter the conclusion that Si_9H_{12} is symmetric (unbuckled).[35] Preliminary calculations, as well as previous ones by Shoemaker et al[5], suggest that the simplest two dimer cluster $Si_{15}H_{16}$ is also symmetric. More recently, Jung et al.[36] have shown that clusters containing up to five dimers are predicted to be symmetric when multi-reference methods are used. So, it seems clear at this point that small silicon dimer clusters are symmetric. The buckling frequency of 190 cm^{-1} (130 cm^{-1} when dynamic correlation is included) is small enough to suggest that any asymmetric attack on the surface, by a substrate or a probe, will be likely to cause buckling. It is also possible that high temperatures or impurities will cause buckling. So, the *chemistry* that occurs on the surface may be somewhat independent of the exact structure of Si_9H_{12}, especially since it is calculated at 0K.

A recent comment[37] criticized the use of multiconfiguration SCF (MCSCF) wave functions to study the structure of Si(100). This comment argued that the dynamic correlation, which is neglected in MCSCF calculations, is essential for reliable predictions of the geometry of Si(100). However, the most recent theoretical studies[38] show that the TCSCF, MP2, and MRMP methods consistently yield symmetric Si_9H_{12} structures. Therefore, dynamic correlation does not play a major role in determining the structure, and DFT incorrectly predicts Si_9H_{12} to be buckled.

4. HYDROGENATION OF THE RECONSTRUCTED SI(100) SURFACE

In view of the many important applications in semiconductor technology, the interaction of hydrogen with silicon surfaces has been intensively studied. Recombinative H_2 desorption from Si(100)-2x1 follows first-order kinetics[39], unusual when compared with the second-order kinetics observed for H_2 desorption from Si(111)-7x7. The measured activation barriers for the desorption of H_2 on Si(100) range from 45 to 66 kcal/mol.[39,40]

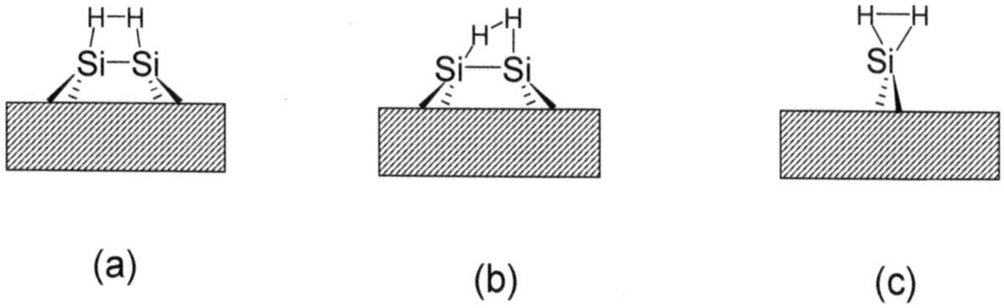

Figure 3. The key intermediates of the Si (100) surface hydrogenation. (a) Symmetric (b) asymmetric approach of the "preparing" mechanism. (c) Dihydride of "defect" mechanism.

Several reaction mechanisms have been proposed for the hydrogenation of Si(100).[41] The two most accepted of these are the "preparing" mechanism and the "surface defect" mechanism. Figures 3a and 3b illustrate the symmetric and asymmetric approaches in the preparing mechanism. This preparing mechanism is based on experimental[42] and theoretical[43] evidence that hydrogen atoms are initially *prepared* on silicon dimers due to the thermodynamic stability of the preparing configuration relative to separated surface

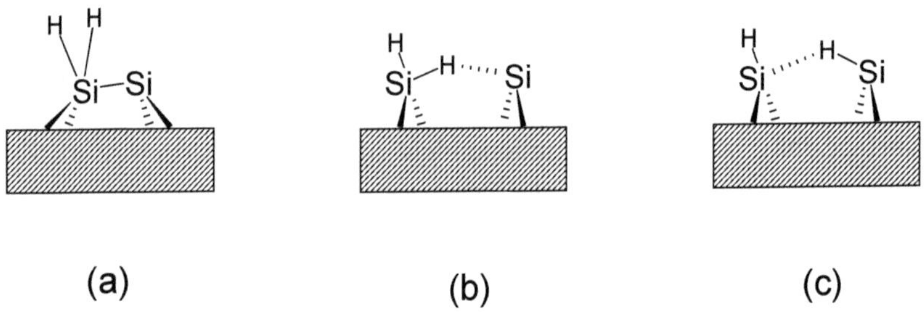

Figure 4. Two step mechanism of the hydrogenation. (a) First transition state that connects H2 adsorbed Si surface and intermediate 4b. (b) intermediate. (c) Second transition state.

and H_2 by about 20 kcal/mol. Using a simple statistical mechanical model, D'Evelyn and co-workers[44] showed that if one assumes the preparing mechanism, desorption follows first-order kinetics. However, many electronic structure calculations with cluster models of the surface have suggested that the activation energy for this mechanism is too high.[43,45]. Using MCSCF and CI levels of theory, Jing and Whitten[46] found both symmetric and asymmetric transition states. For the symmetric transition state, MCSCF is required to obtain the saddle point. At the CI level, they predicted the activation energies for the symmetric and asymmetric pathways to be 86.3 and 85.0 kcal/mol, respectively. Due to these large activation energies compared with experiment (45 ~ 66 kcal/mol), Jing and Whitten suggested a multistep desorption mechanism.

Pai and Doren[47] using DFT and cluster models found only the asymmetric transition state. They predicted the desorption activation energy to be 55 kcal/mol using LSD and 64 kcal/mol using the BLYP exchange-correlation functional. These predictions are in good agreement with the experimental values of 45 ~ 66 kcal/mol. However, similar calculations with the B3LYP functional predict a 77 kcal/mol activation energy for desorption.[45a]

A recent DFT study[48] using 1-dimer and 3-dimer cluster models of the surface suggested that adsorption/desorption occurs in a two-step process through a metastable dihydride-like intermediate (Figure 4b). The transition state **4a** connects H_2 adsorbed on the Si surface and the intermediate **4b**. A second transition state **4c** connects **4b** and the separated surface and H_2 molecule. These authors found an overall adsorption barrier of 15.0 kcal/mol and a desorption barrier of 67.8 kcal/mol. In the intermediate structure, **4b,** the distance between the silicon dimer atoms is 3.21 Å implying a broken dimer bond. Thus the results suggest that during each adsorption or desorption event the dimer bond is broken and reformed providing a picture of the coupling of the adsorption/desorption process to surface vibrations. This interpretation is consistent with the experimentally observed large surface-temperature activation of the adsorption process.[49]

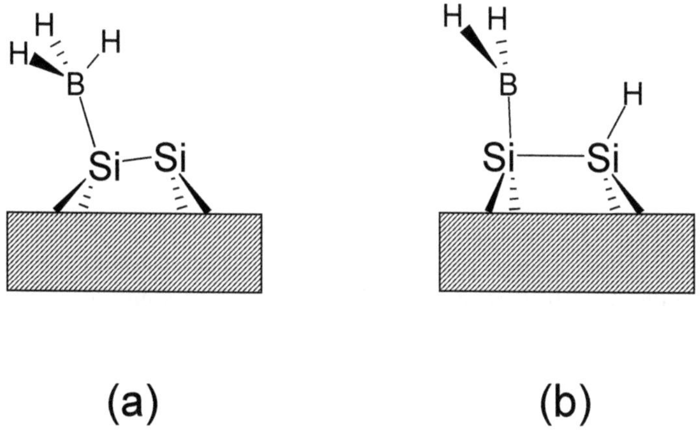

Figure 5. (a) Molecularly adsorbed BH3 on Si(100) surface. (b) Dissociatively adsorbed BH3.

For the surface defect mechanism, using a HF derived kinetic Monte Carlo model, Radeke and Carter[50] showed that H_2 desorption from Si(100)-2x1 via isolated dihydrides (Figure 3c) follows first-order kinetics. They further showed that the concentration of defects on the surface has a profound effect on the desorption rate constant. Recent studies proposed the importance of step sites rather than terraces.[51]

Despite extensive experimental and theoretical studies, it appears that no consensus regarding the mechanism of hydrogen desorption/adsorption reactions exists. Perhaps, better modeling of the systems in combination with accurate *ab initio* methods will help to resolve this long-standing issue.

5. ADSORBATES CONTAINING THE GROUP 3 ELEMENT BORON

Konecny and Doren[52] studied surface reactions of borane on the Si(100) surface with nonlocal, hybrid density functional theory (B3LYP/6-31G**) and a 9-Si model of the surface. The calculations predicted that a Si-B bond is formed via a nucleophilic attack on boron (Fig. 5a), leaving BH_2 and H fragments bound to the surface (Fig 5b) dissociatively. The reaction energy is -43.0 kcal/mol. However, the full potential energy surface including possible transition states was not investigated. Therefore, it is still an open issue as to whether BH_3 can adsorb both molecularly and dissociatively. In any case, it is very interesting that the surface can act as a nucleophile toward electron deficient species. In addition, since the BH_3 monomer does not exist as an independent entity, it would be interesting to study the addition of diborane, B_2H_6, to the Si(100) surface.

6. ADSORBATES CONTAINING GROUP 4 ELEMENTS

6.1. Reactions of alkynes and alkenes with the reconstructed Si(100) surface

$[2_s+2_s]$ cycloadditions (Figure 6a) are formally orbital symmetry forbidden.[53] Thus, a large reaction barrier is expected along the symmetric reaction pathway. In fact, it is known from carbon solution chemistry that

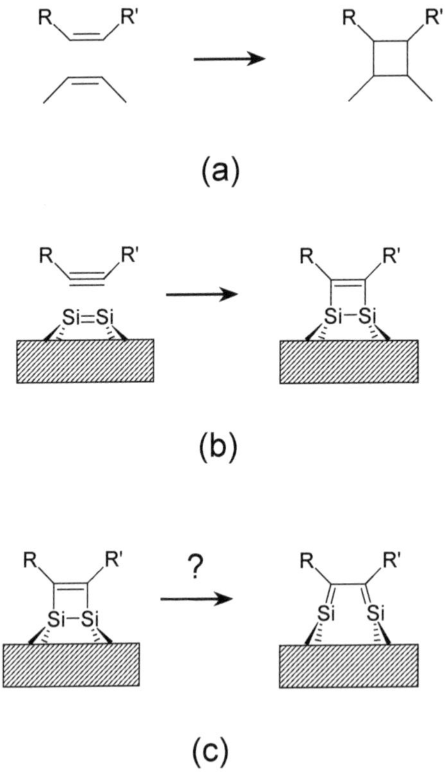

Figure 6. (a) [2+2] cycloaddition reaction. (b) Surface [2+2] cycloaddition between ene and surface dimmer. (c) perycyclic surface reaction.

even the low symmetry reaction path has a high reaction barrier, mainly due to unfavorable geometric configurations along the reaction pathway.[54] The analogous [2+2] reactions of disilenes have been found to be extremely slow[55] suggesting that the same symmetry rule applies in silicon chemistry.

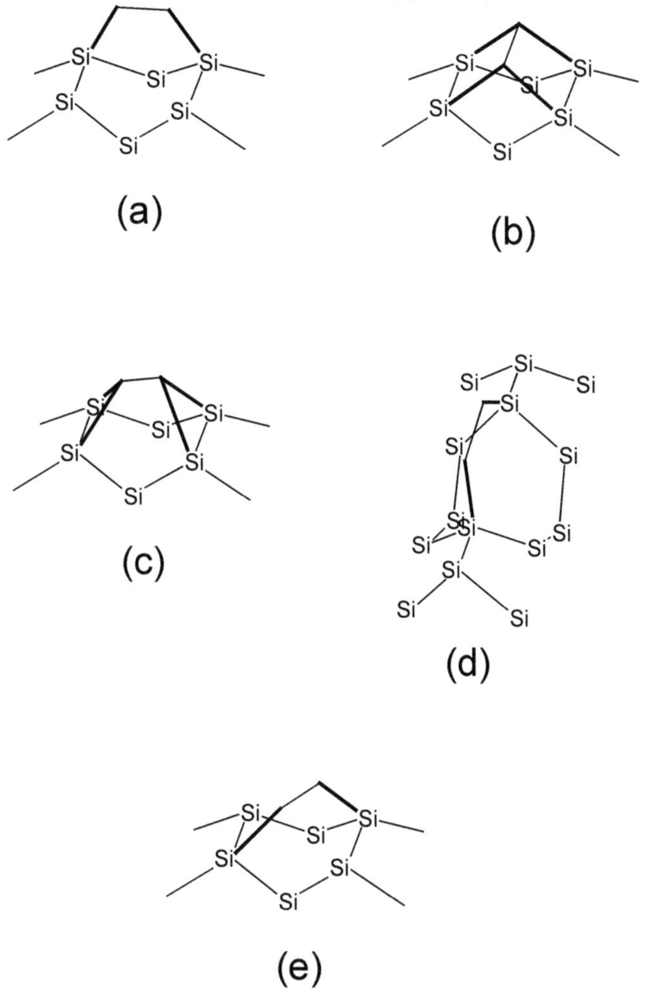

Figure 7. (a) End-bridge. (b) p-bridge. (c) r-bridge. (d) inter-row, (e) cross configuration.

However, the rules governing [2+2] additions on surfaces may be different. Using high-resolution electron energy loss spectroscopy (HREELS) and LEED measurements, Nishijima et al[56] showed that both acetylene and ethylene adsorb molecularly on Si(100). Based on the observed C-H stretching frequencies, they concluded that adsorption takes place on top of the Si dimers with an individual adsorbate molecule being bonded to a single Si-Si surface dimer through formation of two strong Si-C bonds per dimer site. This "di-σ bond structure" results from cleavage of two "π bonds", one each from the alkyne and the Si dimer (see Figure 6b). Yates and co-workers[57] also showed that unsaturated hydrocarbons including ethylene, propylene and acetylene chemisorb on Si(100)-(2x1) surfaces and are subsequently able to resist temperatures up to 600 K. However, there are some

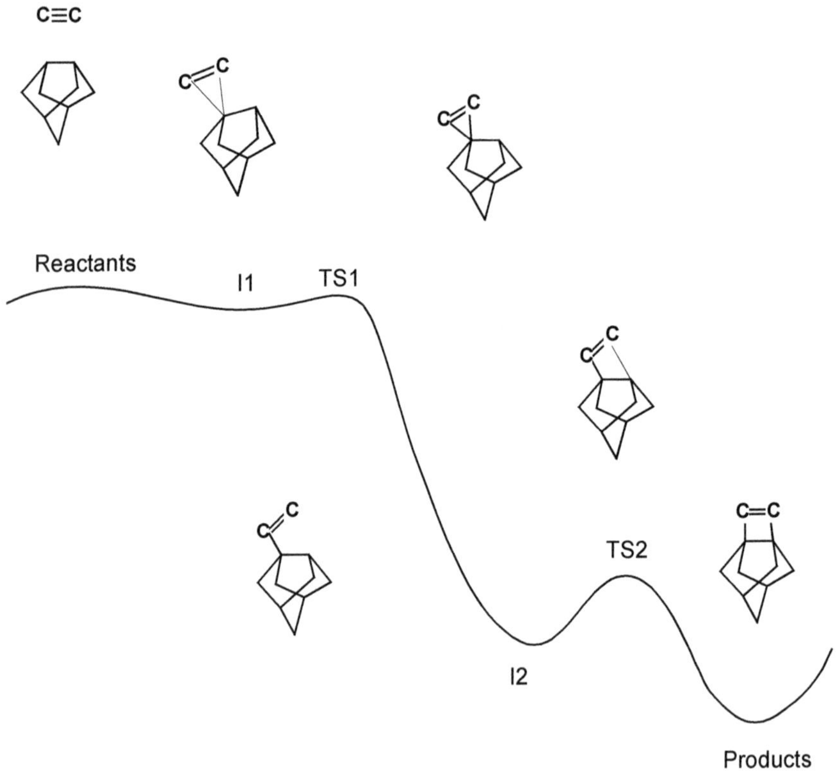

Figure 8. Su rface [2+2] cycloaddition mechanism composed of pi-complex precursor and a biradical intermediate.

disagreements regarding the nature of the C-Si bonds that result from acetylene adsorption. Nishijima et al[56] concluded that the carbon atoms are hybridized to "near sp^3" in the observed structure, while subsequent theoretical[58,59] and experimental[57c,60] studies suggest sp^2 hybridization.

Recent experiments by Wolkow[61] and Mezhenny et al.[62] yielded another adsorption configuration: an end-bridge structure with an acetylene molecule binding to two adjacent dimers and oriented perpendicularly to the dimer rows (Figure 7a). Their[62] calculations also suggest that the p-bridge (Figure 7b) and the r-bridge (Figure 7c) structures are the most probable tetra-coordinated ones. They[62] also showed that the relative populations of the various configurations change with coverage. These experimental results illustrate that the Si(100)/acetylene surface is far more complex than previously assumed.

Liu and Hoffmann[59] studied the chemisorption mechanism of acetylene on Si(100) using Si_9H_{12} models at the UHF/STO-3G level of theory (Figure 6b). They concluded that the symmetric path is forbidden following the gas-phase symmetry rule. However, they found a low symmetry pathway composed of a π-complex precursor and a biradical intermediate that has a low energy barrier to [2+2] cycloaddition products (see Figure 8). The initial intermediate, I1, is predicted to be a π-complex that is 0.25 eV lower in energy than the reactants. The initial transition state, TS1, is only about 0.007 eV above I1. The second intermediate, I2, 3.61 eV lower in energy than I1, now forms one C-Si bond. The second TS, TS2, leads to the formation of another C-Si bond yielding [2+2] products via a small (0.2 ev) energy barrier relative to I2. The final [2+2] product is 1.5 eV more stable than I2. Despite the relatively low level of theory used in this study, it demonstrates that a non-concerted asymmetric low energy pathway to the [2+2] product may exist. This low energy pathway may be attributed to the "pinned-back" cis-bent C_{2v} geometry forced upon the surface Si dimer that reduces the geometric hindrance in the initial ene approach and makes the Si dimer more reactive than simple disilene. The very flat bucking potential may also facilitate the reaction, since buckling is likely to result in a Si-Si charge separation.

Turning to the sp^2 vs. sp^3 carbon issue, acetylene has one C-C π bond left after chemisorption. The remaining Si-Si bond and the C-C π bond may undergo a further pericyclic reaction (Figure 6c) releasing 4-membered ring strain energy and making two new Si-C double bonds. This reaction resembles the electrocyclic ring opening reaction, in which cyclobutene, on heating, gives 1,3-butadiene.[63] Using Auger spectroscopy and temperature-programmed desorption (TPD), Taylor et al[57c] proposed that the Si-Si dimer bond is cleaved when the acetylene molecule adsorbs on top of a dimer. The structure with this complete cleavage of the Si-Si dimer bond was subsequently confirmed[60] by scanning tunneling microscopy (STM) as a major product. However, independent experimental studies concluded that no direct evidence regarding the structure of the Si-Si σ bond has been observed.[64]

Except for the early studies[65], a majority of the theoretical studies support the unbroken Si-Si dimer structure. [58,59,66] The most recent slab model DFT study by Sorescu and Jordan[67] showed that the broken Si-Si dimer structure is about 30 kcal/mol less stable than the unbroken structure. A Si_9H_{12} cluster study of the same system[68] predicts that the Si-Si cleaved structure is not a minimum on the potential energy surface. Since some of the stationary points on the potential energy surface for the addition of acetylene to Si(100) are likely to have significant diradical character, additional calculations that employ multi-configurational wave functions would be useful. Such calculations are currently in progress.

Sorescu and Jordan[67] have performed extensive theoretical studies to explore the recent speculation about tetra-coordinate configurations. Their study included the unbroken and broken Si-Si dimer configurations, end-bridge (Figure 7a), p-bridge (Figure 7b), r-bridge (Figure 7c), inter-row (Figure 7d), and cross configurations (Figure 7e). They predicted that the two most stable species are dicoordinated ones: the unbroken Si-Si dimer and end-bridge structures. These two dicoordinated species are predicted to have adsorption energies of 56.5 and 67.6 kcal/mol, respectively. Of the tetra-coordinate structures, the r-bridge structure is about 19.3 kcal/mol more stable than the p-bridge structure. Their adsorption energies are 49.2 and 29.9 kcal/mol, respectively. The barrier for the r-bridge to p-bridge isomerization is about 39 kcal/mol making this surface inter-conversion unlikely at low temperature. However, the barrier for the r-bridge to end-bridge isomerization is only 4.5 kcal/mol, so the r-bridge may be unimportant. The barrier for isomerization of the p-bridge to one of the dimerized structures is about 19 kcal/mol. This suggests that the p-bridge structure may be a viable candidate for the tetracoordinated species observed in STM measurements.

Density functional calculations using local spin density approximation and BLYP by Silvestrelli et al[69], suggested that among a number of possible adsorption configurations, the lowest-energy structure are di-σ and end-bridge configurations where the C_2H_2 molecule is bonded to two Si atoms. However they also found that tetra-coordinated configurations are much higher in energy and, therefore, can represent only metastable adsorption sites. Similar conclusions are drawn by Kim et al[70] based on scanning tunneling microscopy (STM) and ab initio pseudopotential calculations.

With the help of a gradient corrected density functional method with a pseudopotential technique, Miotto et al[71] predicted that at low temperatures, the di-σ bond configuration is the thermodynamically most stable structure. However, their calculated surface band structure suggests that the end-bridge configuration has metallic character and thus is *Peierls* unstable. They further suggested that other metastable configurations, like the p bridge and the r-bridge are also possible, supporting the existence of the r-bridge model.

Voltage-dependent scanning tunneling microscope images have been calculated by Wang et al[72] using density functional theory (DFT) with the general gradient approximation. (GGA). An image that had been previously attributed to a tetracoordinated acetylene molecule was reassigned to a site with two acetylene molecules bridging two adjacent SiSi surface dimers.

It seems to be the consensus that the majority of the surface product forms two Si-C bonds with a di-σ configuration, an analog of the [2+2] product. This seems sensible, since the orientation of the individual molecules can be controlled, yielding anisotropic physical properties of a chemically modified silicon surface as discussed in recent accounts[73]. However, no general agreement regarding the other possible surface configurations has yet been reached. These studies, however, are mostly based on the relative energies of possible surface products. In order to resolve the current situation, a more complete study of the potential energy surface is needed.

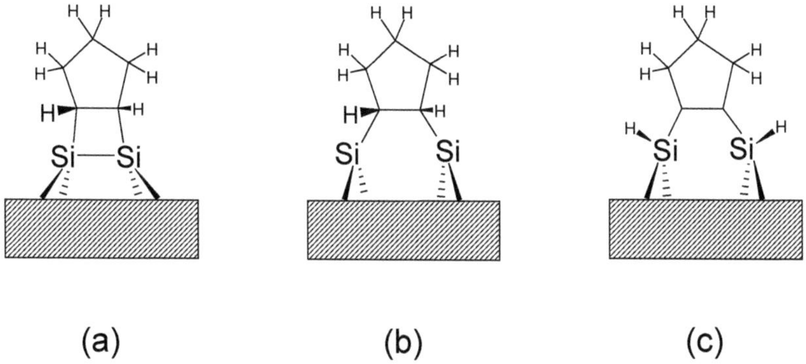

Figure 9. Surface adsorbed configuration of cyclopentene. (a) on-dimer. (b) dimer-bridge. (c) dissociative.

Hamers and co-workers have recently demonstrated that a wide range of complex cyclic olefins react in a manner that is similar to acetylene. They showed, in general, that molecules with high symmetry such as cyclopentene[74], 1,5-cyclooctadiene[75] and 1,3,5,7-cyclooctatetraene[76] chemisorb into a unique [2+2] geometry. More complex molecules such as 3-pyrroline[64] and norbornadiene[64], form multiple bonding geometries. In these cases, attempts to control the surface product distributions lead to molecular fragmentation rather than surface conversions.

In the case of the adsorption of cyclopentene on the Si(100) surface (Figure 9), using DFT methods, Cho and Kleinman[77] recently showed that on-dimer **9a**, dimer-bridge, **9b** and dissociative products, **9c** are possible. Of these, the on-dimer structure is the most stable surface configuration. They further showed that at low coverages cyclopentene molecules favor adsorption on alternate Si dimers rather than on neighboring dimers along a dimer row, because of the repulsive H-H interaction between adsorbed molecules. Once this alternating adsorption process completely fills the surface, further adsorption occurs via a "3-atom" intermediate (the same surface bonding configuration as **I1** in Figure 8) state with some energy barrier. So, they concluded that above 0.5 ML (monolayer) coverage the reaction products may be controlled by both the kinetics and the thermodynamics.

Using plane wave-based density functional theory, Akagi and Tsuneyuki[78] studied the chemisorption of cyclopentene, cyclohexene and 1,4-cyclohexadiene on a clean Si(100) surface. In contrast to Cho and Kleinman[77], they found that conformational isomers such as *cis* (**10a**) and

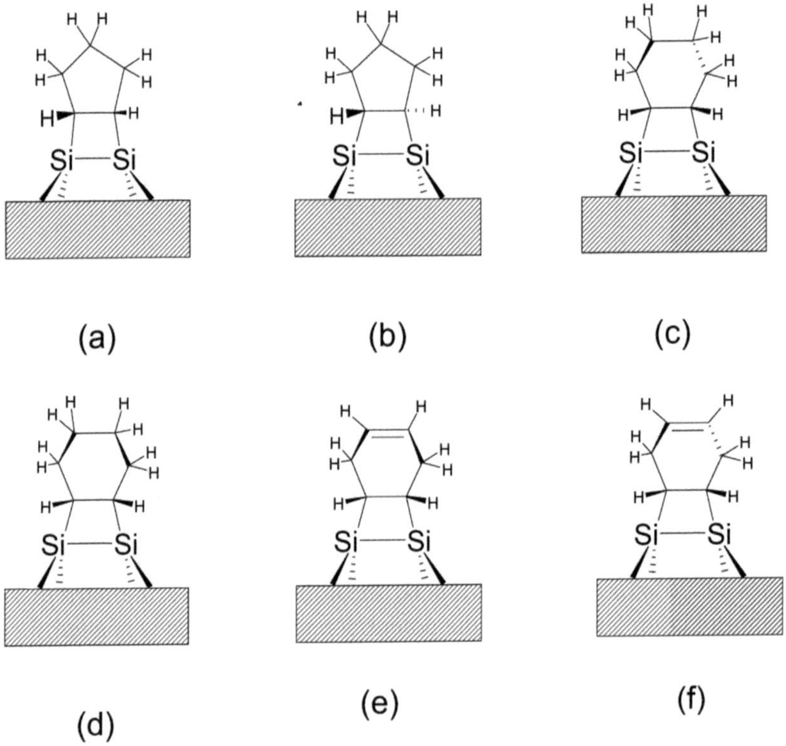

Figure 10. Conformational isomers of (a) cis-, (b) trans-cyclopentene, (c) twisted-, (d) boat-cyclohexene, (e) boat-, (f) plane-1,4-cyclohexadiene.

trans (**10b**) cyclopentene surface products exist depending on the position of the two H atoms, and they interpreted the experiments accordingly. For cyclohexene, two kinds of stable conformers, **10c** and **10d** were determined. These are consistent with two types of characteristic STM images and spectroscopic data. The authors also found two kinds of adsorption structures for 1,4-cyclohexadiene molecules, the boat-type, **10e** and plane-type, **10f** shapes. The structure supported by STM and UPS observations[79] is the plane-type conformation which is predicted by the calculations to be slightly less stable than **10e**. So the situation is still unresolved.

Cho et al. [80] also studied the adsorption of the *non*-conjugated 1,4-cyclohexadiene on the Si(100) surface using GGA-DFT calculations (Figure 11). The "pedestal" structure, **11a**, where the two C=C double bonds react with different Si dimers, is found to be energetically more stable than the "upright" structure, **11b,** in which only one of the two C=C bonds reacts with a Si dimer. They further found that the [2+2] cycloaddition reaction can easily form the upright structure but not the pedestal one. The latter structure can be obtained from the former through a high energy barrier of 0.95 eV, indicating a small reaction rate at room temperature. This theoretical result is consistent with the interpretation of recent LEED and photoelectron spectroscopy data[81] in which the upright structure was observed. More recent

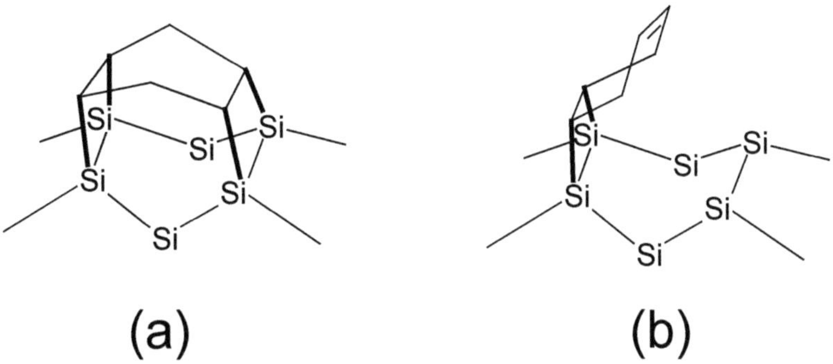

Figure 11. (a) pedestal-, (b) upright-structure of 1,4-cyclohexadiene on Si(100) surface.

STM data by Hamaguchi et al.[82] was also interpreted to be consistent with the upright structure. However, Cho et al.[80] reinterpreted the STM data to be consistent with the pedestal structure by simulating the STM images. Therefore, it is not certain that thermodynamics is the only factor in determining the final surface products. So the correct final products of 1,4-cyclohexadiene adsorption on the Si(100) surface still appears to be an open question. More detailed potential energy surface studies are expected to resolve the current situation.

Another important issue related to surface cycloaddition reactions is the stereochemistry of the products. Advances in this area may open new methods for developing new chemical sensors that may be used for complex molecular recognition tasks. In one study, Wolkow and co-workers[83] showed that a scanning tunneling microscope can be used to determine the absolute chirality of individual molecules of cis- and trans-2-butene adsorbed on silicon. Another study[84] demonstrated that 1S(+)-3-carene, a naturally occurring bicyclic alkene belonging to the terpene family, adsorbs enantiospecifically on silicon. That reaction leads to formation of four chiral centers and a chiral surface.

6.2. Reactions of dienes with reconstructed Si(100) surface

The surface reactions of dienes with the Si(100)-(2x1) surface are of particular interest due to their relation to the [4+2] cycloaddition or "Diels-Alder" reaction of carbon chemistry, in which a conjugated diene reacts with a dienophile to form a six-membered ring (Figure 12a). Since unsaturated hydrocarbons such as acetylene, ethylene, and propylene can readily react with the Si(100)-(2x1) surface yielding [2+2] products, the [2+2] reaction involving only one π bond of a diene (Figure 12c) may compete with the corresponding [4+2] reaction (Figure 12b) in diene chemisorption reactions.

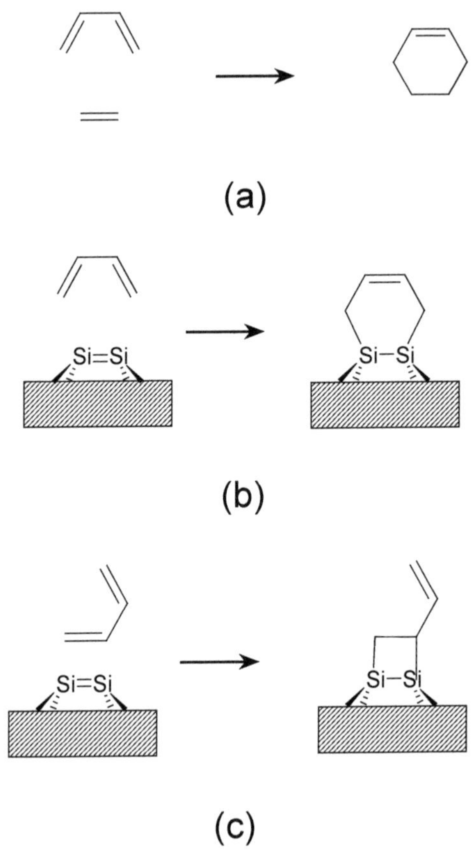

Figure 12. (a) [4+2] cycloaddition. (b) Surface [4+2] cycloaddition. (c) Surface [2+2] cycloaddition.

The [4+2] cycloaddition produces a six-membered ring that is less strained than the four-membered ring produced by [2+2] cycloaddition. Therefore, [4+2] products are expected to be thermodynamically more stable than the [2+2] products. In the case of the adsorption of 1,3-cyclohexadiene, Konecny and Doren[85] predicted that the [4+2] product is more stable than the [2+2] product by 15.2 kcal/mol and that the reaction barrier to the [4+2] product is 0.3 kcal/mol. They further showed that the predicted vibrational spectra of the Diels-Alder products of 1,3-butadiene and 2,3-dimethyl-1,3-butadiene onto the Si(100)-(2x1) surface at room temperature are consistent with the observed spectra and inconsistent with significant formation of [2+2] addition product. These authors did leave room for the possible formation of [2+2] products. The corresponding experimental results by Bent and coworkers[86,87] for the adsorption of these two products showed evidence for only the [4+2] product. These theoretical and experimental investigations established that the Si dimer of the reconstructed Si(100)-(2x1) surface could act as a dienophile in a Diels-Alder cycloaddition fashion, and that the Diels-Alder reaction is facile.

Reinvestigation of the 2,3-dimethyl-1,3-butadiene reaction on the Si(100)-(2x1) surface by Hamers and coworkers[88] supported the observation of a Diels-Alder product for 80% of the surface products. However, they noted a minor (20%) [2+2] product as well. In the case of 1,3-cyclohexadiene, they observed 55% of the [4+2] product, 35% of [2+2] product and 10% unknown product strongly indicating the existence of competition between [4+2] and [2+2] cycloaddition reactions of the diene on the Si(100) surface. Attempts to convert the product distribution to the thermodynamically more stable product by annealing to higher temperatures failed. Surface annealing leads to C-H bond cleavage rather than the [4+2] product. They[88] concluded that the formation of multiple products and the lack of temperature effects indicate that the product distribution is controlled primarily by the kinetics of the adsorption process, not by the thermodynamics.

The primary issue is whether the final products are determined by the initial stage of the surface reactions, or if the initial products are subject to subsequent reactions such as surface isomerization. An understanding of whether and why the [4+2] product is exclusively formed ultimately bears on the chemical selectivity of the Si(100)-(2x1) surface toward conjugated diene systems. By studying the factors that govern the reactivity of these reactions, one hopes to gain control over these surface reactions to an extent that eventually leads to a technique to tailor the reaction selectivity.

Recently, Choi and Gordon[19] have performed systematic investigations on this intriguing system with the MCQDPT2[89] method in combination with CASSCF[90] wavefunctions, performing potential energy surface searches along the possible reaction paths. In the case of chemisorption of 1,3-cyclohexadiene on the Si(100)-(2x1) surface, they showed that both the [2+2] and [4+2] cycloaddition reactions can occur readily, indicating strong competition between these two reactions at least at the initial stage of the chemisorption. Symmetric [4+2] cycloaddition has a negligibly small reaction barrier in agreement with Konecny and Doren's result.[85]

Choi and Gordon[19] showed that the low energy [2+2] cycloaddition pathway on the Si(100) surface constituting a nonsymmetric, nonconcerted reaction path exists. There are at least two reaction paths to the [2+2] product depending on which Si-C bond forms first. Both transition states have small barriers of about 5 kcal/mol, suggesting that [2+2] cycloaddition is facile.

Most importantly, surface isomerization reactions connecting [4+2] and [2+2] products turn out to be very unlikely due to a high (> 40 kcal/mol) energy barrier. In the transition state that connects [2+2] and [4+2] products, one Si-C bond is being broken and another one being formed.

This calculation as well as new experiments by Hamers' group indicate that the final surface reaction products are determined during the initial stage of the surface reactions, and they are not subject to further thermal redistributions or isomerizations among surface products.

Using multiple internal reflection Fourier transform IR (MIR-FTIR) spectroscopy, near-edge X-ray absorption fine structure (NEXAFS), Kong et al[91] confirmed that in the case of 1,3-cyclohexadiene, multiple chemisorption configurations are implicated, consistent with the co-existence of [4+2] and [2+2] cycloaddition products.

The theoretical investigation also demonstrated that methods based on multiconfiguration wavefunction are essential in order to consistently study the entire potential energy surface, since many points on the surface are highly multiconfigurational in nature. The bonding and anti-bonding pair in the active space of the transition state that leads to [2+2] product clearly exhibits singlet diradical character.[19] Recent theoretical studies[92] of the adsorption of 1,3-cyclopentadiene on the Si(100) surface yield similar results, extending the general picture of surface reactions obtained with 1,3-cyclohexadiene.

These recent experimental and theoretical results make it clear that control of selectivity depends on the kinetics rather than thermodynamics. It is interesting to speculate how one might improve the [4+2] addition selectivity of a Si(100)-(2x1) surface toward the diene systems. By replacing hydrogens with other appropriate groups, one may be able to alter the barrier for either the [2+2] reaction or the isomerization reaction. The former may control the initial distribution of surface products, while the latter may change the selectivity of the surface by thermal redistribution.

The growing interest in silicon-germanium ($Si_{1-x}Ge_x$) devices reflects their potential for extending traditional silicon technology, allowing faster field-effect transistors, CMOS and bipolar chips for next-generation wireless telecommunications.[93] A major advantage of $Si_{1-x}Ge_x$ over III-V materials is that it can be integrated with existing silicon CMOS processes, resulting in lower development costs. Mui et al.[94] reported a B3LYP density functional study of the structure and thermochemistry of the products of the chemisorption of 1,3-butadiene on the Ge/Si(100)-2 X1 surface. The surface was modeled using a 2Ge and 7Si cluster which has a Ge surface dimer to study the effects of Ge on the cycloaddition. According to their results, both the [4 + 2] Diels-Alder and the [2 + 2] cycloaddition products are energetically stable on all the Ge/Si(100)-2 **X** 1 surfaces studied. The [4 + 2] Diels-Alder products on the Si-Si, Si-Ge, and Ge-Ge dimers are more stable than the corresponding [2 + 2] cycloaddition product due to ring strain. The binding energies of all cycloaddition reaction products decrease with increasing Ge composition on dimers, which can be explained by differences in bond strength.

However, as discussed earlier, thermodynamic data alone may not be able to correctly describe the surface reactions. So further kinetic studies are expected.

6.3 Reactions of Aromatic Compounds with reconstructed Si(100) surface

Aromatic molecules have a conjugated π system in common with the conjugated dienes discussed above. Due to their unusual stability, however, they do not undergo many of the reactions typical of alkenes and dienes, except electrophilic substitution reactions that preserve the integrity of the π system. It is interesting, therefore, that benzene appears to rather easily undergo addition reactions on the Si(100) surface, especially since such reactions make new C-Si, bonds thereby removing aromatic stability[95,96]. While adsorption of simple alkenes and dienes on Si(100) is essentially irreversible due to the formation of strong C-Si bonds, benzene has been shown to adsorb reversibly[95] and even exhibits redistributions of surface products .[96]

Wolkow and co-workers[96] have performed semi-empirical and ab initio calculations to determine preferred bonding configurations of benzene on Si(100). As shown in Figure 13, at room temperature benzene bonds to silicon dimers in three configurations: 1,4-single dimer (a), tight-bridge (b) and twisted bridge (c). The latter two involve bonds form benzene to two dimers. The STM image shows that the 1,4-dimer configuration converts into the tight-bridge configuration. The 1,2-single dimer **13d** and symmetric-bridge configurations **13e** are not observed. Recently Wolkow and coworkers[97] predicted that the 1,4-single dimer configuration is a metastable chemisorbed state of benzene, while the binding state is the tight bridge across two dimers. This conclusion was drawn by comparing scanning tunneling microscope (STM) measurements on the adsorption of benzene on Si(100) with plane wave-based density functional methods. However, the level of theory used was not sufficient to conclusively determine relative energies among plausible structures. Moreover, it may be kinetics, rather than thermodynamics, that play a major role in determining the product distribution. Therefore, one has to study the relevant potential energy surfaces to fully understand the surface reactions of aromatic systems[98].

For styrene, Hamers and co-workers showed that adsorption occurs almost exclusively through the vinyl group and not through the aromatic ring.[99] This illustrates that due to the aromatic stability, given a choice, adsorption only occurs through the more reactive π bond. In another study with toluene, para-xylene, meta-xylene and ortho-xylene, the same group[100] concluded that the differences in relative peak intensities in the FTIR spectrum as compared with benzene point to the possibility that the methyl substituent groups may steer the ring into different ratios of specific bonding geometries.

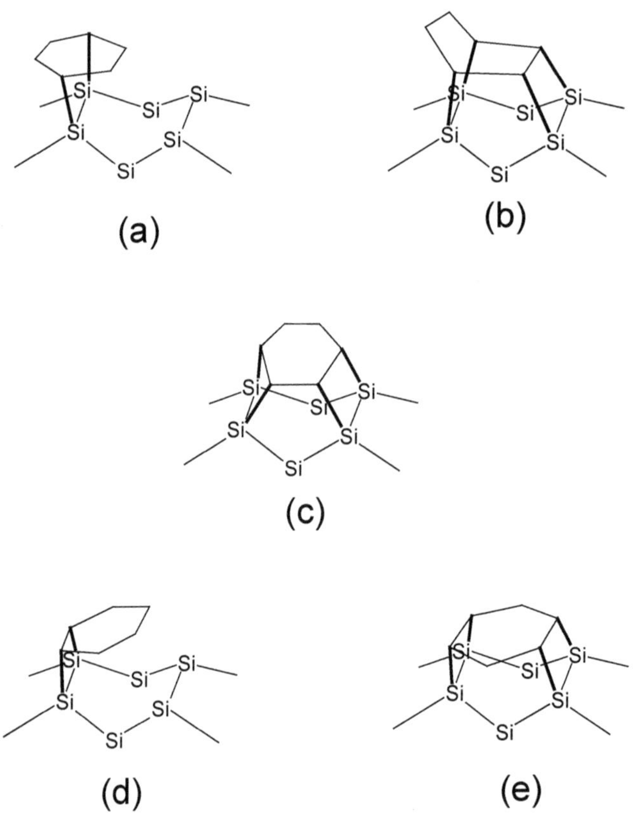

Figure 13. (a) 1,4-single dimer. (b) tight-bridge. (c) twisted bridge. (d) 1,2-single dimer. (e) symmetric-bridge.

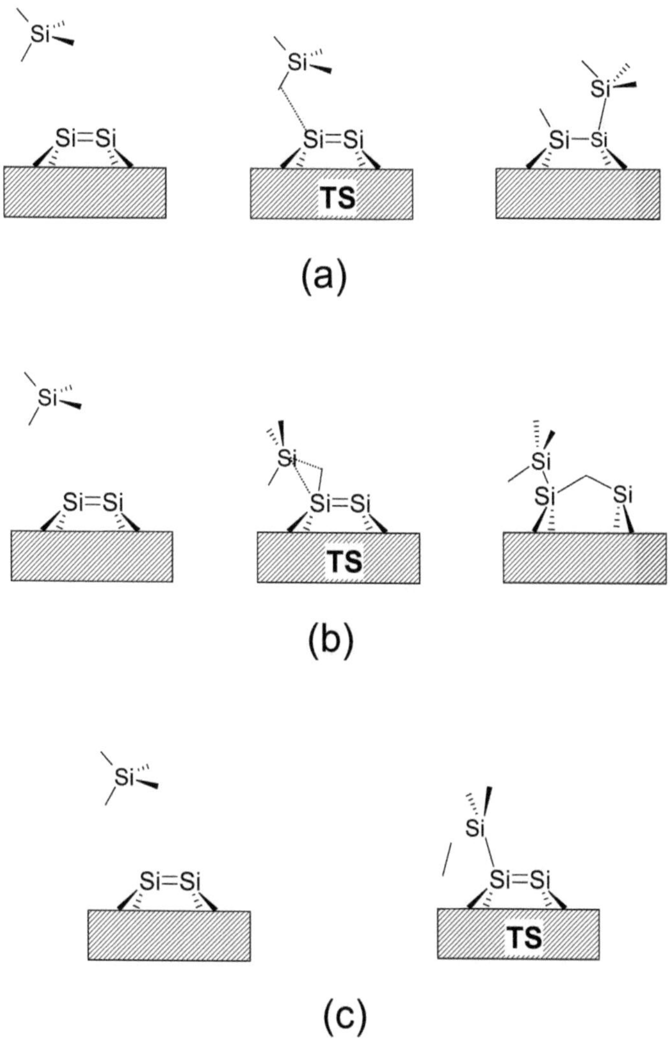

Figure 14. The three dissociative adsorption mechanisms of SiH_4 and surface.

Yamaguchi[101] studied four model clusters of the pentacene molecule adsorbed on dimer rows of the Si(100)-2x1 surface using the DV-Xα-LCAO method. The calculated charge densities of individual occupied and unoccupied orbitals near the Fermi level well reproduced the STM images implying that electrons tunnel from/to the eigenstates of the combined system of the adsorbed organic molecule and the Si substrate where the adsorbate molecule is assumed to bond covalently to dangling bonds of Si dimers.

6.4 Formation of new surface Si-Si bonds

Chemical vapor deposition (CVD) from molecular precursors is a widely used technique for silicon film growth. The CVD growth process is complex with many side products. There have been many kinetic studies of

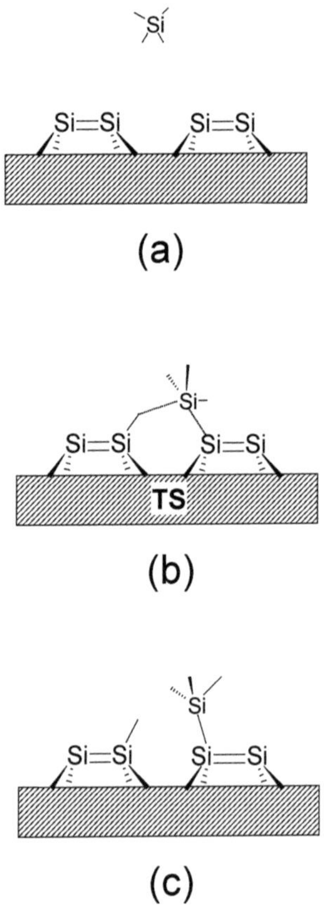

Figure 15. The dissociative adsorption across 2 dimers.

silicon film growth from silane.[102] These studies concluded that the adsorption step on Si(100) occurs dissociatively. Brown and Doren[103] studied the detailed mechanism of dissociative adsorption of silane (SiH_4) on the Si(100)-(2X1) surface using DFT. They found two channels that produce silyl (SiH_3) and hydrogen atom fragments adsorbed on the dimer dangling bonds. The energy barrier on the lowest energy path illustrated in Fig **14a,** is 12-14 kcal/mol (depending on the details of the theoretical method used), while the barrier on the other path, Fig. **14b,** is about 17 kcal/mol. The initial step in both mechanisms is abstraction of a hydrogen atom from silane by an electron-deficient surface atom. They also found another channel, Fig. **14c,** having a prohibitively high barrier (39 kcal/mol), that leads to different products (adsorbed SiH_2 and elimination of H_2). So they concluded that the reaction can be described as electrophilic attack by a surface Si atom to abstract a H atom from silane. However, the predicted 12 ~17 kcal/mol barrier is still too high as compared to the experimental value of 3~4 kcal/mol.[102]

Using DFT/B3LYP, Kang and Musgrave[104] found the barrier to

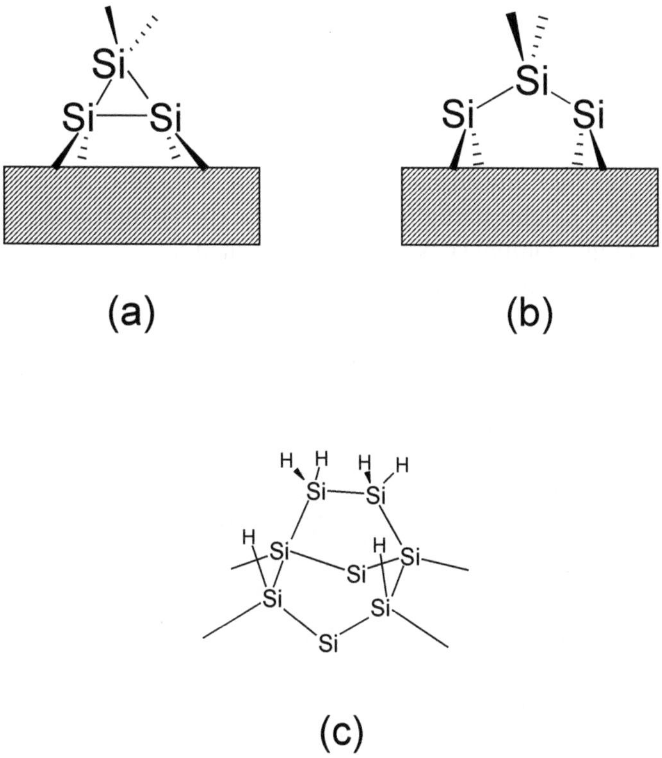

Figure 16. Surface products of disilane adsorption on Si surface. (a) on-dimer. (b) inter-row. (c) 2-dimer bridges.

SiH_4 adsorption on a single dimer, **14a,** to be 7.4 kcal/mol while the barrier across 2 dimers, **15**, is 14.3 kcal/mol. The 7.4 kcal/mol is in qualitative agreement with the TPD value of 3.3 ~ 4 kcal/mol.[102] Subsequently, adsorbed SiH_3 transforms to bridged SiH_2(a) with a barrier of 5.7 kcal/mol relative to SiH_3(a) for the mechanism requiring H(g), while the barrier for the mechanism requiring no H(g) is 32.9 kcal/mol, where (g) and (a) represent gas and adsorbed species, respectively. The dihydride surface transforms to the monohydride surface through two-sequential steps with an overall barrier of 47.0 kcal/mol, in good agreement with the TPD barrier of 43 kcal/mol.

Chlorosilanes are also used as precursors for a wide variety of CVD processes. Chlorosilanes have a number of advantages over normal silane including higher sticking coefficients and superior selectivities. In addition, chlorosilanes are also good candidates for atomic layer epitaxy (ALE) precursors.[105] Using DFT, Hall et al.[106] investigated the adsorption of chlorinated silanes on the Si(100)-2X1 surface. They showed that dissociative adsorption of chlorosilanes via scission of either the Si-H or Si-Cl bond is exothermic by 50-53 kcal/mol. The barrier for SiH_2Cl_2 and $SiHCl_3$ adsorption via scission of the Si-Cl bond is 6-9 kcal/mol lower than the barrier for adsorption via scission of the Si-H bond. This indicates that the reaction pathway is kinetically controlled. Furthermore, the adsorption barrier of SiH_2Cl_2 is 7 kcal/mol lower than the barrier for SiH_4 adsorption. This difference in barrier heights partially explains the experimental observation that SiH_2Cl_2 has a higher initial sticking coefficient than SiH_4.

Adsorption of disilane onto Si(100) has also been investigated extensively, because of the potential of using disilane as the source gas for gas source molecular-beam epitaxy for the growth of Si. It has been demonstrated that the Si(100) growth rate is enhanced using disilane in place of the conventional silane gas, due to higher adsorption rates associated with the ease of cleaving Si-Si bonds, compared to Si-H bonds.[107] Using DFT, Cakmak and Srivastava[108] investigated the dissociative adsorption of Si_2H_6 on the Si(100) surface. However, they did not study the initial adsorption mechanism of Si_2H_6. The cleavage of the Si-Si bond would yield SiH_3 species, which are rather easily converted to SiH_2. Cakmak and Srivastava showed that the on-dimer geometry, **16a**, is energetically more favorable than the inter-row geometry, **16b**, in the case of the SiH_2 species. Recent IR absorption spectroscopy in the multiple internal reflection geometry and hybrid density functional cluster calculations [109] showed that Si_2H_6 dissociatively adsorbs without breaking its Si-Si bond to produce an adatom dimer, -H_2Si-SiH_2-, that bridges 2 adjacent dimers, **16c**. Also they showed that the distribution of the final products is highly depended on the coverage. More detailed studies are needed to resolve current discrepancies.

6.5 Formation of surface Si-Ge bonds

The epitaxial growth of Ge on Si(100) has also been intensively investigated because of its technological importance in the microelectronic industry. The development has in large part been due to the achievement of well controlled experiments based on the molecular beam epitaxy method with the deposition of only a fraction of a monolayer at ultrahigh vacuum conditions. From the chemical point of view, such a mixed surface would add new dimensions to the various surface reactions with organic molecules. At room temperature, the initial deposition of either Ge or Si on these surfaces produces many ad-dimers, which are stable structures formed from two adatoms

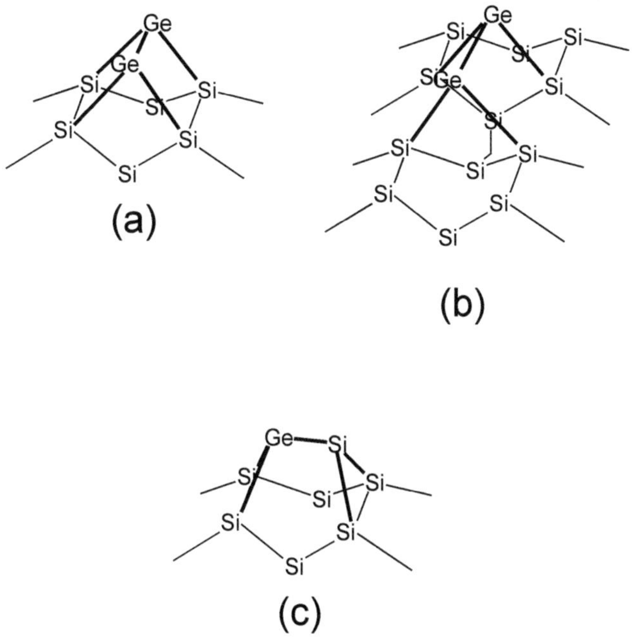

Figure 17. (a) Ad-dimer parallel to substrate dimer. (b) ad-dimer in the trough. (c) Mixed Ge-Si dimer.

that remain bound to each other.

With the help of DFT, Silva et al.[110] suggested two possible ad-dimer configurations. Figure **17(a)** illustrates a Ge ad-dimer located on top of the substrate Si dimer rows with the ad-dimer bond parallel to the substrate dimer bonds, and Figure **17(b)** shows a Ge ad-dimer in the trough between the dimer rows. They further proposed that local buckling of the substrate dimers can give rise to similar structures with very different STM images.

Lu et al.[111] carried out a comparative study of the energetics and dynamics of Si-Si, Ge-Ge, and Ge-Si ad-dimers on top of a Si(100) dimer row, using DFT. In the case of the Ge-Si ad-dimer, they also considered the structure in Figure **17(c)**, which represents an ad-dimer with the ad-dimer bond perpendicular to the substrate dimer bonds. They predicted that the dynamic appearance (dynamic change of STM image) of a Ge-Si dimer is distinctively different from that of a Si-Si or Ge-Ge dimer, providing a unique technique for its identification by scanning tunneling microscopy (STM). The "rocking" motion of the Ge-Si dimer, observed in STM[112], actually reflects a 180° rotation of the dimer. The calculated energy barrier of 0.74 eV for this rotation is in good agreement with the experimental value of 0.82 eV.

7. ADSORBATES CONTAINING GROUP 5 ELEMENTS

7.1 Nucleophilic Surface Reactions with Amine compounds

Silicon nitride thin films have been ubiquitous in the microelectronics industry. Specifically, thin films of Si_3N_4 are being used as insulators, oxidation masks, diffusion barriers, and gate dielectrics.[113] Silicon nitride could be also used as a precursor to oxynitride dielectrics, as an interface between Si and high-K dielectrics[114], and for the creation of ultralarge-scale integrated (ULSI) circuits.[115]. In particular, the NH_3 molecule is an excellent nitridation agent, useful in the growth of ultrathin, sharp silicon nitride.

Widjaja et al.[116] investigated the mechanism of NH_3 adsorption and initial decomposition on the (2**X**1) reconstructed Si(100) surface using B3LYP density functional theory. They predicted that NH_3 adsorbs molecu-

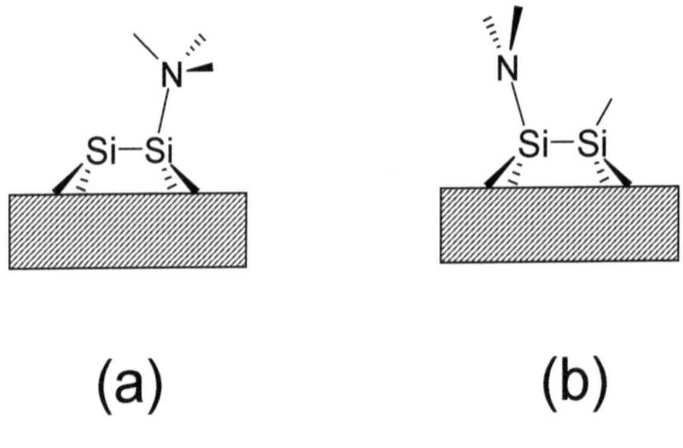

Figure 18. (a) Molecularly adsorbed NH_3. (b) Dissociatively adsorbed NH_3.

larly on the "down" atom of the presumed buckled silicon dimer, Figure **18a,** with no activation barrier. NH_3 adsorption on Si(100)-(2**X**1) is predicted to be exothermic with an adsorption energy of 29 kcal/mol. Dissociation of the molecularly adsorbed NH_3 to form NH_2 and H, Figure **18b,** proceeds with a low activation energy of 5 kcal/mol below the separated NH_3(g) and bare Si(100)-(2X1) energy. The calculated recombination desorption energy of 51 kcal/mol is in good agreement with the temperature-programmed desorption (TPD) experimental result of 47 kcal/mol.

Widjaja and Musgrave further investigated[117] the cluster size effects which they called nonlocal effects in surface reactions. They used the Si_9H_{12} (one-dimer), $Si_{21}H_{20}$ (three-dimer), $Si_{33}H_{28}$ (five-dimer), and two $Si_{23}H_{24}$ trench clusters to determine the nonlocal effects both *along* the dimer row and *across* the trench. The nonlocal effects *along* the dimer row are found to be significant, while nonlocal effects *across* the trench are relatively small. They attributed this difference to charge transfer to nearest neighbor dimers *along* the row, which increases the adsorption energy by

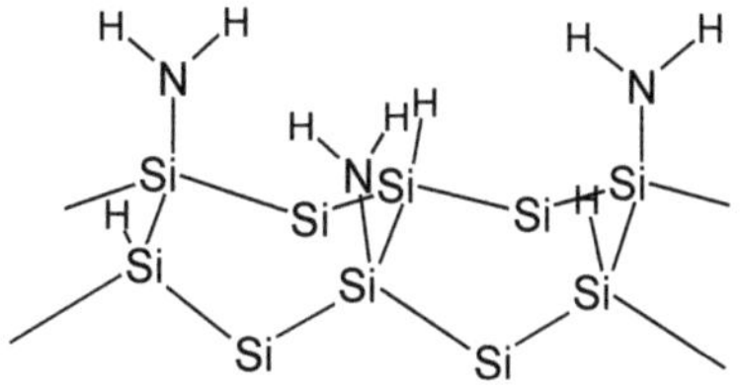

(a)

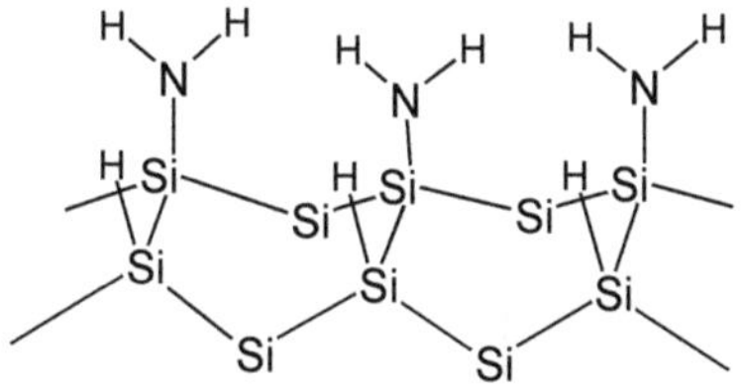

(b)

Figure 19 (a) Alternative adsorption configuration. (b) Ordered adsorption configuration.

26%. However, a complete study of the cluster size effect probably requires the use of larger clusters in the bulk region to prevent the model cluster from artificial bending.

Rignanese and Pasquarello[118] studied N 1s core-level shifts in NH_3 exposed to Si(100)-2X1 surfaces to determine bonding configurations of N atoms using DFT calculations. Their results showed that NH_3 is adsorbed dissociatively as NH_2 and H. They[119] further studied the pattern of NH_3 adsorption. By comparing calculated core-level shifts with measured photoemission spectra, they showed that Figure **19a**, rather than Figure **19b**, dominates, supporting the occurrence of an ordered coverage pattern. Similar conclusions were also drawn by Queeney et al.[120]

Using a pseudopotential DFT method, Miotto and co-workers.[121] have studied the adsorption and dissociation of NH_3, PH_3, and AsH_3 on the Si(100)-(2x1) surface. Phosphine gas is a very common *in situ n*-type dopant source in CVD or gas-source in molecular-beam epitaxy (GSMBE) during epitaxial growth of Si thin films used in ultra-large scale integrated circuits (ULSI). Arsine is a common As source during growth of GaAs on Si, and is

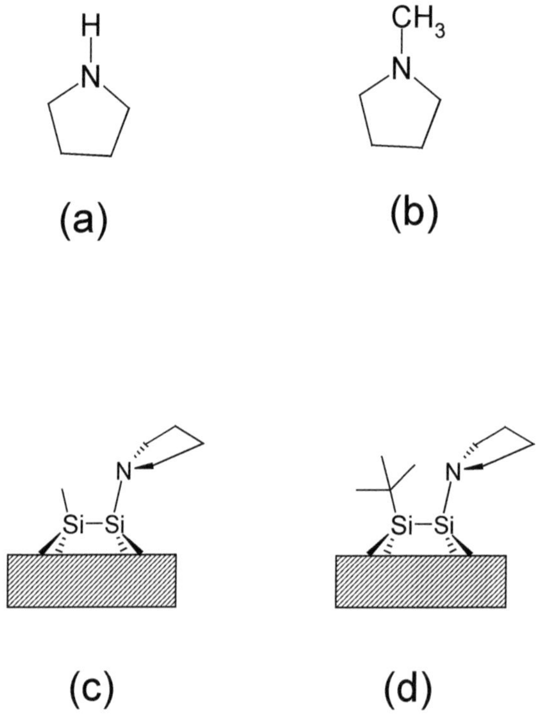

Figure 20. (a) Pyrrolidine. (b) N-methylpyrrolidine. (c) Dissociatively adsorbed pyrrolidine. (d) Dissociatively adsorbed N-methylpyrrolidine.

also a very important dopant used during GSMBE.[122] They showed that apart from the existence of a barrier of about 10 kcal/mol for the adsorption of the initial precursor state for arsine, the general behavior in the chemisorption of the XH_3 moleclues is that they dissociatively adsorb. For low phosphine coverages (0.25 monolayer) phosphine adsorbs molecularly on one side of the Si-Si dimer and, at temperatures around 140 K, fully dissociates into PH_2 and H, with each component attached to one side of the dimer. For higher phosphine coverages (0.5 monolayer) the interaction between adjacent dimers plays a decisive role in the dissociation process. These studies established that ammonia, as well as other group V compounds, dissociate upon adsorption on the Si(100) surface into XH_2 and H, where X=N, P, and As.

The NH_2 fragment is thermally stable and decomposes only at high temperatures around 700 K. Widjaja and Musgrave[123] investigated the NH_2 and H fragments above 700 K using DFT and found that N-containing species prefer to be on the surface, bonded to two surface Si atoms.

In the case of substituted amines, the situation is somewhat different. The room-temperature adsorption of pyrrolidine[124], Figure **20a**, and its methyl-protected analog, N-methylpyrrolidine, Figure **20b**, on the Si(100)-2 X1 surface showed that for both compounds, initial adsorption occurs by barrierless formation of a dative bond between the nitrogen lone pair and an electrophilic atom of the Si dimer. However, while pyrrolidine proceeds to chemisorb dissociatively through N-H bond cleavage, **20c**, methylpyrrolidine is shown to be trapped in its dative-bonded precursor state at room temperature due to a substantial barrier for N-CH_3 cleavage, **20d**. Additionally, the saturation coverage of methylpyrrolidine on Si is seen to be significantly less than that of pyrrolidine, likely due to both steric factors and charge-transfer effects.

Initial molecular adsorption of NH_3 on the Si(100) surface suggests that dative-bonded intermediates play an important role in the decomposition pathway of ammonia and other group V compounds. Under most conditions, however, these intermediate species lead a fleeting existence and are difficult to observe, unless the H atoms are replaced with other protecting groups. In order to observe this rare datively bound species, Cao and

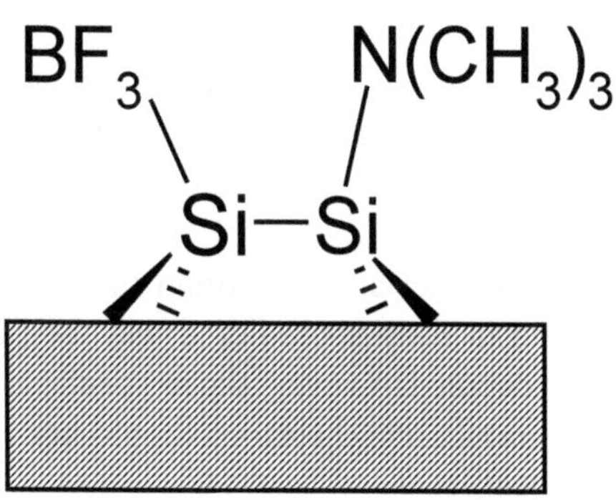

Figure 21. Surface-mediated donor-acceptor complex of TMA-Si-Si-BF_3.

Hamers[125] studied trimethylamine (TMA) and dimethylamine (DMA) with Si surfaces using XPS, Fourier transform IR spectroscopy, and density functional computational methods. They suggested that the dative bond is highly ionic due to the unique geometric structures present on silicon surfaces that permit Si atoms to act as excellent electron acceptors. Following up on this work, they[126] further studied coadsorption of TMA and BF_3 on the Si(100) surface using X-ray photoelectron spectroscopy(XPS). Based on these experiments, they proposed a novel surface-mediated donor-acceptor complex of the structure TMA-Si-Si-BF_3, shown in Figure **21**.

One of the central problems in the field of molecular electronics is the development of well-defined chemical attachment schemes for bonding molecules with delocalized π-electron systems to surfaces without disrupting the π conjugation. In this regard, some functionalized aromatic N-containing molecules may be good candidates. Cao et al. [127] studied pyrrole, aniline, 3-pyrroline, and pyrrolidine on the Si(100)-(2X1) surface using FTIR spectroscopy and XPS. They concluded that both pyrrole and aniline retain their aromatic character after bonding to the surface. Spectroscopic evidence indicates that each of these aromatic molecules can attach to the Si(100) surface via cleavage of one N-H bond, linking the molecules to the surface through a Si-N tether. Isotopic studies of pyrrole showed evidence for additional cleavage of C-H bonds. While strong selectivity favoring bonding through the nitrogen atom is observed for the aromatic molecules, the unsaturated molecule, 3-pyrroline showed evidence for at least two bonding configurations which are either through the nitrogen atom with cleavage of an N-H bond, or through the C=C bond via the surface equivalent of a [2 + 2] cycloaddition reaction. Pyrrolidine appears to bound only through the nitrogen atom.

Using DFT, Luo and Lin[128] predicted the structures and energies for the dissociative adsorption of pyrrole producing C_4H_4N radicals and H atoms on the Si(100)-2**x**1 surface. They concluded that the dissociative adsorption process was found to occur primarily by barrier-less adsorption at an α-C atom followed by N-H dissociation and the isomerization of the radical, leading to the formation of aligned C_4H_4N species on the surface. This result contradicts those of Cao et al. who found that pyrrole retains its aromatic character. Luo and Lin further predicted that the [2+2] and [4+2] cycloadditions also readily occur. However, those species are calculated to be short-lived at room temperature and therefore should not affect the formation of the aligned C_4H_4N radical layer.

7.2 Surface reactions of cyanide group (C≡N) containing compounds

The interaction of CN-containing species with Si surfaces is relevant to the CVD of $SiCN_x$ films that have been shown to exhibit crystalline texture.[129] Bacalzo-Gladden et al.[130] investigated the adsorption, isomerization, and decomposition of HCN on the Si(100)-2 X1 surface with

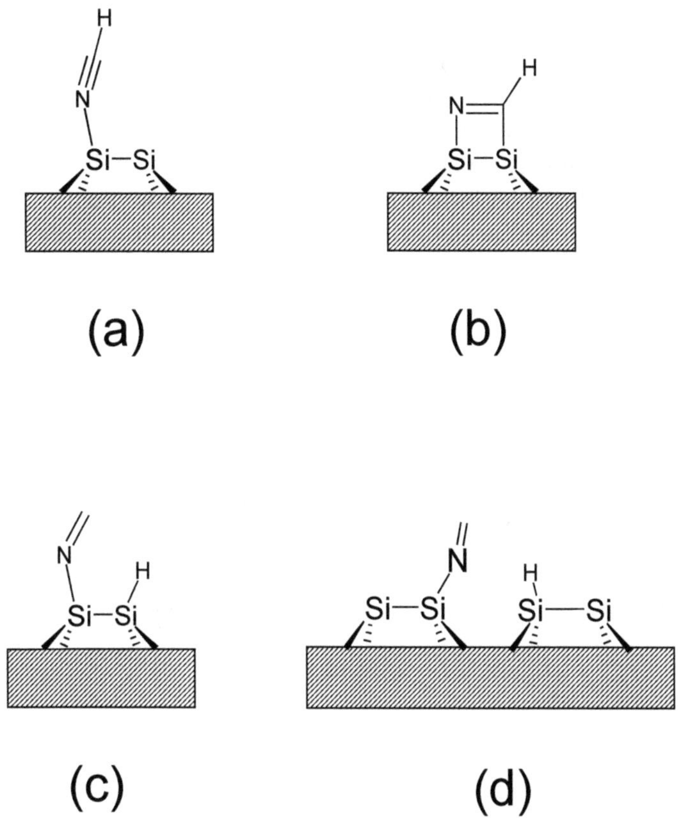

Figure 22.. Molecularly adsorbed configuration of (a) end-on, (b) side-on configuration. Dissociatively adsorbed configuration (c) on the same dimer, (d) across the two dimers.

DFT. They showed that HCN can be readily adsorbed on a Si-Si dimer either dissociatively or molecularly in an end-on, Figure **22a**, and a side-on, Figure **22b**, configuration. They suggested the side-on adsorption, **22b,** occurs by the cycloaddition of the CN triple bond to the Si-Si dimer without barrier. However, the barrier height may be small but nonzero, since it is a thermally forbidden [2+2] reaction. These authors further showed that a transition state connects **22a** and **22c,** yielding a dissociative adsorption on the same dimer or across the two dimers, **22d**. Adsorbate-adsorbate interactions and reactions were also studied with two HCN molecules. For the end-on adsorption, the first HCN exerts a significant effect on the adsorption geometry of the second HCN. In particular, a synergistic effect was observed for the parallel adsorption of two HCNs with their CN groups bridging across the two dimers.

Tao et al.[131,132] studied the covalent attachment and binding configuration of acetonitrile and benzonitrile on Si(100). They concluded that acetonitrile chemisorbs on Si(100) in a side-on di-σ binding configuration, forming Si-C and Si-N bonds. Their TPD measurements reveal the presence of two desorption states, denoted β1 and β2 with desorption energies of 29.8 and 24.6 kcal/mol, respectively. Based on DFT calculations, they proposed that the β1 state may be assigned to di-σ bonded acetonitrile on top of a dimer, Figure **23a**, or an in-row bridging chemisorption, Figure **23b**, while the β2 state is related to acetonitrile bonded in a cross-row bridging configura-

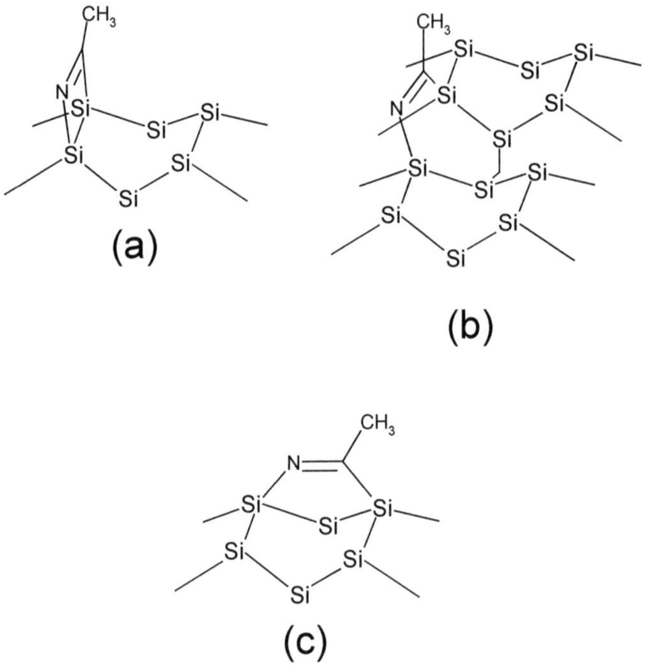

Figure 23. (a) Di-σ bonding configuration. (b) In-row bridging configuration. (c) Cross-row bridging configuration.

tion, Figure **23c**. However, it is also possible that by breaking a C-H bond, acetonitrile can dissociatively chemisorb. In-row and cross-row configurations inevitably create radicals of surface Si atoms. Such low-spin open shell species are not properly studied with single-configurational methods. Further studies are needed using more sophisticated methods.

The reactions of acrylonitrile, which has both a C≡N bond and a C=C bond, with the Si(100)-2x1 surface are also of interest. It has been established[133] that the C=C bond of acrylonitrile acts as a dienophile with the electron-withdrawing C≡N group in traditional solution chemistry. However, it is possible that, on the Si(100)-2x1 surface, acrylonitrile may behave differently, since there is no diene to react with.

Three possible relatively stable reaction products are presented in Figure **24**, where (a) represents the reactants, (b) represents the [4+2] type

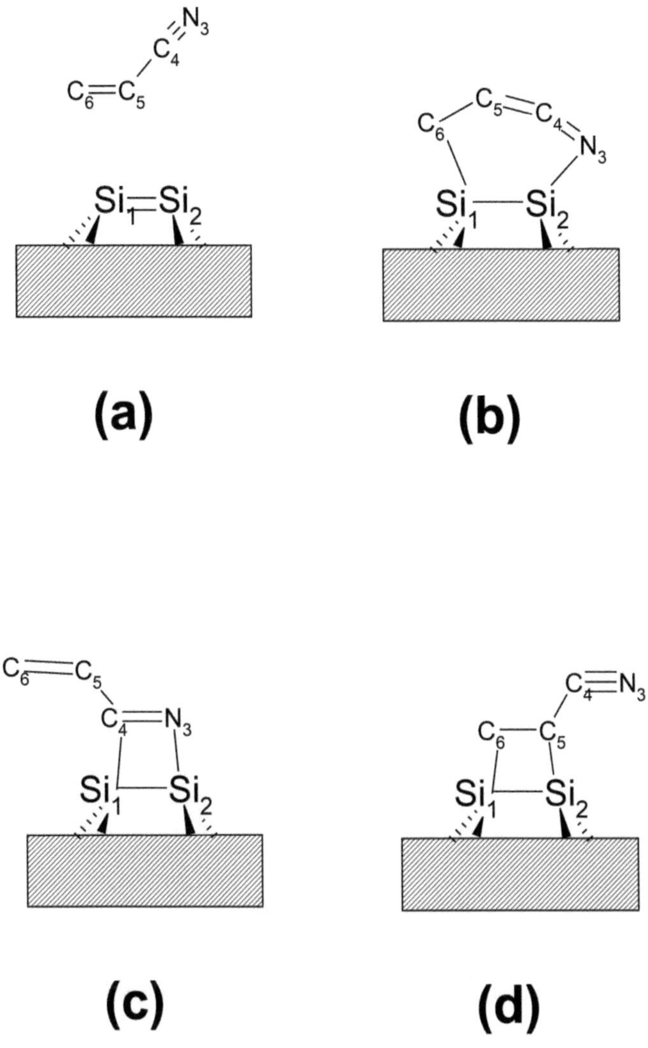

Figure 24. (a) Reactants. (b) [4+2] product. (c) [2+2]CN product. (d) [2+2]CC product.

product, in which Si_1-C_6 and Si_2-N_3 bonds are formed making an allenic N_3=C_4=C_5 configuration, (c) represents the $[2+2]_{CN}$ product with the formation of Si_1-C_4 and Si_2-N_3 bonds, and (d) represents the $[2+2]_{CC}$ product with Si_1-C_6 and Si_2-C_5 bonds being formed. Recent experiments by Tao et al.[134] suggest that acrylonitrile reacts only through the C≡N bond with Si dimers via a [2+2] cycloaddition mechanism yielding exclusively the $[2+2]_{CN}$ surface product. This conclusion is in contrast to the surface reactions of the homonuclear conjugated diene systems considered earlier, for which both [2+2] and [4+2] products were observed.

The question regarding whether and why only one of the many possible products is exclusively formed ultimately bears on the chemical selectivity of the Si(100)-2x1 surface toward conjugated diene systems, homonuclear and heteronuclear alike. By studying the factors that govern the reactivity of these species, one hopes to gain control over these surface reactions to an extent that eventually leads to a technique to tailor the reaction selectivity.

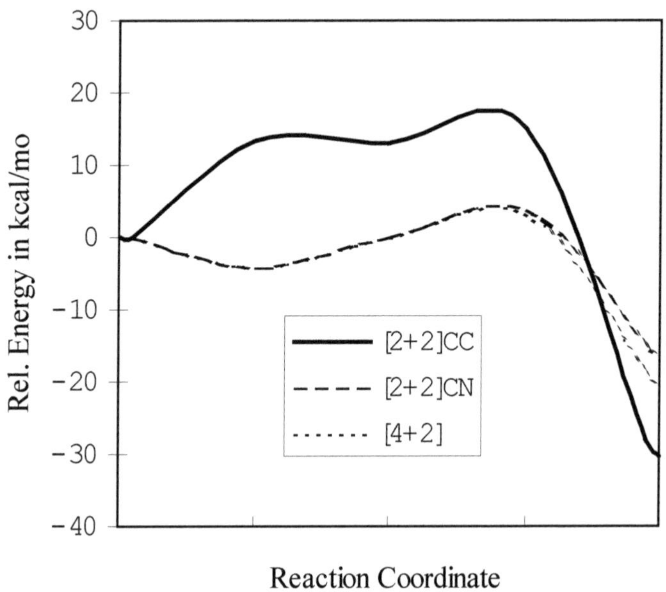

Figure 25. Potential energy surfaces of the three possible surface products.

Using multi-reference as well as single-reference quantum mechanical methods, Choi and Gordon[135] studied the potential energy surface along three possible surface reaction mechanisms of acrylonitrile on the Si(100)-2X1 surface. They showed that all three reactions occur via stepwise radical mechanisms. According to the computed potential energy surfaces presented in Figure **25**, both [4+2] and $[2+2]_{CN}$ cycloaddition products resulting from the reactions of surface dimers with the CN triple bond of acrylonitrile are expected, due to the negligible activation barriers at the surface. Another possible surface product, $[2+2]_{CC}$, requires a 16.7 kcal/mol activation energy barrier. The large barrier makes this route much less favorable kinetically, even though this route produces the thermodynamically most stable products. Isomerization reactions among the surface products are very unlikely due to the predicted large activation barriers preventing thermal redistributions of the surface products. As a result, the distribution of the final surface products is kinetically controlled leading to a reinterpretation of recent experiments. An intermediate Lewis acid-base type complex appears in both the [4+2] and $[2+2]_{CN}$ cycloaddition entrance channels, indicating that the surface may act as an electrophile/Lewis acid toward a strong Lewis base substrate.

8. ADSORBATES CONTAINING GROUP 6 ELEMENTS

8.1 Oxidation reactions on the Si(100) surface

Interactions of atomic or molecular oxygen with a silicon surface can lead to either silicon oxide film growth on the silicon surface (passive oxidation) or to etching of the surface (active oxidation).[136] The outcome of such interactions depends primarily on the surface temperature and also on the oxygen pressure. In the low temperature or high pressure regime, one finds passive oxidation. In the high temperature or low pressure regime, one finds active oxidation or etching of the surface by removal of SiO. The two oxidation processes are:

Passive oxidation: $Si(s) + O_2 \rightarrow SiO_2(s)$
Active oxidation: $Si(s) + 1/2O_2 \rightarrow SiO\ (g)$

Since gate oxide thicknesses of 10 Å or smaller are being grown and will be standard in the near future,[137] understanding the initial oxidation processes and structures of the SiO_2/Si interface is critical. In crystalline SiO_2, oxygen is in a bridging position between two silicon atoms; each silicon atom in turn is surrounded by four oxygen atoms in a tetrahedral configuration in bulk SiO_2, illustrating that di-coordinated oxygen is the most abundant configuration with silicon.

N_2O is a common source of atomic oxygen, as it dissociates at the surface into O and N_2.[138] Many experiments have shown that molecular oxygen (O_2) adsorption on Si(100) is predominantly dissociative.[139] Using silicon clusters containing as many as seven Si atoms and MCSCF wavefunctions, Batra et al[140] have provided theoretical support that the dissociative adsorption of an oxygen molecule is exothermic by 3eV. Conflicting results for the initial oxidation of the Si(100) surface have been reported. Most AES and XPS studies[141] report saturation coverage of 1.0 monolayer (ML) for O_2 adsorption at 300 K. Lower coverage was also reported.[142] An atomic oxygen beam experiment showed no true saturation for atomic oxygen exposure, but the uptake slowed considerably at a coverage of 2-3ML.[141a] Avouris and Lyo[143] concluded that the initial adsorption occurred predominantly at defect sites not at terraces. Many experiments have been devoted to revealing the oxide surface structure. Figure **26** illustrates the possible initial surface configurations of oxygen on the Si(100)- 2x1 surface that have been proposed.

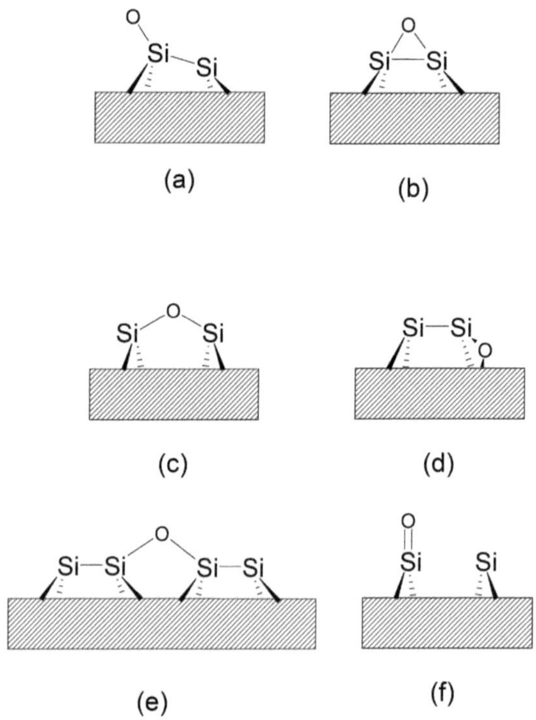

Figure 26. Possible initial oxidized surface structures. (a) on-top. (b) on-dimer. (c) dimer-bridge. (d) backbond. (e) inter-dimer. (f) Silanone.

HREELS[144] and surface extended X-ray-adsorption fine-structure (SEXAFS)[145] measurements reveal that Si-O-Si complexes exist in the early stage of oxidation. Based on the HREELS experiment, it is inferred that the oxygen atom is inserted into the Si-Si "backbond" (Figure **26d**). The SEXAFS experiment suggests that the oxygen atom occupies two different bridge positions, "backbond" and dimer-bridge (Figure **26c**). The Si-O bond length is reported to be 1.65 Å and the Si-O-Si bond angle to be about 120 or 130°. The on-top site (Figure **26a**) has also been proposed.[141a,146] Scanning tunneling microscopy (STM) has been used to study the oxidation. However, the observed features such as position, height, and thermal stability are interpreted differently in these experiments. Avouris and Cahill[147] identified the observed bumps as isolated dimers corresponding to Si atoms ejected from the surface, while Kliese et al[148] have concluded that they are weakly bound species of oxygen atoms or molecules. In addition, small protrusions appear at the very early stage of the oxidation, most frequently in a bridging position between dimer rows with a height of about 0.2 Å.

Theoretically, Smith and Wander[149] have investigated the adsorption of atomic oxygen on Si(100) using a $Si_{18}H_{24}O$ cluster model with HF/STO-3G. These authors predicted that atomic oxygen is adsorbed at the dimer bridge sites (Figure **26c**). The corresponding Si-O bond lengths lie in the range of 1.635 ±0.022 Å. They further found that the dimer and inter-dimer (Figure **26e**) bridge sites become equivalent in the high coverage limit giving rise to a 1x1 pattern. Using a slab model and local-density approximation, Miyamoto and Oshiyama have found three stable sites for the adsorption of atomic oxygen[150]. Of the three, the geometry in which the oxygen atom is inserted into the dimer bond ("dimer-bridge", Figure **26c**) is more stable than the on-dimer (Figure **26b**) and the backbond (Figure **26c**) sites. For molecular oxygen, they found[151] that the backbond site is the most stable. On the other hand, using a molecular dynamics method based on the local density approximation, Uchiyama and Tsukada[152] predicted that the backbond site is more stable than the dimer-bridge site by 0.12 eV even in the case of atomic oxygen. Following calculations of STM images[153], they showed that, in the filled states (negative applied surface bias voltage), the STM images for the backbond and the dimer-bridge site appear to be very similar and almost indistinguishable. The empty-states images (positive applied surface voltage) showed characteristic features of the oxygen site.

There have been some attempts to understand the reaction mechanisms of the initial oxidation. Hoshino et al[154] studied the symmetric mechanism of the direct insertion of molecular oxygen into the dimer bond using silicon cluster models containing two and nine silicon atoms, with the MP2/3-21G method. The activation energy required for this reaction was calculated to be 60.4 kcal/mol. These authors concluded that the reconstructed dimer is barely oxidized at room temperature and that defect sites may be the cause of the natural oxidation of Si(100). This conclusion supported some experiments[143] but contradicted others.[141] Since this symmetric path is formally symmetry forbidden, a low energy asymmetric reaction pathway may exist. Conflicting theoretical results were obtained by Miyamoto and Oshiyama who predicted using LDA slab models that the dissociative chemisorption of O_2 can occur without any activation barrier at all sites they studied.[150,151]

Recently, using scanning reflection electron microscopy (SREM), Watanabe et al[155] obtained strong evidence that the molecular O_2 oxidation of Si(100) proceeds in a layer-by-layer mode and that the first submonolayer

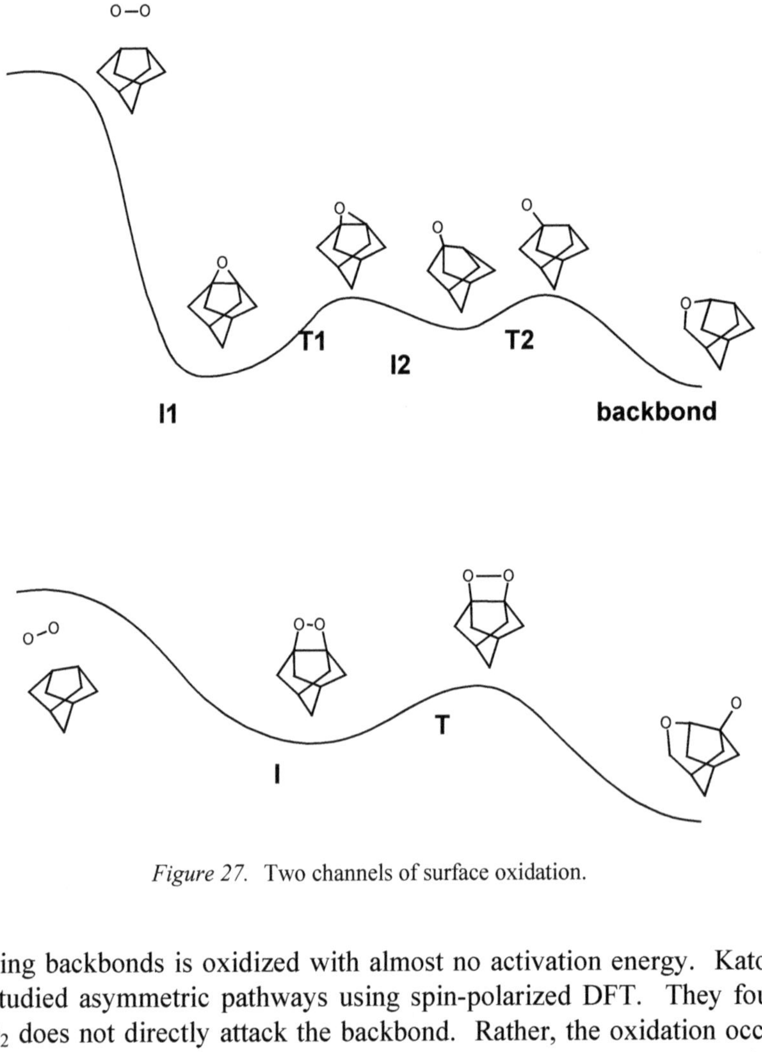

Figure 27. Two channels of surface oxidation.

including backbonds is oxidized with almost no activation energy. Kato et al[156] studied asymmetric pathways using spin-polarized DFT. They found that O_2 does not directly attack the backbond. Rather, the oxidation occurs via metastable chemisorption states along barrierless reaction paths or channels. Figure **27a** shows that the oxygen molecule is proposed to initially dissociate onto two different surface dimers yielding the on-dimer structure, **I1**. The first transition state, **T1** connects **I1** and **I2**, the on-top structure. The second transition state, **T2** connects **I2** and the final "backbond" product. Figure **27b** shows another channel in which the oxygen molecule initially binds on one dimer, **I**. After the transition state, **T,** where the O-O bond is being broken, the final product has a "backbond" oxygen and an on-top oxygen. They argued that a narrowing reaction channel (reduced reaction

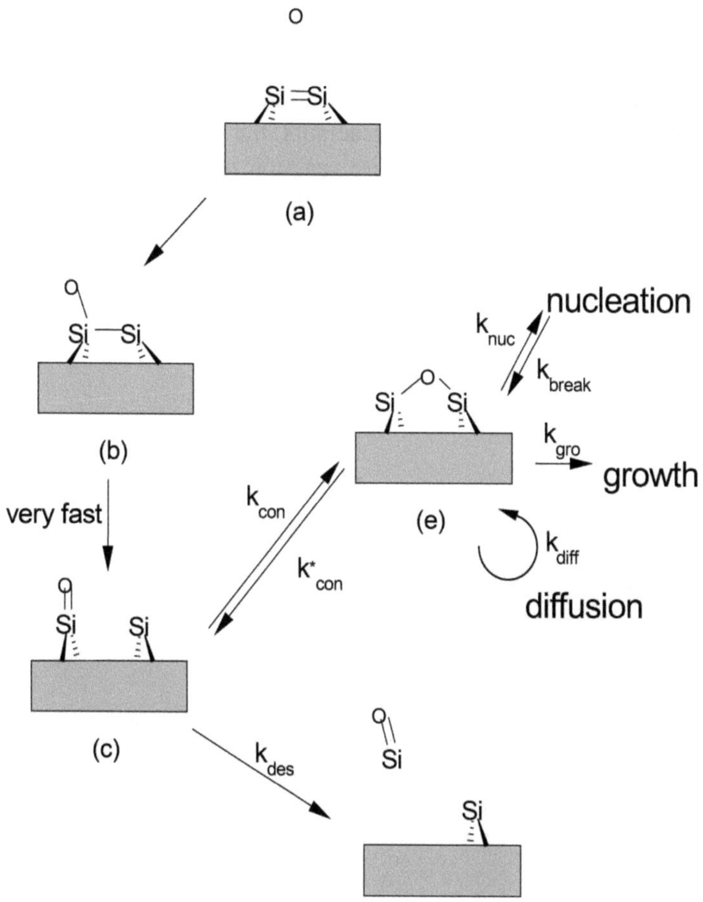

Figure 28. A schematic illustration of a two-species oxidation model. (a) Reactants. (b) on-top. (c) silanone. (d) Product. (e) Dimer-bridge.

probability) yielding a small sticking coefficient with molecular oxygen can be explained by the intersystem crossing (triplet to singlet) that occurs in the middle of the reaction. They estimated the probability of this crossing to be 0.08 and 0.025 for incident kinetic energies of 0.1 and 1.0 eV, respectively.

Choi et al[157] systematically studied the initial O atom oxidations using CASSCF wavefunctions augmented by MRMP2 for dynamic correlation. The SIMOMM method was used in order to study clusters as large as $OSi_{15}H_{20}$ QM atoms embedded in a $OSi_{136}H_{92}$ cluster. It is found that both symmetric and asymmetric approaches of atomic oxygen have no initial reaction barrier. The asymmetric approach initially yields the on-top structure shown in Figure **26a**. The backbond structure is obtained by surmounting a low energy transition state. The symmetric approach yields an on-dimer structure (Figure **26b**), and only a singlet state is found at this geometry. By breaking the remaining Si-Si bond of the dimer, the dimer-bridge structure (Figure **26c**) is obtained. Except for reactants, the singlet surface is lower in energy than the triplet along the entire reaction path. This study illustrates the detailed reaction mechanisms of both "backbond" and dimer-bridge structures. It appears that di-coordinated oxygen configurations such as backbond and dimer-bridge are the most stable forms and play a significant

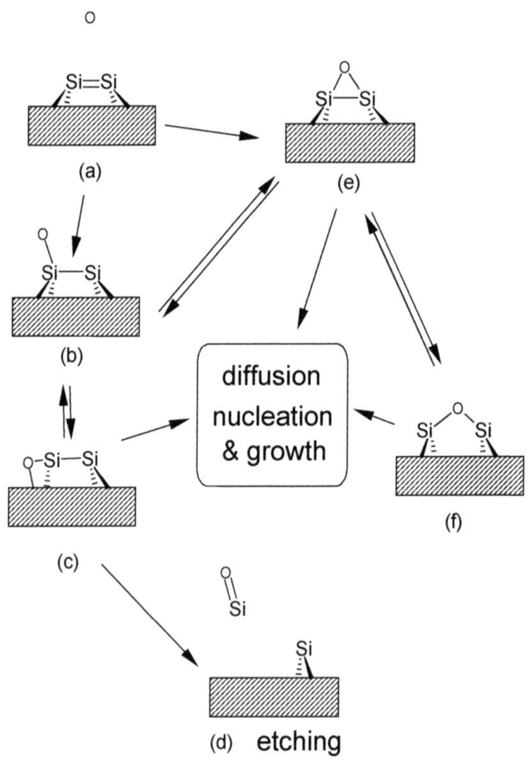

Figure 29. A schematic illustration of oxidation and etching mechanism. (a) Reactants. (b) On-top. (c) Back-bond. (d) Products. (e) on-dimer. (f) Dimer-bridge.

role in the initial growth of silicon oxide film.

For typical oxygen pressures, the transition from "passive" to "active" oxidation occurs at around 600 ~ 750 °C surface temperature. Curves for oxygen uptake vs. time display a transition from a simple Langmuir-Hinshelwood (LH) form for passive oxidation, to a more slowly increasing but sigmoidal form (reflecting autocatalytic aspects to the oxide island formation process) for active oxidation (etching).

Many investigations have been devoted to the elucidation of the reaction mechanisms and the associated energetics of active oxidation of silicon surfaces. Extensive experiments by Engel and coworkers[136,158] that employed modulated molecular beam (MMB) and thermal desorption studies (TDS) at high temperature suggested either the presence of two distinct adsorbed oxygen species (in addition to stable oxide islands), or two distinct desorption behaviors for isolated vs. aggregated species. More recent experimental studies and detailed modeling have provided further insight into the process. Pelz and coworkers[159] performed kinetic Monte Carlo (KMC) simulations on a single species model to reproduce morphological changes during etching observed with STM. Ultimately, they proposed a dual species model which produces an observed decoupling of the etching and nucleation rate.[160]

Figure **28** presents a schematic illustration of a two-species oxidation model. This model assumes the existence of a "desorption precursor" configuration, **28c**, which may be described as a surface silanone species.[161] This silanone can either desorb from the surface as SiO (at rate k_{des}), or convert (at rate k_{con}) to a bridge or other more stable configuration **28e**. The latter can, in turn, diffuse across the surface (at rate k_{diff}) and lead to oxide-cluster nucleation (at rate k_{nuc}) or growth (at rate k_{gro}).

Choi et al[157] found that the surface silanone, **28c** does not appear to exist, at least not when there is a nearby divalent Si atom. One reason is that such a configuration has an unstable diradical Si atom adjacent to the silanone. According to their calculations (summarized in Figure **29**), initial symmetric and asymmetric reaction pathways are both barrierless yielding on-top (Figure **29b**) and on-dimer (Figure **29e**) configurations, respectively. The on-top configuration is converted to the backbond configuration (Figure **29c**) by surmounting a low energy transition state with a barrier of 4.8 kcal/mol. The backbond structure eventually leads to etching (Figure **29d**)

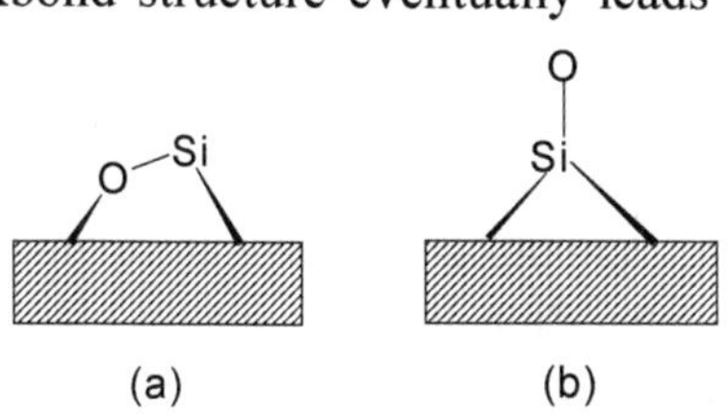

Figure 30. (a) Inserted structure. (b) Silanone structure.

via a complex reaction mechanism with a net barrier of 93 kcal/mol, similar to experimental values of 79 ~ 88[136,158] kcal/mol. They further studied an example secondary active oxidation process, in which the remaining Si atom of the actively oxidized dimer is oxidized. The most stable structure was found to be an inserted configuration (Figure **30a**), similar to the backbond configuration. With a large activation barrier of 39.2 kcal/mol, the inserted configuration is converted into the silanone structure, (Figure **30b**). Subsequently, the SiO leaves the surface without a net reaction barrier beyond the endothermicity. This channel may be compared with the experimental barrier of 46 kcal/mol[162] suggesting that there may be multiple etching channels. Recent KMC studies based on these *ab initio* results successfully describe the transition from passive to active oxidation.[163]

The on-dimer configuration Figure **29e** can be converted into either the on-top species, Figure **29b**, or the dimer-bridge species, Figure **29f**, via transition states with MRMP2 barriers of 66.2 and 12.1 kcal/mol, respectively. The on-top configuration further contributes to the etching process, while the dimer-bridge configuration can be nucleated, grow an oxide film, or diffuse to other sites.

According to this study, the backbond and dimer-bridge configurations are responsible for the etching and oxide film growth, respectively. However, at low temperature, the backbond configuration may be also responsible for film growth. The delicate balance of passive and active oxidation etching, therefore, to some degree depends on the surface conversions between on-dimer and on-top configurations. More elaborate KMC simulations of this process are underway.

Uchiyama et al.[164] have studied the desorption of SiO from the Si(100) surface at small O coverage using DFT dynamics. They concluded that the desorption barrier is 83.0 ~ 87.6 kcal/mol which is quite consistent with the results of Choi et al. Uchiyama et al. suggested the on-dimer and back-bond as the starting structures of active oxidation. However, this suggestion is based on assumptions, rather than a study of the detailed mechanisms.

Using DFT, Widjaja and Musgrave[165] studied the mechanism of O_2 oxidation of the bare Si(100)-(2**X**1) surface. They showed that O_2 adsorbs molecularly on the "up" surface Si atom with no activation barrier and an adsorption energy of 35 kcal/mol. The molecularly adsorbed species further dissociates by first inserting one oxygen atom into the Si-Si dimer bond followed by insertion of the remaining oxygen atom into a Si-Si backbond. They also calculated the activation barriers for SiO_2 formation to be in the range of 65-67 kcal/mol. However, this value is obtained by assuming that the on-top configuration is the starting structure.

8.2. Surface Reactions of O containing compounds

It is known that an oxide layer (SiO_2) grows more easily in the presence of water.[166,167] Therefore, wet oxidation is a preferred choice for the formation of a thick oxide layer. Although there now appears to be consensus that the water undergoes dissociative adsorption on the Si (100) surface, there has been controversy concerning whether the initial adsorption of water is molecular or dissociative. Molecular adsorption (Figure **31a**) leaves one surface Si dangling bond intact, while dissociative adsorption (Figure

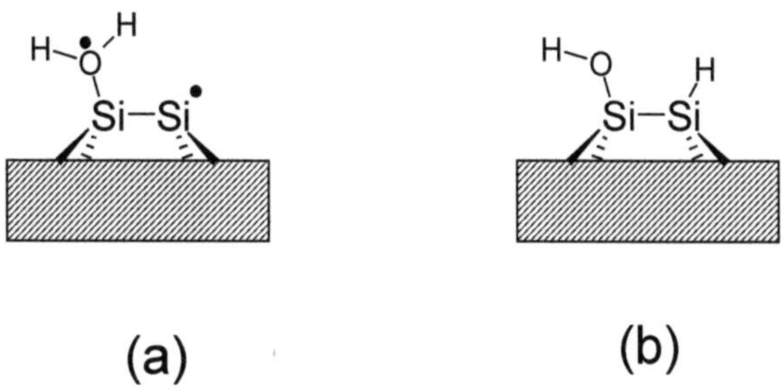

Figure 31. a) Molecularly and (b) Dissociatively adsorbed water on Si surface.

31b) saturates the surface dangling bonds. Initial interpretations of ultraviolet photoelectron spectroscopy (UPS) on Si(100)[168] and Si(111)[169] suggested molecular adsorption, while electron energy loss spectroscopy (EELS)[170] and surface IR[171] studies showed dissociative adsorption on the basis of the Si-H and Si-O-H stretching modes. Later, photoelectron spectroscopy (PES)[172] reinterpretation of the earlier UPS[172a] data and other experiments[173] are all consistent with dissociative adsorption.

At room temperature the sticking coefficient of water on Si(100) is near unity and constant up to saturation.[174] The saturation coverage is reported to be 0.5 ML: one OH and H per Si dimer.[175] These observations suggest a small or zero barrier for the adsorption reaction.

Some structural information is available. An ESDIAD experiment[173a,176] revealed that the O-H bond adsorbed on Si(100) is pointing away from the surface and tilted away from the vertical plane of the dimer bond. TOF-SARS results[177] showed that one hydrogen is close to the surface, while the other one is high above the surface. These were attributed to the hydrogens of Si-H and Si-OH, respectively, and interpreted in favor of the O-H bond pointing away from the surface. The ESDIAD experiment[176b] also revealed that the torsional mode about the Si-O bond in Si-OH is of high amplitude, suggesting a nearly free rotation about the Si-O bond.

Early theoretical studies using extended Hückel[178] or tight binding methods[179] predicted dissociative adsorption, in good agreement with experiment. However, MNDO[180] results suggested that the most stable structure is OH bridging between the two atoms of a dimer. A CNDO[181] study found the O-H bond to be coplanar with the surface dimer bond. These last two theoretical studies are not consistent with experimental evidence.

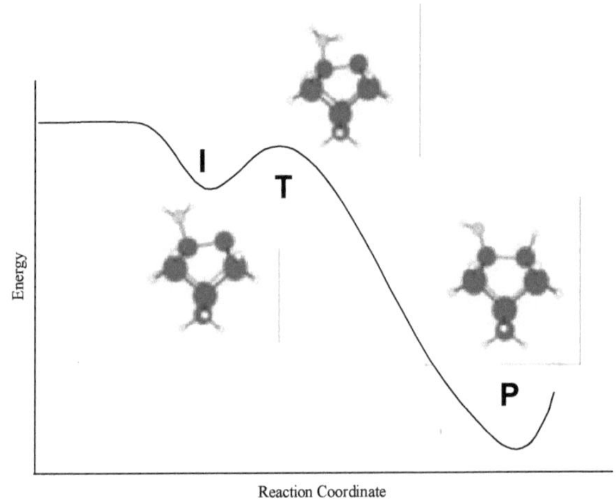

Figure 32. Reaction surface of water on Si surface.

Konecny and Doren[182] detailed the reaction pathway and geometries of the product, the transition state and a molecular precursor state using density functional theory and the Si_9H_{12} cluster model of the surface (see Figure **32**). As a water molecule approaches the surface, it finds a molecular precursor state, **I**, that is about 10 kcal/mol below reactants, with no intervening reaction barrier. Recent, MRMP2//GVB-PP1 results[183] on similar and larger clusters predict **I** to lie 5 kcal/mol below reactants. BLYP/TZ94P[182] and GVB-PP1[183] predict the Si-O distance in **I** to be 2.23 and 1.99 Å, respectively. This rather large difference, consistent with the large difference in stabilization energy, is due to the large decrease in diradical character upon moving from reactants to intermediate. The consequence of this is that a single-reference wavefunction is much less appropriate at reactants than at **I**, so that a multi-reference wavefunction is required for a consistent picture of the mechanism. Since the diradical character is not significant for the rest of the mechanism, single reference methods are adequate to describe subsequent steps, provided that no bare dimers are present.

The transition state **T** that connects the molecular precursor state, **I** with product, **P**, is in the process of breaking one O-H bond and forming a Si-H bond. The reaction barrier at **T** relative to **I** is 1 ~ 8 kcal/mol, depending on basis set and level of theory, providing an easy migration of H from O to Si. The final product, **P** is 45 ~ 55 kcal/mol more stable than **I**. Cho et al.[184] studied this water adsorption using DFT and a slab model and obtained results that are very similar to those obtained with the cluster models. According to MRMP2 results, the net reaction has a small barrier, in contrast with the DFT results. This may be due to the fact that DFT is not able to adequately describe the bare Si dimer of the reactants. In either case, little or no barrier is predicted for the overall process.

Regarding secondary interactions after the chemisorptions of water, Konecny and Doren[182] proposed three types of inter-site interactions that increase the tendency of the O-H bond to lie orthogonal to the dimer bond: Either the oxygen lone pair or the hydroxyl hydrogen may interact with an

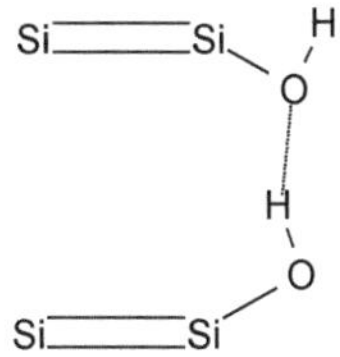

(a)

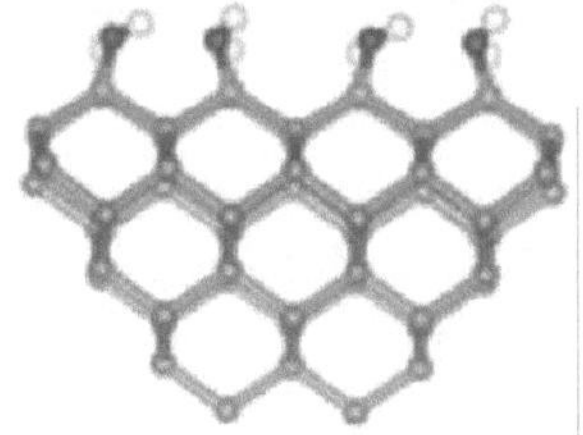

(b)

Figure 33. (a) Surface hydrogen bonding between adsorbed waters on Si surface. (b) Large scale SIMOMM cluster model of four waters on Si surface.

unoccupied adjacent dimer, or two hydroxyls on adjacent dimers may interact as in a weak hydrogen bond.

By combining *ab initio* quantum chemical cluster calculations and IR experiments, Gurevich et al[185] predicted that the initial surface is actually comprised of arrays of isolated and intra-row coupled dimers. The latter are coupled by a hydrogen bonding interaction between OH groups that reside on the same end of adjacent dimers in a dimer row (see Figure **33a**). They further estimated that this inter-dimer bonding increases the stability by ~2 kcal/mol relative to the isolated dimer case. Experiment suggests that the hydroxyl-mediated inter-dimer coupling does not extend to the second adjacent dimer, since there was no further shift of the OH stretch frequency as the coverage was increased. Cho[184] et al also concluded that, while the interaction between water molecules is repulsive, the interaction between dissociated OH species is attractive due to hydrogen bonding.

Large scale SIMOMM cluster calculations[186] depicted in Figure **33b** also show that hydrogen bonding occurs only between the nearest hydroxyls and does not propagate further. The authors[186] concluded that due to the long distance between dimer rows, the formation of hydrogen bonds requires one OH group to be tilted forward and the other OH backward. This pattern makes the extension of hydrogen bonding difficult. A consequence of the coupling is that the Si-Si-O-H torsion potential becomes more rigid in order to direct each H-bonding hydrogen toward its oxygen partner on the neighboring site.

Depending on the surface temperature and water pressure, the ad-

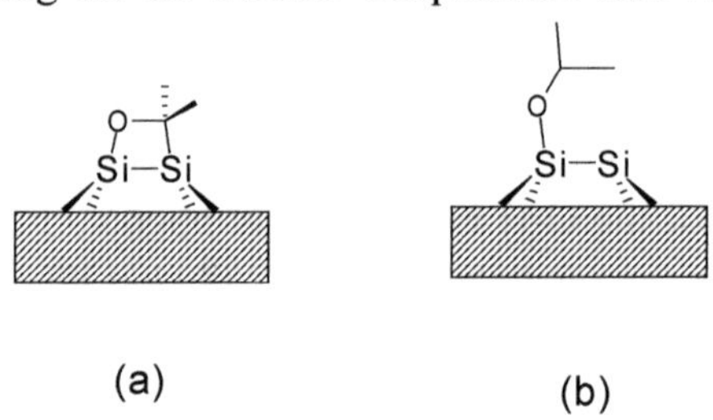

Figure 34. (a) [2+2] and (b) Molecularly adsorbed formaldehyde on Si surface

sorbed water undergoes further reactions yielding an oxide film, decomposition of OH into adsorbed O and H or desorption of SiO.[187] Gurevich et al[185] postulated that the coupling through hydrogen bonding has a significant effect on subsequent oxygen agglomeration, such that single oxygen insertion reactions occur on "isolated" dimers, whereas the coupled dimers lead to the facile production of the doubly inserted dimers.

In the case of methanol[188] and ethanol[189], both experimental and theoretical studies proposed the cleavage of the O-H bond and formation of

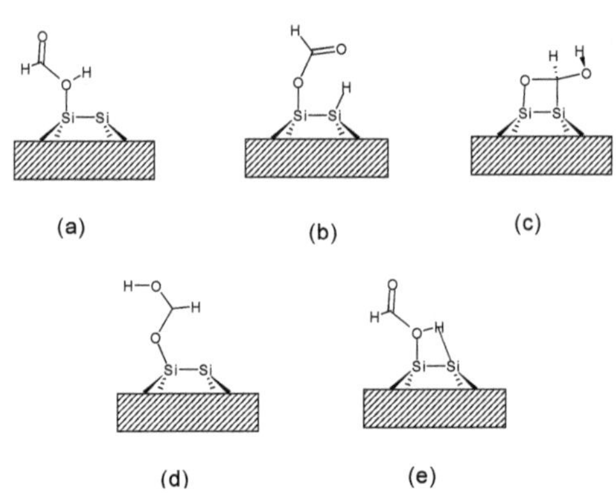

Figure 35. (a), (b), (c) and (d) are the possible surface products of formic acid. (e) A transition state connecting (a) and (b).

Si-OCH_3 and Si-H surface species. According to the results, by making a Si-O bond, the alcohol initially adsorbs molecularly. With the cleavage of the O-H bond, the species dissociates into Si-OR and Si-H.

For formaldehyde[188b], a DFT theoretical study suggested two species. The former, **34a**, contains the Si-O-C-Si 4 membered ring as a result of [2+2] cycloaddition. The latter species, **34b**, adsorbs molecularly.

The adsorption of formic acid on the Si(100) surface was investigated by means of HREELS[190], photo-simulated ion desorption(PSID)[191] and

NEXAFS.[192] These experimental studies suggested that formic acid was partially dissociated to form the unidentate formate (HCOO) and H adatoms on the Si surface. Lin et al. [188b] found 4 adsorption mimima for formic acid on Si(100). They showed that the formation energy of **35b** is -66.3 kcal/mol at B3LYP/6-31G* level of theory, while those of **35a**, **35c**, **35d** are -3.1, -26.7 and -11.8 kcal/mol, respectively. Species **35d** resembles the chemisorption state of CH_2O formed by cycloaddition of the C=O group onto the Si-Si dimer. Theoretical studies predict that **35b** is thermodynamically the most stable. According to the calculations, the reaction barrier of the transition state, **35e** connecting **35a** and **35b** is -0.9 kcal/mol. So they concluded that **35b** is both thermodynamically and kinetically favored.

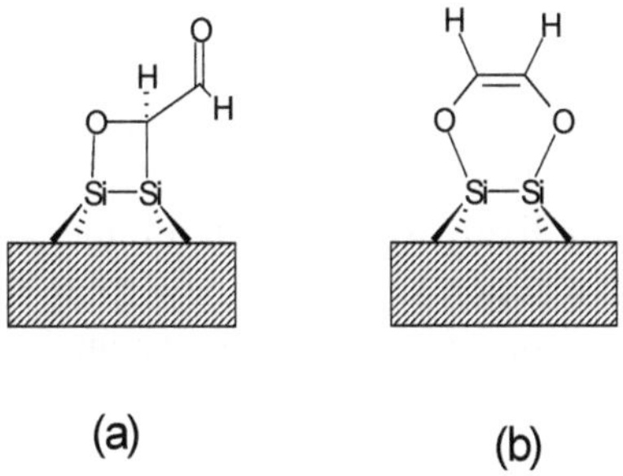

Figure 36. (a) [2+2] cycloaddition product and (b) [4+2] cycloaddition product of dicarbonyl compound.

Using DFT, Barriocanal and Doren[193] studied glyoxal, methylglyoxal and biacetyl carbonyl containing molecules. They concluded that the carbonyl group can adsorb onto the surface by a [2+2] cycloaddition, **36a** with negligible activation barrier. However, they further showed that 1,2-

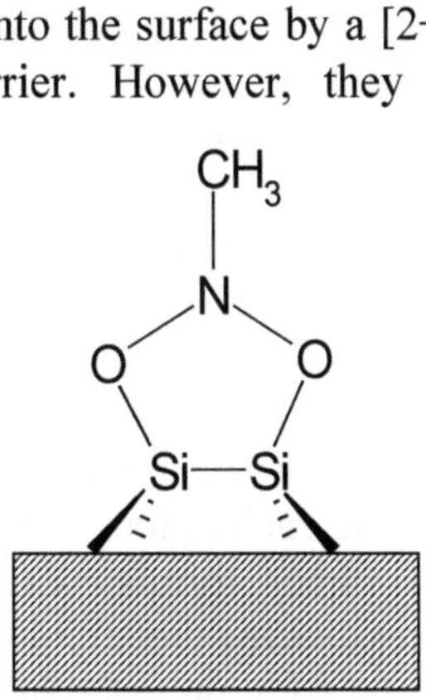

Figure 37. 1,3-dipolar cycloaddition of nitromethane.

dicarbonyls, such as glyoxal, may also react by means of a [4+2] addition to form a hetero-Diels-Alder product, **36b** in which the organic ring stands normal to the surface without significant barrier. In contrast to unsaturated alkyl systems, which must react through the **π** electron system, the reactions of carbonyls may proceed through a very different mechanism, in which the initial surface interaction is through the oxygen lone pair. The presence of lone pairs affects the geometry of the [4+2] adduct, and may alter the competition between [2+2] and [4+2] addition. However, these authors did not study the entire potential energy surface, and did not report a search for possible transition states. Therefore, their mechanistic conclusion does not have supporting kinetic evidence.

The same authors[194] also studied nitromethane with DFT and showed it should strongly chemisorb to the surface dimer by means of a 1,3-dipolar cycloaddition reaction pathway, with no significant activation barrier, see Figure **37**. Though the 1,3-dipolar cycloaddition results in a strongly bound product, this initial cycloadduct is metastable with respect to rearrangement products formed through oxygen migrations between lattice silicons. Multiple products are likely to be present on the surface, so that it will not be well ordered. This reaction suggests a new route for attaching organic monolayers to Si(100) and provides a model system for bonding in silicon oxynitride films.

8.3 S containing compounds

Previous studies using atomic S have used external substituent groups to successfully link π-conjugated sulfur-containg molecules to gold.[195] In particular, sulfur-containing functional groups, such as thiol, sulfide, or disulfide ene groups, have been widely used to selectively tether large multifunctional organic molecules to gold surfaces.[196] In the case of the silicon surface, strong covalent Si-S bonds may increase the robustness of the surface structure. Solution organic reactions have shown that phenyl isothiocyanate (PITC) will undergo a cycloaddition reaction with an alkene via the N=C, C=S and N=C=S linkages.[197] Ellison and Hamers studied the adsorption[198] of PITC on Si(100) surfaces using XPS, Fourier transform IR (FTIR) spectroscopy, scanning tunneling microscopy (STM), and ab initio calculations. They considered the five possible products presented in Fig. **38**, and showed that the adsorption involves the interaction of the N and C atoms of the N=C=S group with the Si=Si dimer, forming a four-member Si_2NC ring at the interface, **38e**. This process leaves the aromatic ring nearly unperturbed and is analogous to a 1,2-dipolar addition reaction.

Coulter et al. [199] studied benzenethiol (C_6H_5SH) and di-phenyl disulfide (C_6H_5S-SC_6H_5) as model systems to compare the interaction of chemically similar π-conjugated molecules with the Si(100)-2X1 surface using Fourier transform IR spectroscopy (FTIR), XPS, scanning tunneling microscopy (STM), and DFT calculations. The results indicate that benzenethiol

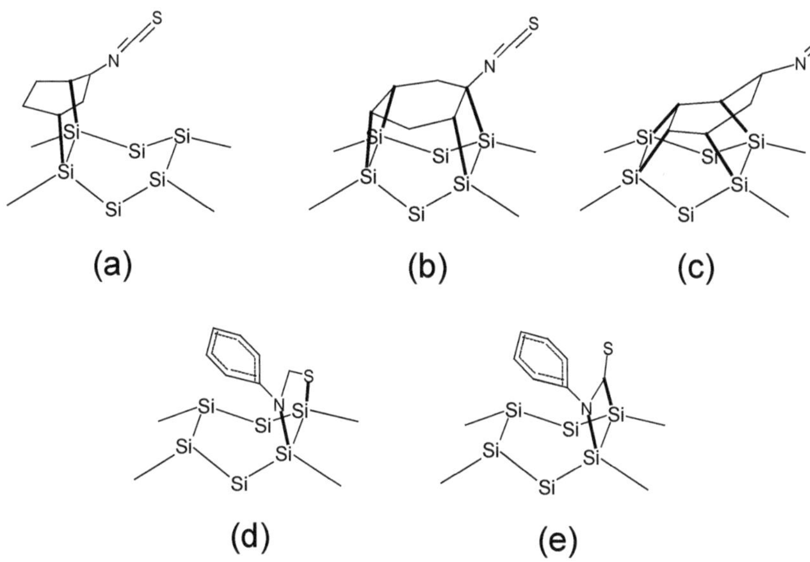

Figure 38. Five possible surface products of phenyl isothiocyanate with Si surface.

molecules chemisorb on the Si(100) surface predominantly through the sulfur atom via deprotonation of the thiol substituent group, **39a**. In the case of di-phenyl disulfide, the S-S bond of di-phenyl disulfide is cleaved and the two sulfur-phenyl moieties are bonded to the silicon surface through the sulfur atoms, **39b**. This new chemistry demonstrated remarkable potential as a means of selectively attaching π-conjugated systems to technologically useful semiconductor surfaces.

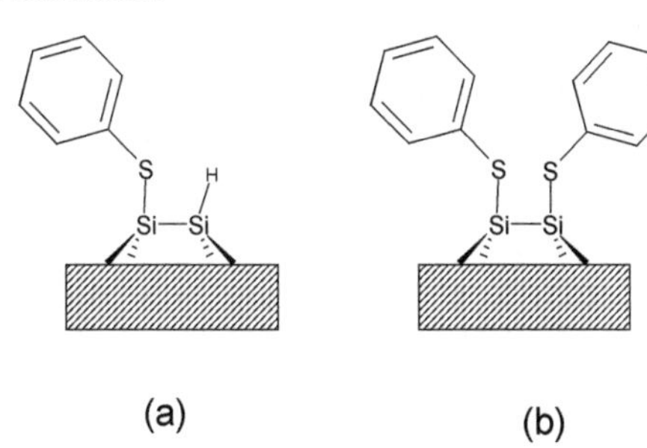

Figure 39. Surface products of (a) benzenethiol, and (b) di-phenyl disulfide on Si surface.

Lu et al.[200] theoretically studied the chemisorption and decomposition of thiophene (C_4H_4S) and furan (C_4H_4O) on the reconstructed Si(100)-2X1 surface. Both thiophene and furan are five-membered ring, π-

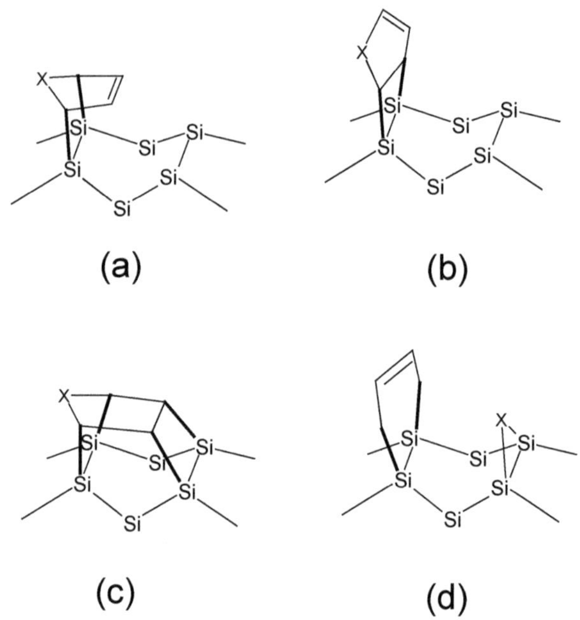

Figure 40. (a) [4+2], (b) [2+2], (c) tetra-□, (d) deoxygenation (desulfurization) of C4H4X (X=S, O)

conjugated compounds with a heteroatom. They differ from common conjugated dienes by showing considerable aromaticity. So they might display somewhat different behavior from that of common dienes and benzene. Lu et al. considered two chemisorption mechanisms, i.e., [4+2], **40a** and [2+2], **40b** cycloadditions of C_4H_4X (X = S,O) onto a surface dimer site. The calculations revealed that the former process is barrierless and favored over the latter, which requires a small activation energy (2.6 kcal/mol for thiophene and 1.2 kcal/mol for furan). The di-σ bonded surface species formed by [4+2] cycloaddition-type chemisorption can either undergo further [2+2] cycloaddition with a neighboring Si:Si dimer site, giving rise to a tetra-□ bonded surface species, **40c**, or undergo deoxygenation (desulfurization) by transferring the heteroatom to a neighboring Si:Si dimer site, leading to a six-member ring metallocyclic $C_4H_4Si_2$ surface species, **40d**. The latter process is slightly more favorable than the former, especially in the case of thiophene.

9. ADSORPTION AND ETCHING OF SI(100) WITH HALOGEN, ADSORBATES CONTAINING GROUP 7 ELEMENTS

Halogen etching of silicon surfaces has been extensively studied.[1,201] Figure 41 shows the proposed surface bonding structures of chlorine on Si(100) that imply dissociative adsorption. The symmetric dimer (Figure 41a) has been suggested by NEXAFS[202], ARPES[203] and EELS experiments.[204] By combining several experimental techniques, Gao et al[205] showed that in addition to the symmetric dimer, there is also a bridging species (Figure 41b). Several theoretical studies[206] have also shown that both the symmetric dimer and bridging species are stable, with the symmetric dimer lower in energy. Different interpretations have been given from other NEXAFS[207] and ESDIAD[208] studies. In these studies, to account for the surface normal Si-Cl bond direction, the asymmetric dimer (Figure 41c) was suggested as the dominant structure. The possibility of a dichloride species[209] (Figure 41d) was also proposed to account for another non-

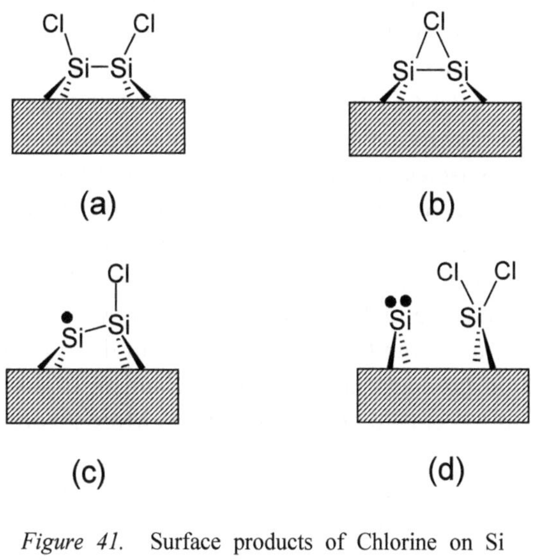

Figure 41. Surface products of Chlorine on Si surface. (a) Symmetric dimer. (b) Bridge. (c) Asymmetric dimer. (d) Dichloride species.

perpendicular bond direction.

In contrast to oxygen, which prefers a di-coordinate configuration, halogen seems to prefer single bond configurations similar to hydrogen and water. This preference may prevent the halogen from diffusing into the silicon bulk, prohibiting growth of a halogen film on the surface.

A majority of the studies[205,209b,210] reported that the thermal etching of Cl occurs exclusively by the desorption of $SiCl_2$ and $SiCl_4$. The former is mainly produced at high temperature (~900 K) and the latter at low and intermediate temperatures (150 ~ 400 K). However, Cl_2[210] and SiCl[211] also have been observed. According to the ion-enhanced etching of silicon, Si, $SiCl_3$, Si^+, and $SiCl^+$ desorption species have been also observed.[212] Of these, Si and SiCl were the major products.

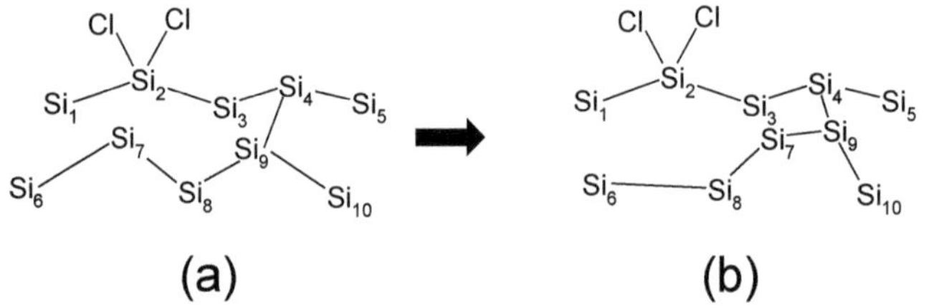

Figure 42. A possible Si migration

To elucidate the reaction mechanisms of $SiCl_2$ desorption, de Wijs et al[213] performed first principle studies using a local density functional in combination with slab models. The study predicted that the two monochlorinated Si atoms (Figure **41a**) of a surface dimer can rearrange into a metastable $SiCl_2$ adsorbed species plus a Cl-free Si atom (Figure **41d**, Figure **42a**). These authors suggested that desorption of $SiCl_2$ occurs via a two-step mechanism, in which the adsorbed species is initially stabilized by diffusion away from the free Si atom (Figure **42b**), followed by the desorption of $SiCl_2$. The estimated activation energy is ~ 3.1eV. Although it was asserted that this result is in reasonable agreement with the experimental value of ~2.4 eV[210a] and comparable with the fluorine etching energy of 3-3.7eV[214], the nearly 1eV difference seems to be too large to have confidence in these results. Due to the multiple desorption products, the surface mechanism of adsorption and desorption of Cl etching seems to be quite complex. More theoretical studies are expected.

Compared with chlorine, studies of other halogens are less common. Iodine[215], fluorine[214] and bromine[216] have been shown to etch the Si(100) surface. ESDIAD experiments[173a,208,217] indicate that F bonds to the dangling bonds of the dimers, similar to the structure shown in Figure **41a**. Using molecular beam techniques, XPS, TPD and LEIS, Engstrom et al[214] showed that molecular fluorine adsorbs dissociatively, and that the saturation coverage of ~1.5 ML is inconsistent with bonding only to the dangling bonds. Ceyer and co-workers[218] have proposed a F atom abstraction mechanism, where by making one Si-F bond with a surface dangling bond, the F_2 molecule ejects a F atom, that can either desorb or chemisorb elsewhere on the surface.

Theoretically, Wu and Carter[219] showed that the 1.5 ML coverage may be explained by the difluoride species. Molecular dynamics simulations by Carter and Carter[220] suggested a stepwise mechanism which yields two pathways involving Si-F bond formation: (a) F atom abstraction, in which one Si-F bond is formed at the expense of the F-F bond, while the remaining F atom is ejected from the surface, and (b) dissociative chemisorption, in which both F atoms in the incident F_2 molecule form Si-F bonds in a consecutive fashion. They showed that depending on the conditions, the preferred reaction channel varies.

Trifluroiodomethane (CF_3I) may find an important use in plasma etching of silicon in semiconductor device fabrication. In this process, electron impact on a halogen containing feed gas creats chemically active ions and radicals that attack silicon to form volatile SiX_y (X=halogen) species. Currently, the most widely used feed gas is CF_4, however, an effort has been mounted to find a replacement owing to the high global warming potential of this species. CF_3I has promise as a substitute for CF_4. Sanabia et al.[221] studied CF_3I adsorption on a silicon surface and the effect of low-energy electron bombardment of a CF_3I-covered silicon surface are relevant to plasma etching using experimental and theoretical methods. They concluded that CF_3X (X=F, Cl, Br, and I) dissociatively adsorbs on Si(100) with X transferred to a silicon atom.

The interaction of alkyl halides with semiconductor surfaces has been of great interest due to the technological applications of the formation of silicon carbide films, CVD of diamond films, and photochemical etching. The high reactivity of methyl halides compared to hydrocarbons makes them potential candidates for diamond and silicon carbide film growth.[222] Lee and Kim[223] studied the adsorption mechanism of CH_3Cl on Si(100)-2**X**1 by experimental and semiempirical calculations. They concluded that the dissociative reaction occurs via a precursor state which yields a nonactivated adsorption process.

10. SUMMARY AND OUTLOOK

It is clear that the chemistry that occurs on silicon surfaces, Si(100) in particular, is not a simple extension of solution chemistry. One of the main reasons is the unique semi-rigid surface structure, as exemplified by the "pinned back" cis-bent structure accompanying very flat buckling potentials. Consequently, such a unique structure reduces the geometric hindrance along the asymmetric approach of substrates, making the surface quite reactive. Cluster models predict both symmetric and buckled static structures, depending on the level of theory. The highest levels of theory, based on multi-reference wave functions, favor the symmetric structure. The most recent experiments find a symmetric species. The discrepancy with some experiments may be due to the experimental conditions, surface defects or insufficient cluster size used in the theoretical studies. In any event, a static picture of the surface structure is likely to be inadequate.

The existence of low symmetry reaction pathways obviates the importance of symmetry rules. The consequence of this is especially apparent in the reactions of enes and dienes with the surface, where formally symmetry forbidden [2+2] reactions occur readily, and the overall reaction mechanisms are kinetically controlled. In the case of reactions with dienes, the surface loses its selectivity for the Diels-Alder product. A unique feature of reactions involving unsaturated hydrocarbons is that more than one Si-C bond are formed, and the integrity of the adsorbate is intact. Furthermore, except for aromatic molecules, surface product inter-conversion is rare.

Surface modification via chemical reactions is currently a very active field due to potential applications in the development of new materials. In addition to the ene and diene systems, other surface reactions including ionic, radical and photochemical reactions with a variety of functional groups are expected to extend the scope of potentially useful surface techniques.

In contrast to the reactions of unsaturated hydrocarbons, hydration, hydrogenation, and oxidation by molecular oxygen seem to be dissociative rather than molecular. In the case of hydration, an asymmetric stepwise low energy pathway to dissociative adsorption is generally accepted. The mechanism initially involves molecular adsorption as a precursor state. A secondary interaction through hydrogen bonding between neighbor dimer OH groups, that seems to affect further surface reactions of adsorbed OH and H, has been illustrated by theoretical and experimental studies.

For hydrogenation, the reaction mechanism is still controversial largely due to discrepancies between theoretically and experimentally determined desorption barriers. Although some alternative mechanisms have been proposed, more studies are needed.

The most stable structures that arise upon initial oxidation may be the backbond and dimer bridge structures. Both structures contain dicoordinated oxygen. These two configurations seem to be the initial building blocks of oxide film growth and the key species for surface etching.

New surface chemical bonds, such as Si-N, Si-S, etc have been shown to form on the Si surface, adding a new dimension to the technological potential. It has been shown that the initial approach of amine and its derivatives occurs via the donation of the N lone pair electron to a surface dimer Si atom, illustrating that surface dimer can act as a Lewis acid/electrophile toward the strong Lewis base. This initial approach usually occurs with no activation barrier.

Due to the multiple desorption products, the etching of the surface by halogen appears to be quite complex. A multi-step reaction mechanism has been suggested to account for the $SiCl_2$ desorption species. In the case of fluorine atom adsorption, F atom abstraction and dissociative chemisorption mechanisms have been suggested. In order to account for the complex surface reactions, more studies are needed.

With the help of new experimental techniques and more sophisticated theoretical methodologies, many fascinating surface structures and mechanisms have been revealed with molecular detail. These combined efforts continue to elucidate new interesting features of surface chemistry. Developments of new theoretical techniques, will facilitate the analysis of much larger, and therefore more realistic clusters. Combined with periodic boundary conditions, sophisticated levels of theory, and dynamics and non-equilibrium statistical mechanics techniques, these efforts will push the convergence of theory and experiment.

ACKNOWLEDGEMENTS

This work was supported by grant No. R05-2002-000-0131-0 to C.H.C. from the Basic Research Program of the Korea Science & Engineering Foundation, and by grant No. F49620-99-1-0063 to M.S.G. from the U.S. Air Force Office of Scientific Research.

REFERENCES

1. H. N. Waltenburg and J. T. Yates, Jr., *Chem. Rev.* **95**, 1589 (1995).
2. M. J. Sailor and E. J. Lee, *Adv. Mater.,* **9**, 783 (1997).
3. J. M. Buriak, *J. Chem. Soc., Chem. Commun.,* 1051 (1999).
4. (a) H. C. Longuet-Higgins and L. Salem, *Prod. Roy. Soc. London A*, **251**, 172 (1959). (b) Y. S. Lee and M. Kertesz, *Int. J. Quantum. Chem. Symp.* **21**, 163 (1987).
5. (a) R. Hoffmann, *Solid and Surfaces: A Chemist's View of Bonding in Extended Structure,* VCH, NY, 1988. (b) S. Y. Hong and D. Marynick, *J. Chem. Phys.* **96**, 5497 (1992).
6. (a) O. H. Nielsen and R. M. Martin, *Phys. Rev. B,* **32**, 3780 (1985). (b) C. Dal, *Lect. Notes Chem.* **67**, 155 (1996). (c) K. Schwarz and P. Blaha, *Lect. Notes Chem.* **67**, 139 (1996). (d) G. Kresse and J. Furthmüller, *Phys. Rev .B* **54**, 11169 (1996).
7. P. J. Feibelman, *Phys. Rev. B.* **44**, 3916 (1991).
8. (a) M. P. Teter, M. C. Payne and D. C. Allan, *Phys. Rev. B.* **40**, 12255 (1989). (b) D. M. Bylander, L. Kleinman and S. Lee, *Phys. Rev. B.* **42**, 1394 (1990).
9. K. N. Kudin and G. E. Scuseria, *Phys. Rev. B.* **61**, 5141 (2000).
10. (a) K. E. Schmidt and M. A. Lee, *J. Stat. Phys.* **63**, 1223 (1991). (b) C. L. Berman and L. Greengard, *J. Math. Phys.* **35**, 6036 (1994). (c) C. G. Lambert, T. A. Darden and J. A. Board, *J. Comput. Phys.* **126**, 274 (1996). (d) M. Challacombe, C. White and M. Head-Gordon, *J. Chem. Phys.* **107**, 10131 (1997). (e) K. N. Kudin and G. E. Scuseria, *Chem. Phys. Lett.* **283**, 61 (1998).
11. D. J. Singh, *Planewaves, Pseudopotentials and the LAPW Method,* Kluwer Academic Pub., Boston, 1994.
12. P. Nachtigall, K. D. Jordan and K. C. Janda, *J. Chem. Phys.* **95**, 8652 (1991).

13. A. Redondo and W. A. Goddard III, *J. Vac. Sci. Technol.* **21**, 344 (1982).
14. J. Shoemaker, L. W. Burggarf and M. S. Gordon, *J. Chem. Phys.* **112**, 2994 (2000).
15. F. W Bobrowicz and W. A. Goddard III, in Modern Theoretical Chemistry, Vol. 3, Schaefer, H. F. III, Ed., Chapter 4. (Plenum, New York, 1977).
16. (a) W. L. Jorgensen, J. Chandrasekhar, J. D. Madura, R. W. Impey, M. L. Klein, J. *Chem. Phys*. **79**, 926 (1983) (b) P. N. Day, J. H. Jensen, M. S. Gordon, S. P. Webb, W. J. Stevens, M. Krauss, D. Garmer, H. Basch, and D. Cohen, *J. Chem. Phys.* **105**, 1968 (1996) (c) H. J. C. Berndsen, J. R. Grigera and T. P. Straatsma, *J. Phys. Chem.* **91**, 6269 (1987) (d) B. Guillot and Y. Guissani, *J. Chem. Phys.* **99**, 8075 (1993) (e) J. Alejandre, D. J. Tildesley and G. A. Chapela, *J. Chem. Phys.* **102**, 4 (1995).
17. (a) B. Weiner, C. S. Carmer and M. Frenklach, *Phys. Rev. B.* **43**, 1678 (1991). (b) C. S. Carmer, B. Weiner and M. Frenklach, *J. Chem. Phys.* **99**, 1356 (1993). (c) F. Maseras and K. Morokuma, *J. Comput. Chem.* **16**, 1170 (1995).
18. J. Shoemaker, L. W. Burggarf and M. S. Gordon *J. Phys. Chem. A*; **103**, 3245 (1999).
19. C. H. Choi and M. S. Gordon, *J. Am. Chem. Soc.* **121**, 11311 (1999).
20. (a) X. Assfeld and J.-L. Rivail, *Chem. Phys. Lett.* **263**, 100 (1996) (b) J. Gao, P. Amara, C. Alhambra and M. J. Field, *J. Phys. Chem. A* **102**, 4714 (1998) (c) D. M. Philipp and R. A. Friesner, *J. Comput. Chem.* **20**, 1468 (1999) (d) V. Kairys and J. H. Jensen, *J. Phys. Chem. A*, **104**, 6656 (2000).
21. (a) M. Challacombe and E. Schwegler, *J. Chem. Phys.* **106,** 5526 (1997) (b) C. A. White, B. G. Johnson, P. M. W. Gill and M. Head-Gordon, *Chem. Phys. Lett.* **253,** 268 (1996). (c) T.-S. Lee, D. M. York and W. Yang, *J. Chem. Phys.* **105**, 2744 (1996). (d) M. C. Strain, G. E. Scuseria and M. J. Frisch, *Science* **271**, 51 (1996). (e) C. H. Choi, J. Ivanic, M. S. Gordon and K. Ruedenberg, *J. Chem. Phys.* **111**, 8825 (1999). (f) G. Hetzer, M. Schutz, H. Stoll, and H. Werner, *J. Chem. Phys.* **113**, 9443 (2000) (g) S. Saebo and P. Pulay, *J. Chem. Phys.* **115**, 3975 (2001) (h) M. Schutz, and H.-J. Werner, *J. Chem. Phys.* **114**, 661 (2001)
22. R. E. Schlier and H. E. Farnsworth, *J. Chem. Phys.* **30**, 917 (1959).
23. J. D. Levine *Surf. Sci.* **34**, 90 (1973)
24. D. J. Chadi, *Phys. Rev. Lett.* **43**, 43 (1979)
25. (a) N. Jedrecy, M. Sauvage-Simkin, R. Pinchaux, J. Massies, N. Greiser and V. H. Etgens *Surf. Sci.* **230**, 197 (1990) (b)D.-S. Lin, T. Miller and T.-C. Chiang *Phys. Rev. Lett.* **67**, 2187 (1991) (c) E. Landemark, C. J. Karlsson, Y.-C. Chao and R. I. G. Uhrberg, *Phys. Rev. Lett.* **69**, 1588 (1992) (d) E. Landemark, C. J. Karlsson, Y.-C. Chao, R. I. G. Uhrberg, *Surf. Sci.* **287/288**, 529 (1993) (e) R. A. Wolkow, *Phys. Rev. Lett.* **68**, 2636 (1992).
26. (a) P. C. Weakliem, G. W. Smith and E. A. Carter, *Surf. Sci.* **232**, L219 (1990) (b) J. Dabrowski and M. Scheffler, *Appl. Surf. Sci.* **56-58**, 15 (1992) (c) N. Roberts and R. J. Needs, *Surf. Sci.* **236**, 112 (1990) (d) P. Kruger and J. Pollmann, *Phys. Rev. B.* **47**, 1898 (1993) (e) J. E. Northrup, *Phys. Rev. B.* **47**, 10032 (1993) (f) A. Ramstad, G. Brocks and P. J. Kelly, *Phys. Rev. B.* **51**,

14504 (1995) (g) C. Yang, S. Y. Lee and H. C. Kang, *J. Chem. Phys.* **107**, 3295 (1997). (h) R. Konecny and D. J. Doren, *J. Chem. Phys.* **106**, 2426 (1997).
27. (a) R. M. Tromp, R. J. Hamers and J. E. Demuth, *Phys. Rev. Lett.* **55**, 1303 (1985) (b) R. J. Hamers, R. M. Tromp, and J. E. Demuth, *Phys. Rev. B.* 34, 5343 (1986) (c) R. J. Hamers, P. Avouris and F. Bozso, *Phys. Rev. Lett.* **59**, 2071 (1987) (d) R. J. Hamers, P. Avouris and F. Bozso, *J. Vac. Sci.* Technol. A, **6**, 508 (1988) (e) R. Wiesendanger, D. Burgler, G. Tarrach, and H. J. Guntherodt, *Surf. Sci.* **232**, 1 (1990) (f). Z. Jing and J. L. Whitten, *Surf. Sci.* **274**, 106 (1992) (g) M. Tsuda, T. Hoshino, S. Oikawa and I. Ohdomari, *Phys. Rev. B.* **44**, 11241 (1991) (h) T. Hoshino, S. Oikawa, M. Tsuda and I. Ohdomari, *Phys. Rev. B.* **44**, 11248 (1991) (i) T. Hoshino, M. Hata, S. Oikawa and M. Tsuda, *Phys. Rev. B.* **54**, 11331 (1996) (j) I. P. Batra, *Phys. Rev. B* **41**, 5048 (1990).
28. A. Redondo and W. A. Goddard III, *J. Vac. Sci. Technol.*, **21**, 344 (1982).
29. B.Paulus, *Surf. Sci.* **408**, 195 (1998).
30. P. C. Weakiem, G. W. Smith, and E. A. Carter, *Surf. Sci. Rep.* **232**, L219 (1990).
31. T. Yokoyama, and K. Takayanagi, *Phys. Rev. B* **61**, R5078, (2000)
32. M.W. Schmidt, P. Truong, and M.S. Gordon, *J. Am. Chem. Soc.,* **109**, 5217 (1987).
33. F. W. Bobrowicz and W. A. Goddard, III in Modern Theoretical Chemistry, V 3, H. F. Schaefer III, Ed., Chapter 4 (Plenum, New York, 1977).
34. M. W. Schmidt and M. S. Gordon, *Ann. Rev. Phys. Chem.* **49**, 233 (1998).
35. M. S. Gordon, J. Shoemaker and L. W. Burggraf, *J. Chem. Phys.* (in press)
36. Y. Jung, Y. Akinaga, K.D. Jordan, and M.S. Gordon, *Theor. Chem. Accts.*, in press
37. J. S. Hess and D. J. Doren, *J. Chem. Phys.* **113**, 9353 (2000).
38. M. S. Gordon, J. R. Shoemaker, and Burggraf, *J. Chem. Phys.* **113**, 9355 (2000).
39. (a) K. Sinniah, M. G. Sherman, L. B. Lewis, W. H. Weinberg, J. T. Yates, Jr., and K. C. Janda, *Phys. Rev. Lett.* **62**, 567 (1989) (b) K. Sinniah, M. G. Sherman, L. B. Lewis, W. H. Weinberg, J. T. Yates, Jr., and K. C. Janda, *J. Chem. Phys.* **92**, 5700 (1990) (c) M. L. Wise, B. G. Koehler, P. Gupta, P. A. Coon, and S. M. George, *Surf. Sci.* **258**, 5482 (1991) (d) U. Höfer, L. Li and T. F. Heinz, *Phys. Rev. B,* **45**, 9485 (1992) (e) M. C. Flowers, N. B. H. Honathan, Y. Liu and A. Morris, *J. Chem. Phys.* **99**, 7038 (1993)
40. (a) R. J. Hamers, R. M. Tromp and J. E. Demuth, *Phys. Rev. B.* **34**, 5343 (1986) (b) G. Schulze and M. Henzler, *Surf. Sci.* **124**, 336 (1983) (c) P. Gupta, V. L. Colvin and S. M. George, *Phys. Rev. B.* **37**, 8234 (1988) (d) M. L. Wise, B. G. Koehler, P. Gupta, P. A. Coon and S. M. George, *Mater. Res. Soc. Symp. Proc.* **204**, 319 (1991).
41. M. R. Radeke and E. A. Carter, *Phys. Rev. B.* **54**, 11803 (1996).
42. J. J. Boland, *Phys. Rev. Lett.* **67**, 1539 (1991).
43. (a) P. Nachtigall, K. D. Jordan and K. C. Janda, *J. Chem. Phys.* **95**, 8652 (1991). (b) C. J. Wu and E. A. Carter, *Chem. Phys. Lett.* **185,** 172 (1991).
44. (a) M. P. D'Evelyn, Y. L. Yang and L. F. Sutcu, *J. Chem. Phys.* **96**, 852 (1992). (b) Y. L. Yang and M. P. D'Evelyn, *J. Vac. Sci. Techno. A* **11**, 2200 (1993).
45. (a) P. Nachtigall and K. D. Jordan, *J. Chem. Phys.* **101**, 2648 (1994). (b) Z. Jing and J. L. Whitten, *Phys. Rev. B.* **50**, 5506 (1994).

46. Z. Jing and J. L. Whitten, *J. Chem. Phys.* **102**, 3867 (1995).
47. S. Pai and D. Doren, *J. Chem. Phys.* **103**, 1232 (1995).
48. E. S. Tok and H. C. Kang, *J. Chem. Phys.* **115**, 6550 (2001).
49. M. Durr, M. B. Raschke, and U. Hofer, *J. Chem. Phys.* **111**, 10411 (1999).
50. M. R. Radeke and E. A. Carter, *Phys. Rev. B.* **55**, 4649 (1997).
51. (a) P. Kratzer, E. Pehlke, M. Scheffler, M. B. Raschke and U. Köfer, *Phys. Rev. Lett.* **81**, 5596 (1998) (b) E. Pehlke and P. Kratzer, *Phys. Rev. B.* **59**, 2790 (1999).
52. R. Konecny, D. J. Doren, *J. Phys. Chem.* B **101**, 10983 (1997).
53. R. B. Woodward and R. Hoffmann, *The Conservation of Orbital Symmetry*, Verlag Chemie, Weinheim, 1970.
54. See p. 90, I. Fleming, *Frontier Orbitals and Organic Chemical Reactions*, Wiley, NY, 1976.
55. R. West, *Angew. Chem., Int. Ed. Engl.* **26**, 1201 (1987); P. J. Domaille, *J. Am. Chem. Soc.* **106**, 7677 (1984).
56. (a) M. Nishijima and J. Yoshinobu, H. Tsuda and M. Onchi, *Surf. Sci.* **192**, 383 (1987). (b) J. Yoshinobu, H. Tsuda, M. Onchi and M. Nishijima, *J. Chem. Phys.* **87**, 7332 (1987).
57. (a) J. T. Yates, *J. Phys.: Condens. Matter,* **3**, S43 (1991) (b) M. J.Bozack, P. A. Taylor, W. J. Choyke and J. T. Yates, *Surf. Sci.*, **177**, L933 (1986) (c) P. A. Taylor, R. M. Wallace, C. C. Cheng, W. H. Weinberg, M. J. Dresser, W. J. Choyke and J. T. J. Yates, *J. Am. Chem. Soc.*, **114**, 6754 (1992) (d) L. Clemen, R. M. Wallace, P. A. Taylor, M. J. Dresser, W. J. Choyke, W. H. Weinberg and J. T. Yates, *Surf. Sci.,* **268**, 205 (1992) (e) C. C. Cheng, R. M. Wallace, P. A. Taylor, W. J. Choyke and J. T. J. Yates, *J. Appl. Phys.*, **67**, 3693 (1990) (f) C. C. Cheng, W. J. Choyke and J. T. J. Yates, *Surf. Sci.*, **231**, 289 (1990).
58. Y.Imamura, Y. Morikawa, T. Yamasaki, H. Nakatsuji, *Surf. Sci.* **341**, L1091 (1995).
59. Q. Liu and R. Hoffmann, *J. Am. Chem. Soc.*, **117**, 4082 (1995).
60. L. Li, C. Tindall, O. Ttakaoka, Y. Hasegawa and T. Sakurai, *Phys. Rev. B.* **56**, 4648 (1997).
61. R. A. Wolkow, *Ann. Rev. Phys. Chem.* **50**, 413 (1999).
62. S. Mezhenny, L. Lyubinetsky, W. J. Choyke, R. A. Wolkow and J. T. Yates, Jr. *Chem. Phys. Lett.* **344**, 7 (2001).
63. I. Fleming, *Frontier Orbitals and Organic Chemical Reactions*, John Wiley & Sons, New York, 1976.
64. J. Hovis, S. Lee, H. Liu and R. J. Hamers, *J. Vac. Sci. Techno. B*, **15**, 1153 (1997)
65. (a) B. I. Craig, P. V. Smith, *Surf. Sci.* **276**, 174 (1992) (b) C. S. Cramer, B. Weiner and M. Frenklach, *J. Chem. Phys.* **99**, 1356 (1993).
66. (a) A. J. Fisher, P. E. Blochl, G. A. D. Briggs, *Surf. Sci.* **374**, 298 (1997) (b) W. Pan, T. Zhu, W. Yang, *J. Chem. Phys.* **107**, 3981 (1997).
67. D. C. Sorescu and K. D. Jordan, *J. Phys. Chem. A*, **104**, 8259 (2000).
68. R. Konecny and D. J. Doren, *Surf. Sci.* **417**, 169 (1998).
69. Silvestrelli, Pier Luigi; Toigo, Flavio; Ancilotto, Francesco. *J. Chem. Phys.* 114, 8539 (2001).
70. W. Kim, H. Kim, G. Lee, Y.-K. Hong, K, Lee, C. Hwang, D.-H. Kim, J.-Y. Koo, *Phys. Rev. B* **64**, 193313/1 (2001).

71. R. Miotto, A. C. Ferraz, and G. P. Srivastava, *Phys. Rev. B* **65**, 075401/1 (2002).
72. F. Wang, D. C. Sorescu, and K. D. Jordan, *J. Phys. Chem. B,* **106**, 1316 (2002)
73. R. J. Hamers, S. K. Coulter, M. D. Ellison, J. S. Hovis, D. F. Padowitz, and M. P. Schwartz , C. M. Greenlief , J. N. Russell, Jr. *Acc. Chem. Res.* **33**, 617 (2000).
74. R. J. Hamers, J. S. Hovis, S. Lee, H. Liu and J. Shan, *J. Phys. Chem. B*, **101**, 1489 (1997).
75. J. S. Hovis and R. J. Hamers, *J. Phys. Chem. B*, **101**, 9581 (1997).
76. J. S. Hovis and R. J. Hamers, *J. Phys. Chem. B.*, **102**, 687 (1998).
77. J.-H. Cho and L. Kleinman *Phys. Rev. B* **64**, 235420/1 (2001).
78. K. Akagi and S. Tsuneyuki, *Surf. Sci.* **493**, 131 (2001).
79. K. Hamaguchi, S. Machida, M. Nagao, F. Nagao, F. Yasui, K. Mukai, Y. Yamashita, J. Yoshinobu, H. S. Kato, H. Okuyama, M. Kawa, T. Sato, and M. Iwatsuki, *J. Phys. Chem. B* **105** 3718 (2001).
80. J.-H Cho, D.-H Oh, K. S. Kim, and L. Kleinman, *J. Chem. Phys.* **116**, 3800 (2002).
81. K. Hamaguchi, S. Machida, K. Mukai, Y. Yamashita and J. Yoshinobu, *Phys. Rev. B.* **62**, 7576 (2000).
82. K. Hamaguchi, S. Machida, M. Nagao, *J. Phys. Chem.* **105**, 3718 (2001).
83. G. P. Lopinski, D. J. Moffatt, D. D. M. Wayner, and R. A. Wolkow, *J. Am. Chem. Soc.*, **122**, 3548 (2000).
84. G. P. Lopinski, D. J. Moffatt, D. D. M. Wayner, M. Z. Zgierski, and R. A. Wolkow, *J. Am. Chem. Soc.* **121**, 4532 (1999).
85. R. Konecny and D. Doren, *J. Am. Chem. Soc.*, **119**, 11098 (1997).
86. A. V. Teplyakov, M. J. Kong and S. F. Bent, *J. Am. Chem. Soc.,* **119**, 11100 (1997).
87. A. V. Teplyakov, M. J. Kong and S. F. Bent, *J. Chem. Phys.* **108**, 4599 (1998).
88. J. S. Hovis, H. B. Liu and R. J. Hamers, *J. Phys. Chem. B*, **102**, 6873 (1998).
89. (a) H. Nakano, *J. Chem. Phys.* **99**, 7983 (1993) (b) H. Nakano, *Chem. Phys. Lett.* **207**, 372 (1993).
90. (a) K. R. Sunberg and K. Ruedenberg, in *Quantum Science;* J. L. Calais, O. Goscinski, J. Linderberg and Y. Ohrn, Eds.; Plenum: New York, 1976 (b) L. M. Cheung, K. R. Sunberg and K. Ruedenberg, *Int. J. Quan. Chem.* **16**, 1103 (1979) (c) K. Ruedenberg, M. Schmidt, M. M. Gilbert and S. T. Elbert, *Chem. Phys.* **71**, 41 (1982) (d) B. O. Roos, P. Taylor and P. E. Siegbahn, *Chem. Phys.* **48**, 157 (1980) (e) M. W. Schmidt and M. S. Gordon, *Annu. Rev. Phys. Chem.* **49**, 233 (1998).
91. M. J. Kong, A. V. Teplyakov, J. Jagmohan, J. G. Lyubovitsky, C. Mui and S. F. Bent, *J. Phys. Chem. B* **104**, 3000 (2000).
92. Y. Jung, C. H. Choi and M. S. Gordon, *J. Phys. Chem. B.* **105**, 4039 (2001).
93. (a) D. J. Paul, *Thin Solid Films*, **321**, 172 (1998), (b) L. E. Larson, *J. Vac. Sci. Technol. B.* **16**, 1541 (1998).
94. C. Mui, S. F. Bent, and C. B. Musgrave, *J. Phys. Chem. A* **104**, 2457 (2000).
95. Y. Taguchi, M. Fujisawa, T. Takaoka; T. Okasa and M. Nishijima, *J. Chem. Phys.*, **95**, 6870 (1991).
96. (a) G. P. Lopinski, T. M. Fortier, D. J. Moffatt and R. A. Wolkow, *J. Vac. Sci. Tech. A*, **16**, 1937 (1998). (b) G. P. Lopinski, D. J. Moffatt and R. A. Wolkow, *Chem. Phys. Lett.,* **282**, 305 (1998).

97. Hofer, W. A.; Fisher, A. J.; Lopinski, G. P.; Wolkow, R. A. *Surf. Sci.* **482-485(Pt. 2)**, 1181 (2001).
98. Y. Jung, M. S. Gordon, in preparation
99. M. Schwartz, S. Coulter, Hovis, J. and R. J. Hamers, *J. Phys. Chem. B.*, (submitted)
100. S. K. Coulter, J. S. Hovis, M. D. Ellison and R. J. Hamers, *J. Vac. Sci. Technol. A* **18**, 1965 (2000).
101. Yamaguchi, Tsuyoshi. *J. Phys. Soc. Japan* **68**, 1321 (1999).
102. (a) S. M. Gates, C. M. Greenlief, D. B. Beach and P. A. Holbert, *J. Chem. Phys.* **92**, 3144 (1990) (b) S. M. Gates, C. M. Greenlief and D. B. Beach, *J. Chem. Phys.* **93**, 7493 (1990) (c) S. M. Gates, *J. Cryst. Growth* **120,** 269 (1992).
103. A. R. Brown and D. J. Doren, *J. Chem. Phys.* **110**, 2643 (1999).
104. J. K Kang, C. B. Musgrave, *Phys. Rev. B* **64**, 245330 (2001).
105. S. Sugahara, E. Hasunuma, S. Imai, and M. Matsumura, *Appl. Surf. Sci.*, **107**, 161 (1996).
106. M. A. Hall, C. Mui, C. B. Musgrave, *J. Phys. Chem. B* **105**, 12068 (2001).
107. T. R. Bramblett, Q. Lu, T. Karasawa, M. A. Hasan, S. K. Jo and J. E. Greene, *J. Appl. Phys.* **76**, 1884 (1994).
108. M. Cakmak and G. P. Srivastava, *Phys. Rev. B* **61**, 10216 (2000).
109. M. Shinohara, A. Seyama, Y. Kimura, M. Niwano, M. Saito, *Phys. Rev. B* **65**, 075319 (2002).
110. A. J. R. da Silva, G. M. Dalpian, A. Janotti, A. Fazzio, *Phys. Rev. Lett.* **87**, 036104/1 (2001).
111. Z.-Y. Lu, F. Liu, C-Z. Wang, X. R. Qin, B. S. Swartzentruber, M. G. Lagally, and K.-M. Ho, *Phys. Rev. Lett.* **85**, 5603 (2000).
112. X. R. Qin, B. S. Swartzentruber and M .G. Lagally, *Phys. Rev. Lett.* **85**, 3660 (2000).
113. Y. Ma, T. Yashda, G. Lucovsky, *J. Vac. Sci. Technol., B* **11**, 1533 (1993).
114. G. W. Yoon, A. B. Joshi, J. Kim, G.Q. Lo, D. L. Kwong, *IEEE Electron Device Lett.* **13**, 606 (1992).
115. A. Izumi and H. Matsumura, *Appl. Phys. Lett.* **71**, 1371, (1997).
116. Y. Widjaja, M. M. Mysinger, and C. B. Musgrave, *J. Phys. Chem. B* **104**, 2527 (2000),
117. Y. Widjaja and C. B. Musgrave, *Surf. Sci.* **469**, 9 (2000).
118. G.-M. Rignanese and A. Pasquarello, *Appl. Phys. Lett.* **76**, 553 (2000).
119. G.-M. Rignanese and A. Pasquarello, *Surf. Sci.* **490**, L614 (2001).
120. K. T. Queeney, Y. J. Chabal and K. Raghavachari, *Phys. Rev. Lett.* **86**, 1046 (2001).
121. (a) R. Miotto, G. P. Srivastava, R. H. Miwa, and A. C. Ferraz, *J. Chem. Phys.* (2001), 114(21), 9549-9556. (b) R. Miotto, G. P. Srivastava, and A. C. Ferraz, *Phys. Rev. B* **63**, 125321/1 (2001) (c) R.Miotto, G. P. Srivastava, and A. C. Ferraz, *Surf. Sci.* **482-485**, 160 (2001).
122. J. M. Hartmann, B. Gallas, R. Ferguson, J. Fernandesz, J. Zang, and J. J. Harris, *Semicond. Sci. Technol.* **15**, 362 (2000).

123. Y. Widjaja and C. B. Musgrave, *Phys. Rev. B* **64**, 205303/1 (2001).
124. G. T. Wang, C. Mui, C. B. Musgrave, and S. F. Bent, *J. Phys. Chem. B* **105**, 3295 (2001).
125. X. Cao and R. J. Hamers, *J. Am. Chem. Soc.* **123**, 10988 (2001).
126. X. Cao and R. J. Hamers, *J. Phys. Chem. B* **106**, 1840 (2002).
127. X. Cao, S. K. Coulter, Sarah K.; M. D. Ellison, H. Liu, J. Liu, Jianming, R. J. Hamers, *J. Phys. Chem. B* **105**, 3759 (2001).
128. H.Luo and M. C. Lin, *Chem. Phys. Lett.* **343**, 219 (2001).
129. D. M. Bhusari, C. K. Chen, K. H. Chen, T. J. Chuang, L. C. Chen, and M. C. Lin, *J. Mater. Res.* **12**, 1 (1997)
130. Bacalzo-Gladden, F.; Lu, Xin; Lin, M. C. *J. Phys. Chem. B* **105**, 4368 (2001).
131. F. Tao, Z.H. Wang, M. H. Qiao, Q. Liu, W. S. Sim and G. Q. Xu, *J. Chem. Phys.* **115**, 8563 (2001).
132. F. Tao, Z.-H. Wang, G. Q. Xu, *J. Chem. Phys.* **106**, 3557 (2002).
133. I.Fleming, *Frontier Orbitals and Organic Chemical Reactions*, Wiley, London 1976.
134. F. Tao, W. S. Sim, G. Q. Xu, and M. H. Qiao, *J. Am. Chem. Soc.* **123**, 9397 (2001).
135. C. H. Choi and M. S. Gordon, *J. Am. Chem. Soc.* **124**, 6162 (2002).
136. T. Engel, *Surf. Sci. Rep.* **18**, 91 (1993).
137. (a) M. S. Krishnan, L. Chang, T. King, J. Bokor and C. Hu, *Tech. Dig. - Int. Electron Devices Meet.* 241 (1999). (b) Y. Ma, Y. Ono, L. Stecker, D. R. Evans and S. T. Hsu, *Tech. Dig. - Int. Electron Devices Meet.* 149, (1999).
138. H. Wormeester, E. G. Keim, A. van Silfhout, *Surf. Sci.* **271**, 340 (1992)
139. (a) X. M. Zheng, P. V. Smith, *Surf. Sci.* **232**, 6 (1990) (b) U. Höfer, P. Mogen and W. Wurth, *Phys. Rev. B.* **40**, 1130 (1989) (c) P. Gupta, C. H. Mak, P. A. Coon and S. M. George, *Phys. Rev. B.* **40**, 7739 (1989) (d) B. Schubert, P. Avouris and R. Hoffmann, *J. Chem. Phys.* **98**, 7593 (1993) (e) K. Sakamoto, K. Nakatsuji, H. Daimon, T. Yonezawa and S. Suga, *Surf. Sci.* **306**, 93 (1994).
140. I. P .Batra, P. S. Bagus and K. Hermann, *Phys. Rev. Lett.* **52**, 384 (1984).
141. (a) J. R. Engstrom, D. J. Bonser and T. Engel, *Surf. Sci.* **268**, 238 (1992) (b) J. Westermann, H. Nienhous and W. Mönch, *Surf. Sci.* **311**, 101 (1994) (c) K. E. Johnson, P. K. Wu, M. Sander and T. Engel, *Surf. Sci.* **290**, 213 (1993).
142. H. Yaguchi, K. Fujita, S. Fukatsu, Y. Shiraki, R. Ito, T. Igarashi and T. Hattori, *Surf. Sci.* **275**, 395 (1992).
143. P. Avouris, I.-W. Lyo, *App. Surf. Sci.* **60/61**, 426 (1992).
144. J. A. Schaefer, F. Stucki, D. J. Frankel, W. Göpel and G. L. Lapeyre, *J. Vac. Sci. Technol. B.* **2**, 359 (1984).
145. L. Incoccia, A. Balerna, S. Cramm, C. Kunz, F. Senf and I. Storjohann, *Surf. Sci.* **189/190**, 453 (1987).
146. (a) A. Namiki, K. Tanimoto, T. Nakamura, N. Ohtake and T. Suzaki, *Surf. Sci.* **222**, 530 (1989).
147. (a) P. Avouris and D. G. Cahill, *Ultramicrosc.* **42-44**, 838 (1992). (b) D. G. Cahill and P. Avouris, *Appl. Phys. Lett.* **60**, 326 (1992).
148. P. Kliese, B. Röttger, D. Badt and H. Neddermeyer, *Ultramicrosc.* **42-44**, 824 (1992).

149. P. V. Smith and A. Wander, *Surf. Sci.* **219**, 77 (1989).
150. Y. Miyamoto and A. Oshiyama, *Phys. Rev. B.* **41**, 12680 (1990).
151. Y. Miyamoto and A. Oshiyama, *Phys. Rev. B.* **43**, 9287 (1991).
152. T. Uchiyama and M. Tsukada, *Phys. Rev. B.* **53**, 7917 (1996).
153. T. Uchiyama and M. Tsukada, *Phys. Rev. B.* **55**, 9356 (1997).
154. T. Hoshino, M. Tsuda, S. Oikawa and I. Ohdomari, *Phys. Rev. B.* **50**, 14999 (1994).
155. H. Watanabe, K. Kato, T. Uda, K. Fujita, M. Ichikawa, T. Kawamura and K. Terakura, *Phys. Rev. Lett.* **80**, 345 (1998).
156. K. Kato, T. Uda and K. Terakura, *Phys. Rev. Lett.* **80**, 2000 (1998).
157. C. H. Choi, D.-J. Liu, J. W. Evans and M. S. Gordon, (in preparation).
158. J. R. Engstrom, D. J. Bonser, M. M. Nelson and T. Engel, *Surf. Sci.* **256**, 317 (1991).
159. (a) C. Ebner, J. V. Seiple and J. P. Pelz, *Phys. Rev. B.* **52**, 16651 (1996). (b) J. V. Seiple and J. P. Pelz, *Phys. Rev. Lett.* **73**, 999 (1994).
160. J. V. Seiple, C. Ebner and J. P. Pelz, *Phys. Rev. B.* **53**, 15432 (1996).
161. R. Ludeke and A. Koma, *Phys. Rev. Lett.* **34**, 1170 (1975)
162. (a) J. B. Hannon, M. C. Bartelt, N. C. Bartelt, and G. L. Kellog, *Phys. Rev. Lett.* **81**, 4676 (1998). (b) M. C. Bartelt, J. B. Hannon, A. K. Schmid, C. R. Stoldt and J. W. Evans, *Colloids and Surfaces A* **165**, 373 (2000).
163. D.-J. Liu, C. H. Choi, M. S. Gordon and J. W. Evans, *MRS Proceedings,* **619** (2000).
164. T. Uchiyama, T. Uda, and K. Terakura, *Surf. Sci.* 474, 21 (2001).
165. Y. Widjaja and C. B. Musgrave, *J. Chem. Phys.* 116, 5774 (2002)
166. E. A. Irene, *J. Electrochem. Soc.* **125** 1708 (1978).
167. P. A. Thiel and T. E. Madey, *Surf. Sci. Rep.* **1**, 211 (1987).
168. D. Schmeisser, F. J. Himpsel, and G. Hollinger, *Phys. Rev. B.* **27**, 7813 (1983).
169. K. Fujiwara, *Surf. Sci.* **108**, 124 (1981).
170. M. Nishijima, K. Edamoto, Y. Kubota, S. Tanaka, and M. Onchi, *J. Chem. Phys.* **84**, 6458 (1986).
171. Y. J. Chabal and S. B. Christman, *Phys. Rev. B.* **29**, 6974 (1984).
172. (a) E. M. Oellig, R. Butz, H. Wagner and H. Ibach, *Solid State Commun.* **51**, 7 (1984) (b) C. U. S. Larsson, A. S. Flodstrom, R. Nyholm, L. Incoccia, and F. Senf, *J. Vac. Sci. Technol. A.* **5**, 3321 (1987) (c) K. Fives, R. McGrath, C. Stephens, I. T. McGovern, R. Cimino, D. S.-L. Law, A. L. Johnson, and G. Thornton, *J. Phys.: Condens. Matter.* **1**, SB105 (1989).
173. (a) A. L. Johnson, M. M. Walczak and T. E. Madey, *Langmuir,* **4**, 277 (1988). (b) Q. Gao, Z. Dohnalek, C. C. Cheng, W. J. Choyke and J. T. Yates, Jr., *Surf. Sci.* **312**, 261 (1994). (c) H. Bu and J. W. Rabalais, *Surf. Sci.* **301**, 285 (1994). (d) A. T. S. Wee, C. H. A. Huan, P. S. P. Thong and K .L. Tan, *Corrosion Sci.* **36**, 9 (1994).
174. W. Ranke and Y. R. Xing, *Surf. Sci.* **157**, 339 (1985).
175. (a) E. Schröder-Bergen and W. Ranke, *Surf. Sci.* **236**, 103 (1990). (b) M. Chander, Y. Z. Li, J. C. Patrin, and J. H. Weaver, *Phys. Rev. B.* **48**, 2493 (1993).
176. (a) C. U. S. Larsson, A. L. Johnson, A. S. Flodstrom, and T. E. Madey, *J. Vac. Sci. Technol. A.* **5**, 842 (1987). (b) Q. Gao, Z. Dohnalek, C. C. Cheng, W. J. Choyke, and J. T. Yates, Jr. *Surf. Sci.* **312**, 261 (1994).

177. H. Bu and J. W. Rabalais, *Surf. Sci.* **301**, 285 (1994).
178. S. Ciraci and H. Wagner, *Phys. Rev. B.* **27**, 5180 (1983).
179. S. Katircioglu, *Surf. Sci.* **187**, 569 (1987).
180. (a) N. Russo, M. Toscano, V. Barone and F. Lelj, *Surf. Sci.* **180**, 599 (1987). (b) V. Barone, *Surf. Sci.* **189/190**, 106 (1987).
181. C. K. Ong, *Solid State Commun.* **72,** 1141 (1989).
182. R. Konecny and D. J. Doren, *J. Chem. Phys.* **106**, 2426 (1997).
183. Y. Jung and M. S. Gordon, (in preparation).
184. J. Cho, K. S. Kim, S. Lee and M. Kang, *Phys. Rev. B.* **61**, 4503 (2000).
185. A. B. Gurevich, B. B. Stefanov, M. K. Weldon, Y. J. Chabal and K. Raghavachari, *Phys. Rev. B.* **58**, R13434 (1998).
186. Jung, Y.; Choi, C. H.; Gordon, M. S.; *J. Phys. Chem. B.* ***105***; 4039 **(**2001**).**
187. (a) X.-L. Zhou, C. R. Flores and J. M. Whilte, *App. Surf. Sci.* **62**, 223 (1992). (b) R. K. Schulze and J. F. Evans, *App. Surf. Sci.* **81**, 449 (1994). (c) L. Andersohn and U. Köhler, *Surf. Sci.* **284**, 77 (1993).
188. (a) J. A. Glass, E. A. Wovchko and J. T. Yates, Jr., *Surf. Sci.* **338**, 125 (1995). (b) X. Lu, Q. Zhang and M. C. Lin, *Phys. Chem. Chem. Phys.* **3**, 2156 (2100).
189. (a) J. Eng, Jr., K. Raghavachari, L. M. Struck, Y. J. Chabal, B. E. Bent, G. W. Flynn, S. B. Christman, E. E. Chaban, G. P. Wiliams, K. Rademacher and S. Mantl, *J. Chem. Phys.* **106**, 9889 (1997). (b) M. P. Casaletto, R. Zanoni, and M. Carbone, M. N. Piancastelli, L. Abelle, K. Weiss and K. Horn, *Surf. Sci.* **447**, 237 (2000).
190. S. Tanaka, M. Onchi and M. Nishijima, *J. Chem. Phys.* **91**, 2712 (1989).
191. H. Ikeura-Sekiguchi and T. Sekiguchi, *Surf. Sci.* **390**, 214 (1997).
192. H. Ikeura-Sekiguchi and T. Sekiguchi, *Surf. Sci.* **433-435**, 549 (1999).
193. J. A. Barriocanal and D. J. Doren, *J. Am. Chem. Soc.* **123**, 7340 (2001).
194. J. A. Barriocanal and D. J. Doren, *J. Phys. Chem. B* 104, 12269 (2000)
195. R. P. Andres, J. D. Bielefiled, J. I. Henderson, and D. B. Janes, V. R. Kolagunta, C. P. Kubiak, W. J. Mahoney, and R. G. Osifchin, *Science* **273**, 1690 (1996).
196. (a) K. L. Prime and G. M. Whitesides, *Science* **252**, 1164 (1991) (b) C. Jung, O. Dannenberger, Y. Xu, M. Buck, and M. Grunze, *Langmuir* **14**, 1103 (1998).
197. (a) C. K. Reddy, P.S.N. Reddy, and C. V. Ratnam, *Ind. J. Chem.* **26B**, 882 (1987) (b) E. Schaumann, H.-G. Bauch, S. Sieveking, and G. Adiwidjaja, *Chem. Ber.* 1982, 115, 3340.
198. M. D. Ellison and R. J. Hamers, *J. Phys. Chem. B* **103**, 6243 (1999).
199. S. K. Coulter, M. P. Schwartz, and R. J. Hamers, *J. Phys. Chem. B* **105**, 3079 (2001).
200. X. Lu, X Xu, N. Wang, Q. Zhang, and M. C. Lin, *J. Phys. Chem. B* **105**, 10069 (2001).
201. H. F. Winters, and J. W. Coburn, *Surf. Sci. Rep.* **14**, 161 (1992).
202. G. Thornton, P. L. Wincott, R. McGrath, I. T. McGovern, F. M. Quinn, D. Norman and D. D. Vvedensky, *Surf. Sci.* **211/212**, 959 (1989).
203. L. S. O. Johansson, R. I. G. Uhrberg, R. Lindsay, P. L. Wincott, and G. Thornton, *Phys. Rev. B* **42**, 9534 (1990).
204. N. Aoto, E. Ikawa and Y. Kurogi, *Surf. Sci.* **199**¸408 (1988).

205. Q. Gao, C. C. Cheng, P. J. Chen, W. J. Choyke, and J. T. Yates, Jr. *J. Chem. Phys.* **98**, 8308 (1993).
206. (a) B. I. Craig and P. V. Smith, *Surf. Sci.* **290**, L662 (1993). (b) L.-Q. Lee and P.-L. Gao, *J. Phys. Condens. Matter* **6**, 6169 (1994). (c) P. Kruger and J. Pollmann, *Phys. Rev. B* **47**, 10032 (1993).
207. D. Purdie, N. S. Prakash, K. G. Purcell, P. L. Wincott, G. Thornton, and D. S.-L. Law, *Phys. Rev. B* **48**, 2275 (1993).
208. S. L. Bennett, C. L. Greenwood, E. M. Williams, *Surf. Sci.* **290**, 267 (1993).
209. (a) J. Matsuo, K. Karahashi, A. Sato, and S. Hijiya, *Jpn. J. Appl. Phys.* **31**, 2025 (1992). (b) A. Szabó, P. D. Farrall, and T. Engel, *Surf. Sci.* **312**, 284 (1994).
210. (a) R. B. jackman, H. Ebert, and J. S. Foord, *Surf. Sci.* **176**, 183 (1986). (b) M. A. Mendicino, and E. G. Seebauer, *Appl. Surf. Sci.* **68**, 285 (1993).
211. K. Karahashi, J. Matsuo, and S. Hijiya, *Appl. Surf. Sci.* **60/61**, 126 (1992).
212. (a) D. J. Oostra, A. Haring, R. P. v. Ingen, A. E. de Vries, and G. N. A. v. Veen, *J. Appl. Phys.* **64**, 315 (1988) (b) N. Materer, R. S. Goodman, and S. R. Leone, *J. Phys. Chem. B* **104**, 3261 (2000).
213. (a) G. A. de Wijs, A. De Vita and A. Selloni, *Phys. Rev. Lett.* **78**, 4877 (1997) (b) G. A. de Wijs, A. De Vita and A. Selloni, *Phys. Rev. B* **57**, 10021 (1998)
214. J. R. Engstrom, M. M. Nelson, T. Engel, *Surf. Sci.* **215(3)**, 437 (1989).
215. E. G. Michel, T. Pauly, V. Eteläniemi, and O. Materlik, *Surf. Sci.* **241**, 111 (1991).
216. D. D. Koleske, and S. M. Gates, *J. Chem. Phys.* **99**, 8218 (1993).
217. M. J. Bozack, M. J. Dresser, W. J. Choyke, P. A. Taylor, J. T. Yates, Jr. *Surf. Sci.* **184**, L332 (1987).
218. Y. L. Li, D. P. Pullman, J. J. Yang, A. A. Tsekouras, D. B. Gosalvez, K. B. Laughlin, Z. Zhang, M. T. Schulberg, D. J. Gladstone, M. McGonigal, and S. T. Ceyer, *Phys. Rev. Lett.* **74**, 2603 (1995).
219. C. J. Wu and E. A. Carter, *Phys. Rev. B* **45**, 9065 (1992).
220. L. E. Carter and E. A. Carter, *J. Phys. Chem.* **100**, 873 (1996).
221. J. E. Sanabia, J. H. Moore, and J. A. Tossell, *J. Chem. Phys.* **116**, 10402 (2002).
222. K. A. Brown and W. Ho, *Surf. Sci.* **228**, 111 (1995).
223. J. Y. Lee and S. Kim, *Surf. Sci.* **482-485(Pt. 1)**, 196 (2001).

Chapter 5

QUANTUM-CHEMICAL STUDIES OF MOLECULAR REACTIVITY IN NANOPOROUS MATERIALS

Stanislaus A. Zygmunt[a] and Larry A. Curtiss[b]

[a] *Department of Physics and Astronomy, Valparaiso University, Valparaiso, Indiana 46383 USA*

[b] *Materials Science and Chemistry Divisions, Argonne National Laboratory, Argonne, Illinois 60439 U.S.A.*

1. INTRODUCTION

Broadly understood, chemical reactions in nanoporous solids occur via four sequential physical processes. First, the reactant molecules must diffuse through the pores and channels of the material to reach sites at which reactions can favorably occur. Second, the gas phase reactants adsorb on the interior surface of the nanopore. The third step is the chemical reaction proper, during which reactants are transformed into products through the breaking and formation of chemical bonds. In the fourth step, which is the reverse of the first two, product molecules desorb from the nanopore surface and diffuse out of the solid. The classic example of this kind of reaction is the catalytic transformation of hydrocarbons in zeolites, which are strong Brønsted acids in widespread use in the petrochemical industry.

As the techniques of computational quantum chemistry became more sophisticated and computer processing speed increased at a dramatic rate through the 1990s, theoretical studies of the interaction of simple molecules with the active sites of zeolites were carried out and published with increasing frequency. Due to the limitation of available computational resources, these studies were initially restricted to small cluster models of the zeolite catalyst that were generic and did not represent a unique zeolite framework. This made it impossible to explore and explain the molecular basis for the

L.A. Curtiss and M.S. Gordon (eds.),
Computational Materials Chemistry: Methods and Applications, 191–245.

differences in activity and selectivity among the variety of zeolite framework types. However, in the last ten years, the development of fast and accurate computational methods based on density functional theory (DFT) has helped to overcome this limitation. The chemical literature now contains a small but growing number of computational studies based either on periodic unit cells or on cluster models large enough to represent a particular zeolite framework. This chapter will review these exciting developments and the results of their application to problems of chemical interest. Although the large majority of such applications deal with the reactivity of zeolites and related aluminosilicate or aluminophosphate materials, a small number of computational studies have been carried out for carbon nanotube systems as well.

Probe molecules have been used along with a variety of spectroscopic techniques to characterize the acidity of nanoporous solid acids such as zeolites. Theoretical calculations of the interaction of small probe molecules with the active sites of these materials provide very useful complementary information. In many cases quantum-chemical calculations have been used to help interpret the results of spectroscopic measurements. In a 1994 review of theoretical calculations of van der Waals complexes of molecules with surfaces, Sauer et al. devoted considerable attention to studies of the interaction of molecules with the Brønsted acid site in zeolites.[1] A year later, another useful survey of computational studies of the properties of acidic zeolites and their interaction with various probe molecules was published in a review of Brønsted acidity in zeolites by van Santen et al.[2] A 1998 review by Bates and van Santen updated the earlier survey and extended it to discuss molecular simulations of diffusion and adsorption as well as quantum-mechanical studies of chemical reactivity for a few representative hydrocarbon reactions.[3] Much briefer reviews of quantum-chemical modeling of a few hydrocarbon reactions in zeolites also appeared in 1997[4] and 1999.[5] Since that time a significant effort has been made in applying both ab initio quantum chemistry and DFT methods to study the role of nanoporous materials in catalyzing hydrocarbon reactions. On the basis of these calculations, pathways for cracking, dehydrogenation, H/D exchange, alkene chemisorption, hydride transfer, skeletal isomerization, and alkylation reactions have been proposed, including the identification of transition states and computation of activation energies. Preliminary efforts have also been made to model key steps of the methanol-to-gasoline (MTG) and methanol-to-olefins (MTO) reactions that are of widespread industrial importance. In addition, quantum chemistry has been applied to the study of the industrially significant hydrodesulfurization (HDS) and NO_x decomposition reactions. Theoretical models and computational resources have progressed to the point where we can realistically compare computed reaction pathways and activation energies to experimentally obtained results. In the first decade of the 21st century, computational chemistry has the great potential to enhance our understanding of catalytic reactions by assisting in the interpretation of experimental results from spectroscopy and reaction kinetics and by fostering a more detailed molecular understanding of catalysis.

In Section 2 we will briefly describe the various structural models and computational methods used in these studies. Some of these methods are reviewed more extensively in other chapters of this book. Section 3 will review at some length a broad range of significant computational reaction studies published since the mid-1990s. In Section 4 we will identify common themes that emerge from the growing body of computational reactivity studies and discuss lessons they have provided. Hopefully this conclusion will suggest avenues of profitable future computational research in the molecular reactivity of nanoporous materials.

2. STRUCTURAL MODELS AND COMPUTATIONAL METHODS

2.1 Structural Models of Active Sites in Nanoporous Materials

The first step in applying computational chemistry to study a chemical system is to choose a representative model for the system being considered. For chemical reactions in nanoporous materials, the system of interest is bulk-like, and there are three main categories of models. (1) A finite cluster of atoms containing the important chemical constituents for a particular process is defined and and terminated with suitable boundary conditions. (2) A finite cluster, within which quantum-chemical methods are employed, is embedded in a much larger model of the bulk, which is treated with more approximate theoretical methods. (3) By contrast, the system can be represented by a periodic unit cell, which is repeated in all directions in space.

The cluster model includes a finite number of atoms from a surface in the nanoporous material that are terminated in a suitable manner. One common approach is to terminate using hydrogens. The number of nonhydrogen atoms used in such a cluster is usually less than 20. This permits full optimization of all atomic coordinates to be carried out. The $Si_4AlO_4H_{13}$ cluster shown in Fig. 1 is a typical cluster used to model a Brønsted acid site on the surface of a zeolite pore. Such a model is referred to as a 5T cluster in the literature, denoting the presence of five tetrahedrally-coordinated atoms. These types of clusters have been used to study reactions of molecules with surfaces.

While for molecular reactions full geometry relaxations are preferable, in the case of surface or bulk studies this can sometimes be unrealistic because the edge atoms move more than if they were constrained by the missing bulk atoms. Thus, the outer atoms may be constrained at the positions they occupy in the crystal structure. Various embedding procedures have also been developed and used that take into account the constraint of the edge atoms as well as the effect of the bulk on the reactions at a surface site. The embedding approaches basically include a cluster that is treated quantum mechanically while the influence of the rest of the crystal on the electrons in the cluster is treated by an embedding potential.[6] Alternatively, part of the environment around the cluster can partially treated at a lower level of theory. An example of this type of approach is to carry out a study on a small cluster at a high level of theory and then embed it in a much larger cluster that is treated at a lower quantum mechanical level to obtain an approximate energy correction for the bulk. The edge atoms of the small cluster are held rigid at their positions in the crystal to facilitate embedding in the larger cluster as shown in Fig. 2.[7] The SIMONN (Surface Integrated Molecular Orbital/Molecular Mechanics) [8] is a similar embedding approach

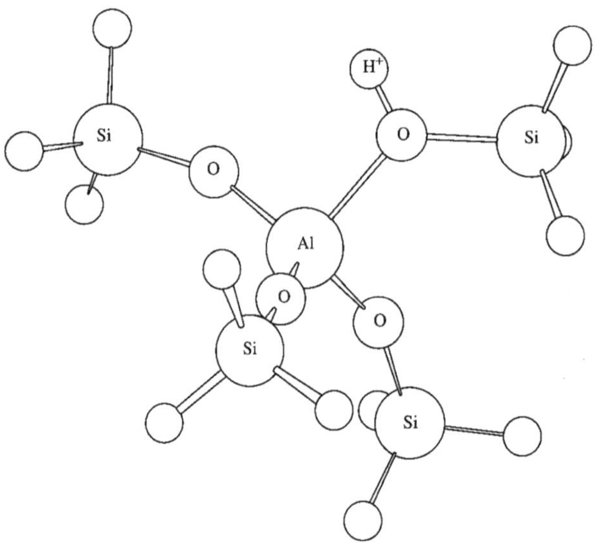

Figure 1. Illustration of typical cluster ($Si_4AlO_4H_{13}$, denoted 5T) used to model a Brønsted acid site on the surface of a zeolite pore.

except it uses molecular mechanics for the outer region. A more detailed discussion of the latter approach is given in Chapter 4.

Recently, periodic models have also been used for reactivity studies in nanoporous materials.[9] They perform best in studies under the conditions of high surface coverage as they permit the study of periodic adsorbate structure using relatively small unit cells. Adsorption of isolated molecules at the surface represents a challenge for periodic models because a sufficiently large unit cell must be chosen to avoid interactions between the molecules in the neighbouring cells. The expansion of valence electron wavefunctions in

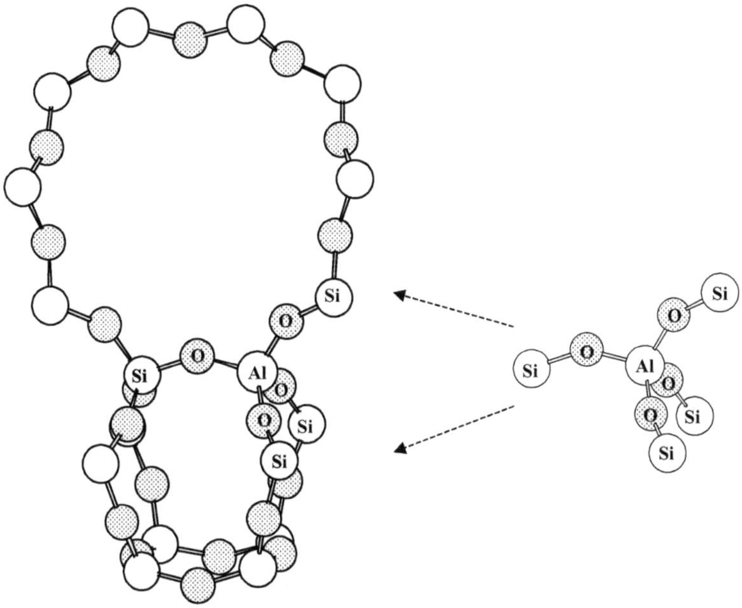

Figure 2. Illustration of embedding procedure used to estimate effect of zeolite lattice on reactions at an acid site.

a plane wave basis and the use of pseudopotentials to replace core electrons are tools that are ideally suited to periodic calculations, such as in the VASP computer code.[10]

2.2 Computational Methods

2.2.1 Ab Initio Molecular Orbital Theory

The simplest form of ab initio molecular orbital theory is Hartree-Fock (HF) or self-consistent-field (SCF) theory. The accuracy obtained from HF level calculations depends on the size of the basis set used in the calculation. The most commonly used basis sets are split-valence plus polarization such as 6-31G(d). The HF level calculations are most commonly used for geometries and vibrational frequencies (with scaling). They are not very reliable for reaction energies or barriers, where correlation effects are important and need to be included.

Correlation effects can be included by several methods. One of the more commonly used approach is to start with an HF calculation and use perturbation theory (e.g. MP2, MP3, MP4) to evaluate the electron correlation energy. Other methods include configuration interaction (CI) or coupled cluster theory (e.g., CCSD(T), QCISD(T)).[11,12]

2.2.2 Density Functional Theory (DFT) Methods

Over the past 30 years, density functional theory (DFT) has been widely used by physicists to study the electronic structure of solids. More recently chemists have been using the Kohn-Sham version[13] of DFT as a cost-effective method to study properties of molecular systems. They have become very popular for computational studies of reaction mechanisms because they are quite reliable and scale as ~N**3. We briefly review some of the DFT methods commonly used for computational studies of reaction mechanisms in nanoporous materials.

The simplest density functional method is the local spin density functional, which treats the electronic environment of a given position in a molecule as if it were a uniform gas of the same density at that point. One of these is the SVWN functional that uses the Slater functional[14] for exchange and the uniform gas approximate correlation functional of Vosko, Wilk and Nusair.[15] The more sophisticated functional BPW91 combines the 1988 exchange functional of Becke[16] with the correlation functional of Perdew and Wang.[17] Both components involve local density gradients as well as densities. The BLYP functional uses the Becke 1988 part for exchange, together with the correlation part of Lee, Yang and Parr.[18] A number of other functionals use parameters that are fitted to energies in the G2 molecular test set.[19] These give a functional that is a linear combination of Hartree-Fock exchange, 1988 Becke exchange, and various correlation parts. This idea was introduced by Becke[20] who obtained parameters by fitting to the molecular data. This is the basis of the B3PW91 functional. The others (B3P86 and B3LYP) are constructed in a similar manner, although the parameters are the same as in B3PW91.[21]

In several validation studies for molecular geometries and frequencies, DFT has given results of quality similar to that of MP2 theory.[22,23] It has also been examined for use in calculation of thermochemical data.[24] While DFT is not as accurate as very high level ab initio molecular orbital methods such as G3 theory or CCSD(T), it is more cost effective and is more accurate for reaction energies than MP2 or HF methods. One problem with DFT methods is that they do poorly for van der Waals interactions such as those of nonpolar hydrocarbons with a zeolite pore surface.

2.2.3 Reaction Path Computation

Determining the stable equilibrium configuration of a molecular complex or periodic unit cell is a straightforward task involving the minimization of forces on the atomic constituents of the system. Algorithmic implementation of this procedure is standard in all widely used quantum chemistry computer programs. However, fully characterizing a reaction path connecting desired reactant and product structures requires the location of transition states (TS) along the minimum energy path (MEP). This is a much more difficult procedure, and automatic algorithms for finding TS structures are generally only reliable if one has a very good initial guess for the TS structure.

However, the recent development of the nudged elastic band (NEB) method has provided an efficient means for determining the MEP for a chemical reaction.[25] The MEP is found by constructing a series of images (generally between 5 and 20) of the system that interpolate between the reactant and product structures. A spring interaction between adjacent images is added to guarantee continuity along the reaction path, and this defines a so-called "elastic band". Optimization of the MEP involves the minimization of the forces acting on each image structure. A special force projection ensures that the spring forces neither interfere with convergence to the MEP nor affect the distribution of the images along the path. This force projection is known as "nudging". The NEB method, which can easily be implemented in standard electronic structure programs, has been successfully applied to studies of atomic diffusion in Si, crystal growth on Al (100), and dissociative adsorption of CH_4 on the Ir (111) surface.[26] Recent computational studies of molecular reactivity in nanoporous solids have utilized the NEB method, as will be described in Section 3.

3. SURVEY OF REACTIVITY STUDIES

The vast majority of studies of chemical reactivity in nanoporous materials has dealt with hydrocarbon conversion reactions of relatively short alkanes. Only recently have computational resources allowed researchers to study reactions involving aromatic molecules, but the success of these efforts is encouraging. We begin this survey by considering those reactions that have been the subject of the most numerous computational studies. These are the three simple monomolecular reactions of alkanes catalyzed by solid acids: cracking, dehydrogenation, and H/D exchange. After a rather thorough review of various other alkane transformation reactions, including those involved in alkene chemisorption, skeletal isomerization, and alkylation reactions, we turn to even more ambitious studies of methanol-to-gasoline (MTG) chemistry, NO_x decomposition reactions, and hydrodesulfurization (HDS) reactions. This section will close with a brief discussion of recent computational studies of reactivity in carbon nanotubes.

3.1 Cracking

Two groundbreaking papers by Kazansky et al. appeared in 1994 reporting the first application of ab initio quantum chemistry to catalytic reactions of hydrocarbons with model zeolitic clusters.[27,28] Although the cluster model used to represent the Brønsted acid site was $H_2OAl(OH)_3$ and thus contained only one tetrahedral atom (1T), and the HF/3-21G level of theory was used, the results suggested a plausible pathway for protolytic attack on ethane and the cleavage of the C-C bond to form methane and a surface-stabilized methyl alkoxide. The apparent activation energy (with respect to the gas phase reactants) obtained for this model at this level of theory was found to be 93.4 kcal/mol, which is much higher than typical experimental

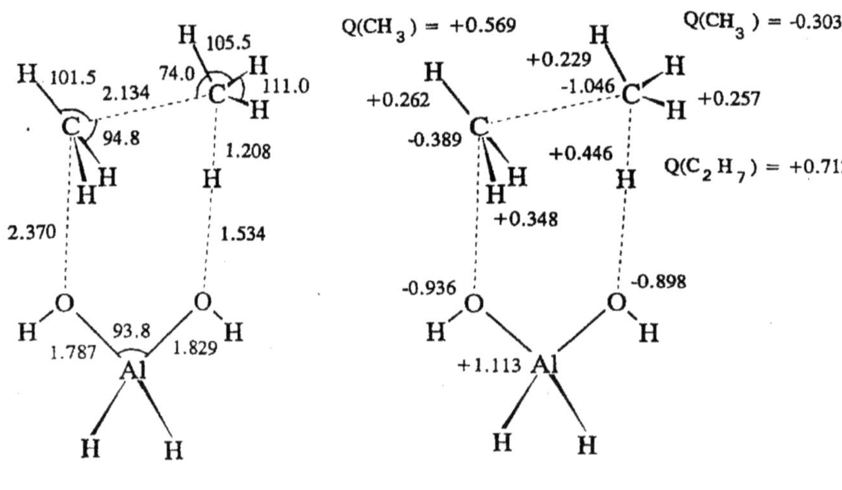

Figure 3. HF/3-21G geometry (a) and charge distribution (b) for transition state for protolytic cracking of ethane using 1T cluster model of zeolite acid site. Reprinted by permission from Ref. 27.

values for short-chain alkanes. However, the computational results offered important qualitative insights about the pathway for this reaction. In particular, these calculations revealed that the protolytic attack on ethane did *not* form a stable intermediate carbonium ion ($C_2H_7^+$) as analogies to liquid superacid chemistry had suggested, but rather formed a transition state (TS) complex containing a nearly planar CH_3 fragment, as shown in Figure 3. The reaction coordinate involved motion of the acidic proton directly toward one C atom, with a concomitant elongation of the C-C bond. On the product side of the barrier, the C-C bond was cleaved, with one C forming methane and the other inverting through the plane of its three H atoms and forming a C-O bond with a surface oxygen atom in the zeolite model cluster. Thus the reaction resulted in a surface-bound alkoxide species. Although no stable carbocation was formed, the calculations revealed a strongly ionic TS complex, in which the total Mulliken charge on the protonated ethane fragment was +0.71e.

Soon afterward Collins and O'Malley carried out a similar computational study of protolytic cracking of n-butane by an $H_3SiOAl(OH)SiH_3$ (3T) zeolitic cluster model.[29] Use of the semi-empirical AM1 method predicted the formation of a stable protonated intermediate ($C_4H_{11}^+$), but the BLYP/3-21G* DFT method produced a pathway through a TS formed by the protonation of one of the C-C bonds in butane. In this complex, shown in Figure 4, the zeolitic proton was essentially equidistant from the two C atoms, which distinguishes this pathway from that found by Kazansky et al.[27]

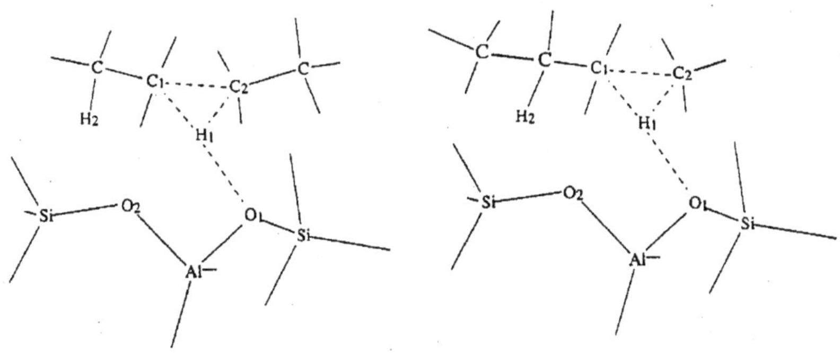

Figure 4. Schematic BLYP/3-21G* transition states for secondary and primary protolytic cracking of n-butane using a 3T cluster model for zeolite acid site. Reprinted by permission from Ref. 29.

Moreover, upon C-C bond cleavage and formation of a shorter alkane, a proton was *directly* donated back to one of the framework oxygen atoms to regenerate the acid site and form the product alkene. Thus the reaction was predicted to occur *without* the formation of a surface alkoxide species. The apparent activation energies for primary and secondary C-C cracking of n-butane were found to be 52.7 and 49.6 kcal/mol, respectively, which are still significantly higher than representative experimental values of 34.0 and 32.1 kcal/mol[30] in zeolite H-ZSM-5.

In 1996 Blaszkowski et al. published an ambitious study of protolytic cracking, dehydrogenation, and H/D exchange involving ethane over a 3T cluster model of a zeolitic acid site.[31] This effort relied on geometry optimization at the SVWN/DZVP level of (local) DFT, followed by single-point energy calculations using the (nonlocal) Becke-Perdew (BP) functional. The authors investigated two pathways for the cracking reaction, proceeding through two structurally different TS complexes. In the first pathway the reaction coordinate was nearly identical to that of Kazansky's 1994 paper, but the apparent activation energy was a much smaller value of 69.8 kcal/mol, due to the larger cluster size, greater accuracy of a DFT treatment of electron correlation, and the larger basis set. The second pathway was considerably more complex, involving an initial rotation of one $-CH_3$ group about the C-C axis of ethane, followed by motion of the acidic proton toward the other C atom, resulting in C-C cleavage. However, the authors did not find all of the stationary points along this second pathway, so it was not fully characterized. Incidentally, both pathways have nearly the same activation energies, and the product structures (methane and a surface-bound methoxy species) are identical, so the two reaction channels would appear to be difficult to distinguish experimentally. The fact that two TS were found with very different structures but nearly identical energies underscores the complexity of the potential energy surfaces for such systems, and indicates the difficulty of searching through all possible reaction pathways in configuration space. No experimental value of the activation energy is available for comparison, since ethane cracking does not form a catalytic cycle, but the value of 37 kcal/mol measured for propane in H-ZSM-5[32] indicates that the computed result for ethane must still be far too high.

In a subsequent paper, Kazansky et al. re-determined the activation energy for ethane cracking using a 3T cluster model of the acid site and more accurate computational methods.[33] The geometry optimization was performed at the HF/6-31G(d) level without electron correlation, and then a single-point energy correction was carried out with the MP2/6-31++G(d,p) method. The final activation energy of 75.4 kcal/mol was much lower than the 1994 result, but still much higher than experiment. In this paper the authors also determined the TS structure and activation energy for protolytic cracking of i-butane using a 1T cluster model and the same technique described above. The lower activation energy of 57.5 kcal/mol compared to ethane reflects the relative ease with which cracking can proceed at a tertiary C atom. Still, this value is still a serious overestimate of the reported experimental values of 29.9,[34] 31,[35] and 29.1[36] kcal/mol. However, one experimental study reported the much higher value of 57 kcal/mol,[37] highlighting the difficulty in measuring activation energies reproducibility for these reactions, especially when different types of reactors are utilized.

Very similar results were reported by Rigby et al., who carried out a computational survey of reaction paths for protolytic cracking of alkanes along with several other hydrocarbon reactions.[38] This work, based on geometry optimizations at the HF/3-21G level of theory with a 3T zeolite cluster, showed the importance of full optimization of structures, rather than imposing artificial symmetry constraints that may be computationally helpful but will provide inaccurate results. Using single-point MP2/6-31G(d) energy corrections, the authors calculated true activation energies (relative to the adsorbed reactant rather than the gas phase species) for ethane, propane, and n-butane of 78, 68, and 67 kcal/mol, respectively. As one would expect, the activation energy for i-butane cracking (60 kcal/mol) was found to be somewhat lower than for n-butane. The essentially identical value of activation energy for propane and n-butane cracking was qualitatively consistent with experimental results showing that the activation energy for n-alkane cracking is independent of chain length.[32] However, quantitative agreement between theory and experiment was still lacking.

In an attempt to better reconcile theory and experiment, Zygmunt et al. carried out a study of ethane cracking using both ab initio and DFT methods applied to a larger 5T cluster model of the zeolite acid site.[39] Geometry optimizations were carried out using the Hartree-Fock method, as well as the B3LYP and MP2 correlated methods, all with the 6-31G(d) basis set. For the correlated methods, single-point energies were subsequently calculated with the very large 6-311+G(3df,2p) basis set used in the G2 method of theoretical thermochemistry. All three methods indicate the formation of a surface alkoxide intermediate, in agreement with the results of Kazansky et al.[27] and Blaszkowski et al.[31] However, the importance of electron correlation was illustrated by the reaction pathways obtained by these three methods. The uncorrelated HF/6-31G(d) method identified three different TS structures connecting two local energy minima through which the C-C bond cleavage proceeds, with the TS energies roughly 0.5 kcal/mol above those of the minima. The two energy minima were not found with the B3LYP and MP2 methods, which identified only one TS between the adsorbed ethane complex and the surface-bound methoxy complex. The authors concluded that the high-energy local minima found with the HF method were spurious artifacts of an uncorrelated method, which reinforces the need for a critical attitude toward the results of such calculations.

The use of the larger 5T cluster model, compared to the 3T model from the study by Kazansky,[33] reduced the calculated activation energy by roughly 8 kcal/mol. The 5T cluster contains a complete shell of four O atoms around the central Al atom, while the 3T cluster included only two such O atoms. The TS for the cracking reaction, shown in Figure 5, is stabilized in the 5T model due to interaction with three of the four O atoms, leading to the reduced activation energy. This clearly shows a potential cluster size dependence for reactivity studies that use finite cluster models. However, even with the 5T cluster model of the zeolite, Zygmunt et al. obtained true activation energies (relative to the adsorbed ethane) of 66 kcal/mol (B3LYP) and 69 kcal/mol (MP2), which are still much higher than the essentially constant experimental value of 47 kcal/mol for propane through n-hexane.[32]

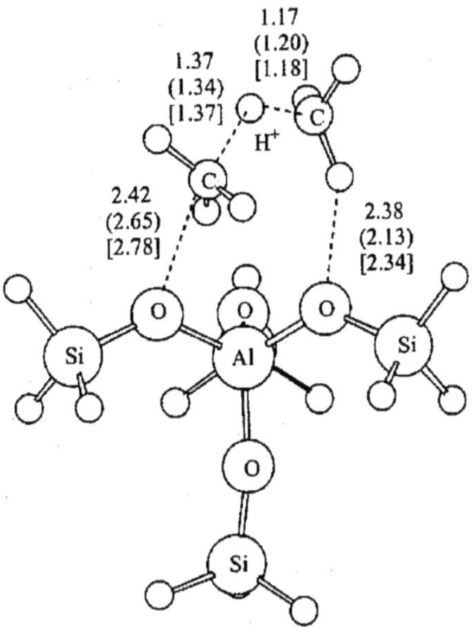

Figure 5. Transition state for protolytic cracking of ethane using a 5T cluster model, with MP2(FC)/6-31G(d), B3LYP/6-31G(d) (in parentheses), and HF/6-31G(d) (in brackets) distances shown. Reprinted by permission from Ref. 39.

However, for the first time this study also attempted to estimate the long-range effect of the zeolite lattice on the activation energy of a chemical reaction. The HF/6-31G(d) optimized geometries of the adsorbed and TS complexes obtained in the 5T cluster were inserted into successively larger 18T, 28T, 38T, 46T, and 58T cluster models of the ZSM-5 framework. Without further optimization, single-point energies were calculated using the HF/6-31G(d) method. The results, shown in Table 1, demonstrated that the electric field produced by the zeolite lattice in the micropore stabilized the ionic TS with respect to the adsorbed ethane complex by approximately 15 kcal/mol with the largest 58T cluster. This led to an approximate activation energy of 51 kcal/mol (B3LYP) and 54 kcal/mol (MP2). Since cleavage of a C-C bond between two primary C atoms is more difficult than when one C atom is secondary, the activation energy for ethane is expected to be higher than for propane (47 kcal/mol), so the calculated results, for the first time, were in reasonable agreement with experiment. This study identified the long-range stabilizing effect of the zeolite lattice, which had been neglected in earlier computational reactivity studies, as the most important reason for the large discrepancies between earlier theoretical and experimental activation energies.

Table 1. Calculated HF/6-31G(d) correction energy to activation barrier for ethane cracking in H-ZSM-5 produced by embedding reactant and TS structures in successively larger models of the zeolite. Adapted from Ref. 39.

Cluster size of H-ZSM-5 Model	Correction energy (kcal/mol)
18T ($Si_{17}AlO_{22}H_{29}$)	-7.6
28T ($Si_{27}AlO_{40}H_{33}$)	-7.9
38T ($Si_{37}AlO_{54}H_{45}$)	-12.7
46T ($Si_{45}AlO_{68}H_{49}$)	-14.0
58T ($Si_{57}AlO_{88}H_{57}$)	-14.5

Of course all studies discussed so far in this section have dealt only with protolytic (i.e., monomolecular) cracking of hydrocarbons, and while this may be the important initiation step in industrial cracking processes, it is widely believed that once protolytic cracking has produced a sufficiently high concentration of surface alkoxide species, cracking proceeds much more readily through a β–scission mechanism. Two computational studies have examined this process in some detail. Frash et al. located stationary point structures along the reaction pathway for initial but-1-oxy and pent-2-oxy species using both 1T and 3T zeolite model clusters.[40] Geometry optimizations were performed at both the HF/6-31G(d) and B3LYP/6-31G(d) levels of theory.

The results indicated a very complex potential energy surface for these reactions, with HF/6-31G(d) and B3LYP/6-31G(d) optimizations leading to different reaction pathways involving two distinct TS structures. Though the pathways differ, in each case the surface-bound but-1-oxy reactant species undergoes β–scission and forms an ethoxy surface species along with ethene. Final activation energies, based on single-point B3LYP/6-31++G(d,p) calculations, were found to be 57.4 kcal/mol for but-1-oxy and 52-53 kcal/mol for pent-2-oxy. Due to the complexity of the β–scission process and the number of attendant side reactions, the activation energy is difficult to measure directly. However, the calculated values are significantly higher than effective β–scission activation energies obtained from kinetic modeling studies of hydrocarbon cracking, which are about 30 kcal/mol.[41,42] Another computational study subsequently published by Hay et al. for β–scission of the pent-2-oxy species using a 3T zeolite cluster and the B3LYP/6-31G(d) method gave a somewhat different reaction pathway, but yielded a very similar activation energy of 56 kcal/mol.[43] Based on the results of Zygmunt et al.,[39] it seems likely that the remaining discrepancy between theory and experiment is at least in part due to the long-range electrostatic effect of the zeolite crystal, which significantly stabilizes an ionic TS relative to a reactant complex that has a much lower dipole moment.

3.2 Dehydrogenation

3.2.1 In Unmodified Acidic Zeolites

In contrast to the protolytic cracking reaction, which proceeds via attack on the C-C bond, protolytic dehydrogenation of alkanes occurs as a result of attack on the C-H bond and subsequent hydride abstraction, leading to the formation of molecular hydrogen and an alkene molecule. Two early computational studies of this reaction were published in 1994. Kazansky et al., in a paper paralleling their earlier study of ethane cracking, used a 1T cluster model and the HF/3-21G method to trace the dehydrogenation reaction paths for methane and ethane.[28] For each reactant, the TS complex shown in Figure 6 is characterized by transfer of the acidic proton from the cluster to the alkane and incipient cleavage of one of its C-H bonds.

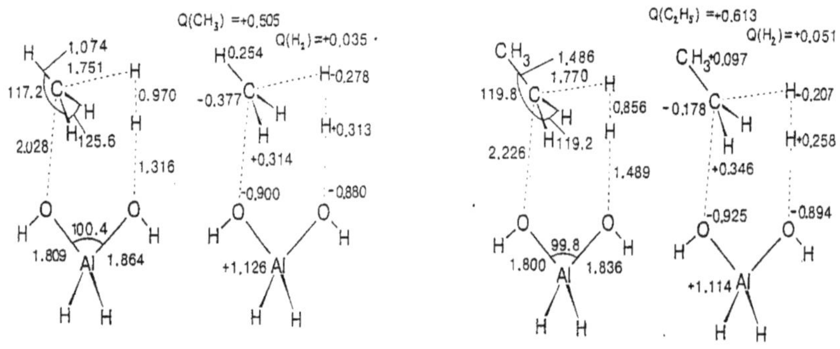

Figure 6. Geometry and charge distribution of HF/3-21G transition states for protolytic dehydrogenation of methane and ethane using 1T cluster model. Reprinted by permission from Ref. 28.

As was true for the cracking reaction, the dehydrogenation TS was quite ionic, with the Mulliken charge on the protonated alkane equal to +0.54e (methane) and +0.66e (ethane). The reaction products were molecular hydrogen and a surface-bound alkoxy species arising from formation of a C-O bond between the alkane and zeolite framework. This alkoxy species is properly a stable reaction intermediate, and a subsequent proton transfer from one of its methyl groups to a nearby framework oxygen would be required to form the final product alkene and close the catalytic cycle.

The apparent activation energies obtained from these calculations were 104.5 kcal/mol (methane) and 94.8 kcal/mol (ethane), reflecting the fact that removal of the hydride ion from ethane is easier than from methane. The activation energy for ethane dehydrogenation was thus found to be nearly the same as for ethane cracking (see Section 3.2), although the TS structures differed greatly. It is noteworthy that the cracking and dehydrogenation reactions proceed via transition states that strongly resemble the two structurally distinct forms of the ethyl carbonium ion $C_2H_7^+$. The TS analogous to the classical (C-H protonated) form leads to dehydrogenation, while the nonclassical (C-C protonated) structure leads to cracking.

At about the same time a similar study of methane dehydrogenation was published by Blaszkowski et al.[44] Using a larger 3T zeolite cluster model and the BP/DZVP nonlocal DFT approach, they obtained a TS very similar in structure to that of Kazansky but a significantly lower activation energy of 82.0 kcal/mol. The Mulliken charge on the CH_5^+ fragment in the TS, +0.54e, was the same in the two studies, despite the different computational methods. Clearly the larger cluster model, larger basis set, and treatment of electron correlation lowers the calculated activation energy due to two factors. First, the deprotonation energy of the 3T cluster model is less than for the 1T model, and secondly, there is an increased stabilization of the ionic TS relative to the gas phase reactants. Blaszkowski et al.[31] followed up with a similar study of ethane dehydrogenation, again finding a TS structure quite similar to that of Ref. 28 but with a smaller apparent activation energy of 71.0 kcal/mol. Although experimental activation energies for methane and ethane dehydrogenation are not available for comparison, values of 22.7 and 15.5 kcal/mol were reported for propane in H-ZSM-5 and H-Y zeolites, respectively.[34] It is expected that the activation energy for propane dehydrogenation should be somewhat less than that of ethane, since removal of the hydride ion from a secondary carbon is energetically easier. However, the overall heat of reaction for the gas phase dehydrogenation reaction $C_3H_8 \rightarrow C_3H_6 + H_2$ is known experimentally to be about 30 kcal/mol, which is difficult to reconcile with the reported experimental barriers. In any case, it seems clear that the computational results are too high at this level of theory. This parallels the results for ethane cracking from Section 3.1.

An interesting issue was raised by Kazansky et al. in their computational study of i-butane.[33] The *two-step* dehydrogenation reaction path described above proceeds through a surface-bound alkoxy intermediate, which must subsequently return a proton to the zeolite to re-form the acid site and produce an alkene. However, it is quite conceivable that an alternate path could form the alkene *directly* in a concerted process that returns a proton to the zeolite simultaneous with the formation of molecular hydrogen. Indeed, TS structures for these two pathways were found using the HF/6-31G(d) method and a 1T cluster model. Single-point MP2/6-31++G(d,p) energies revealed that the activation energies were 66.8 kcal/mol for the two-step reaction and 74.7 kcal/mol for the direct, concerted reaction. However, the "transition state" for the direct reaction was optimized using symmetry constraints and actually had *three* imaginary frequencies instead of *one*, as is the case for a true TS. An effort to remove this constraint caused the structure to collapse into the TS for the two-step pathway. So if a true TS exists for the direct reaction, it would be lower in energy than this artificially constrained structure, which would in turn lower the activation energy for the process. Thus this study did not definitively establish whether two pathways are possible and which of the two is favored.

The pathway for i-butane dehydrogenation was recently re-examined in a more sophisticated computational study using both 3T and 5T cluster models of the zeolite and the B3LYP/6-31G(d,p) and B3LYP/6-311G(d,p) methods for geometry optimization.[45,46] Based on calculated TS structures shown in Figure 7 and intrinsic reaction coordinate (IRC) calculations that optimize the structure of the zeolite-hydrocarbon complex along the reaction coordinate, the authors concluded that dehydrogenation of i-butane leads *directly* to the formation of molecular hydrogen and i-butene, *without* the formation of the tert-butyl alkoxide intermediate found in Kazansky's work. The apparent activation energies were found to be 58.5 kcal/mol for the 3T cluster with the 6-31G(d,p) basis set, and 53.4 kcal/mol for the 5T cluster with the

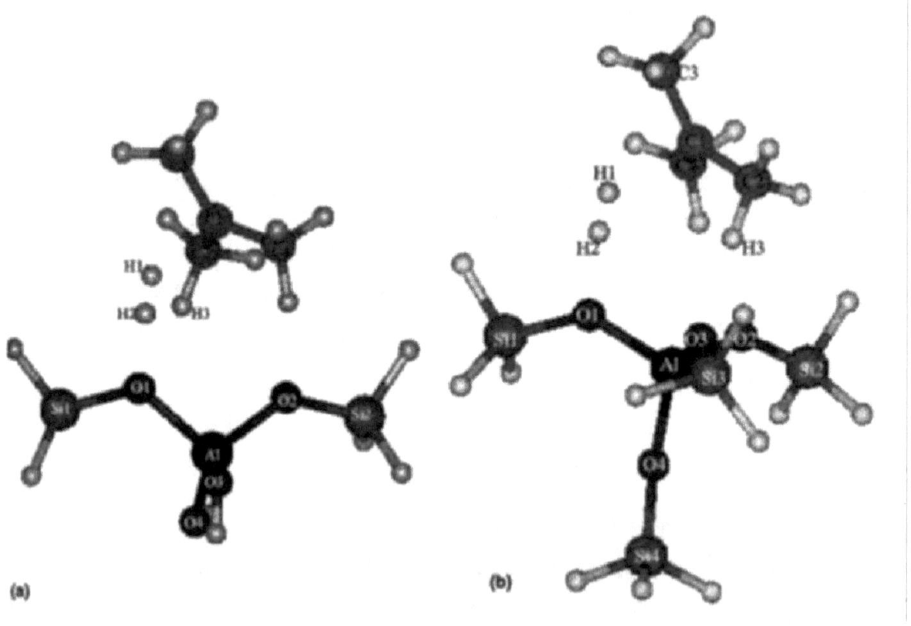

Figure 7. Geometries of B3LYP/6-311G(d,p) transition states for protolytic dehydrogenation of isobutane using (a) 3T and (b) 5T cluster models. Reprinted by permission from Ref. 45.

6-311G(d,p) basis. These energies, which do not account for the stabilizing influence of the zeolite lattice, are clearly much larger than the experimental values of 23.9[34] and 29.5[36] kcal/mol in H-ZSM-5 and 38.7 kcal/mol in H-Y.[47] However, the reaction coordinate found in these calculations is somewhat surprising. As the zeolitic proton moves away from a framework O atoms and attacks the C-H bond on the tertiary C atom, a proton from a primary C atom simultaneously moves toward *the same framework O atom* to which the acidic proton was originally bound. As a result, the two protons involved in the reaction must have a certain degree of repulsion, which increases the energy of the TS complex. It might be more energetically favorable for the proton to return to a *different framework O atom* in a concerted process that minimizes the repulsive interaction between the protons. This was the kind of pathway found (with, however, some symmetry constraints) in Ref. 33.

We recently investigated the possibility of such a single-step dehydrogenation pathway for ethane, propane, and n-butane using a 1T cluster model and the HF/3-21G method. For each reactant we were able to locate TS complexes corresponding to both the single-step and two-step pathways. Comparing the activation energies for the two pathways is quite interesting: for ethane the single-step activation energy is 10.6 kcal/mol *lower* in energy, while for propane and n-butane the single-step energies are 2.1 and 3.5 kcal/mol *higher*, respectively. We also studied these two pathways for ethane dehydrogenation using a 5T cluster model and the B3LYP/6-31G(d) method, followed by single-point B3LYP/6-311+G(3df,2p) energy calculations and zero-point energy corrections. As before, we located TS complexes for both the single-step and two-step reaction pathways described above. These structures are shown in Figure 8. The single-step apparent activation energy was 68.6 kcal/mol, compared to 71.6 kcal/mol for the two-step process, suggesting that the "ring-like" TS for the single-step, concerted process is more strongly bound to the zeolite than the TS leading to the formation of the alkoxide intermediate. However, if the HF/3-21G 1T cluster results yield qualitatively correct trends, then this situation may be reversed for propane, n-butane, and (perhaps) longer alkanes. The reason for this behavior remains to be clearly explained.

When we embedded the ethane TS for the two-step process into a 64T cluster model of H-ZSM-5, as described in Section 3.1, the activation energy was reduced by 18 kcal/mol, leading to an estimated activation energy in H-ZSM-5 of 53.6 kcal/mol. Calculating the TS for propane dehydrogenation, followed by this embedding procedure, leads to an apparent activation energy of about 43 kcal/mol, still far above the experimental value of 22.7 kcal/mol from Ref. 34. Certainly some of the discrepancy is caused by the failure of DFT to account for van der Waals interactions between the protonated propane TS structure and the zeolite framework. But in light of the heat of reaction value mentioned earlier, it is also quite possible that the experimental value is too low.

3.2.2 In Ga- and Zn-Modified Zeolites

In addition to the aforementioned studies of alkane dehydrogenation over acidic zeolites, several authors have recently used computational chemistry to model ethane dehydrogenation over Ga- and Zn-exchanged zeolites. Ga-exchanged zeolite catalysts are in wide industrial use in the commercial Cy-

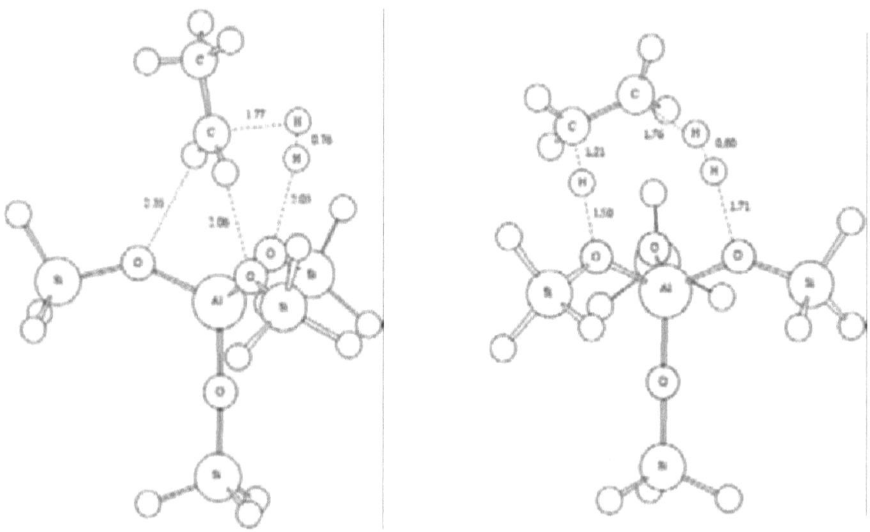

(a) (b)

Figure 8. B3LYP/6-31G(d) transition states corresponding to two different pathways for protolytic dehydrogenation of ethane using a 5T cluster model of the zeolite acid site. (a) two-step pathway; (b) single-step pathway.

clar process for the transformation of alkanes to aromatics,[48] and dehydrogenation is thought to be the first step in the aromatization process. Use of a Zn modifier in place of Ga may be advantageous from an environmental perspective. Introduction of these ionic species into the zeolite greatly complicates a theoretical analysis, because the structure of the adsorbed ion complexes are not well-known. Thus a detailed reaction path study must consider not only different candidate structural models for the active site, but also several possible reaction mechanisms for each. Frash and van Santen considered two possible structures for active sites in a Ga-exchanged zeolite.[49] One is based on an adsorbed gallium dihydride ion (GaH_2^+) and the other includes an adsorbed gallyl ion ($Ga{=}O^+$). Each of these ions adsorbs in a bidentate structure that straddles two of the O atoms bound to the substitutional Al atom. For each structure the authors calculated stable intermediates and TS complexes along two proposed reaction paths. One was based on an "alkyl" activation of a C-H bond via a $C_2H_5^{\delta-}\cdots H^{\delta+}$ TS structure, while the other involved the "carbenium-like" activation of a C-H bond via a $C_2H_5^{\delta+}\cdots H^{\delta-}$ TS structure, similar to the reaction described above for acidic zeolites. A 3T zeolite cluster model was used, and geometries were optimized at both the HF/6-31G(d) and B3LYP/6-31G(d) levels of theory, with the Ga basis set recommended by Binning and Curtiss.[50] Based on the four distinct mechanisms investigated in this study, the authors concluded that the "alkyl" route over the active site formed from GaH_2^+ is the main reaction channel for ethane dehydrogenation in Ga-doped zeolites. The apparent activation energies for the three elementary steps of the reaction are 36.9, ≈0, and 57.9 kcal/mol, calculated using the B3LYP/6-31G(d) structures and the MP2(FC)/6-311++G(2df,p) method. The authors went on to calculate reaction rates for each step and obtained the somewhat surprising result that the first step, with activation energy 36.9 kcal/mol, was rate-limiting for temperatures 700-900 K due to its low Arrhenius pre-exponential factor. The agreement with the experimental activation energy of 39.0 kcal/mol for this reaction in Ga-exchanged zeolite H-ZSM-5[51] is remarkable and perhaps somewhat fortuitous. However, the TS complex for the rate-limiting step appears to be less ionic than those found for cracking and dehydrogenation in the proton form of zeolites, which would indicate that the electrostatic influence of the extended zeolite framework on the activation energy would be smaller.

The same authors carried out a similar study of ethane dehydrogenation using a small cluster model of a Zn-exchanged zeolite.[52] The catalytic site consisted of a charge-balancing Zn^{+2} ion atop a four-membered silicate ring with two Al and two Si atoms, each terminated by two H atoms. The Zn^{+2} cation is coordinated by the four O atoms in the ring at a distance of just over 2.0Å. Such a cluster is a reasonable model for the local geometry of the active site in a zeolite like faujasite (Y). Using the B3LYP/6-31G(d) method for geometry optimization, followed by single-point energy calculations at the MP2/6-31+G(d,p) level of theory (with the Wachters-Hay basis set for Zn[53]), both the "alkyl" and "carbenium" routes to dehydrogenation were analyzed, just as for the Ga-modified zeolite.

As in the case of the Ga-exchanged system, the authors concluded that the "alkyl" mechanism was the likely reaction pathway for ethane dehydrogenation. This pathway consisted of three steps: (1) cleavage of the C-H bond by the Zn-O pair at the active site (ethane dissociation); (2) formation of ethene from the $C_2H_5^+$ alkyl group bound to Zn, and (3) formation of H_2 from two H atoms bound to Zn and O atoms in the cluster. Although in the case of the Ga cluster the activation energies were different, the major qualitative difference was the reversal of the order of steps (2) and (3) above.

A similar study of ethane dissociation (the first step in dehydrogenation) on a Zn-exchanged zeolite took a different cluster model as its starting point. Shubin et al. noted that Zn/ZSM-5 is much more active than zeolite Zn/Y for aromatization of hydrocarbons and thus represented the active site by a Zn^{+2} ion atop a five-membered silicate ring, again with two substitutional Al atoms.[54] This cluster model represents a possible active site in the ZSM-5 structure. They also considered four- and six-membered ring structures as models for active sites in faujasite for comparative purposes. Geometries of the stable intermediate structures for the dehydrogenation reaction were partially optimized (in order to reflect lattice constraints) with the HF/LANL2DZ method, which utilizes pseudopotentials rather than an all-electron basis set. Since TS structures connecting the reaction intermediates were *not* calculated, no conclusions could be drawn regarding relative activation energies for the different active site models. However, some useful information emerged about the relative stability of different active site models.

The same authors went on to publish a more detailed study using two different models for the active site in Zn/ZSM-5.[55] One was similar to the model presented above, with a single Zn^{+2} cation atop a cluster composed of *two* fused five-rings. In the new model, an oxygen-bridged binuclear $[Zn\text{-}O\text{-}Zn]^{+2}$ moeity was attached to the fused five-ring. The advantage of the binuclear complex is that it allows the two Al atoms to sit in *different* five-rings rather than forcing them to be in *the same* five-ring, as they must in order to stabilize the single Zn^{+2} ion. Since the Si/Al ratio is typically quite high in the ZSM-5 structure, the likelihood of two Al atoms in the same five-ring is small.

Restricted geometry optimizations, constrained by fixing the positions of the terminating H atoms, were carried out for the ethane dehydrogenation reaction pathway using the PW method of DFT and the TZVP basis set. Once again only the structures of the various stable reaction intermediates were computed, so activation barriers cannot be compared. However the authors concluded that the $[Zn\text{-}O\text{-}Zn]^{+2}$ bridged binuclear active site is likely to be more active than the single Zn^{+2} site for this reaction because of the much higher basicity of the non-framework O atom compared to that of the framework O atoms in the cluster. These kinds of calculations can certainly be extended in the future to probe the entire dehydrogenation reaction pathway by locating the TS complexes along the reaction coordinate. This will provide activation energies for the different active site models and allow for the interpretation and rationalization of experimental reactivity studies for different zeolites.

3.3 H/D Exchange

The adsorbate-mediated proton exchange "reaction" seems almost too trivial to study, since it appears that little of consequence occurs. However, this reaction *is* important, because its simple reaction pathway is readily amenable to computational study and its activation energy can be easily measured using isotopic substitution. The reaction is also not as trivial as it might first appear, since the proton is bound to different framework O atoms before and after the exchange, and the basicity of the framework is known to vary widely throughout the zeolite. Such variation has been found to be up to 10 kcal/mol even among the four O atoms bound to a single substitutional Al atom (see below)! As a result, this reaction has a most prominent role in the computational zeolite chemistry literature (at least ten different articles from 1993-2001) because it can be used to test the quantitative reliability of different theoretical methods. In fact, one of the earliest computational studies of hydrocarbon reactivity in zeolites was carried out for hydrogen-deuterium (H/D) exchange between deuterated methane and a 3T cluster model of an acidic zeolite.[56]

In this work Kramer et al. found HF/6-31G(d,p) optimized geometries for the adsorbed methane complex and for the TS for H/D exchange. This transition state, shown in Figure 9, differs significantly from those found for the

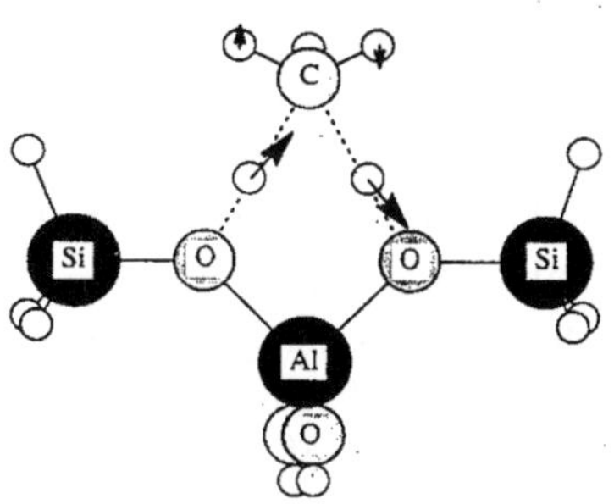

Figure 9. HF/6-31G(d,p) transition state for H/D exchange between methane and 3T cluster model for zeolite acid site. Reprinted with permission from Ref. 56.

cracking and dehydrogenation reactions since it contains a proton and a deuteron approximately equidistant from two different framework O atoms in the zeolite cluster. The hydrocarbon fragment in the TS structure is similar to the gas phase CH_5^+ ion. The reaction coordinate involves proton transfer from the zeolite to methane and simultaneous deuteron transfer to an adjacent O atom, so the degree of charge transfer is much less than for typical hydrocarbon reactions initiated by proton transfer. Single-point energies for each structure were calculated with the approximate CISD configuration interaction technique, which includes some of the effects of electron correlation. This led to a final estimate of 35.9 kcal/mol for the true activation energy. The authors also used theoretical calculations of the deprotonation energies of oxygen atoms in the zeolites ZSM-5 and Y to rationalize the observed differences in H/D exchange rate between these two structures.

Evleth et al. performed a similar calculation of the activation energy for this reaction, but using the smaller 1T cluster model of the zeolite.[57] Reactant and TS geometries were optimized using the HF/6-31G(d) method with subsequent energies obtained at the MP2/6-31++G(d,p) level. The resulting activation energy of 39.9 kcal/mol reflects the lower acidity of the 1T cluster compared to the 3T model used in Ref. 56. More importantly, Evleth et al. performed a detailed structural and charge analysis of the TS structure for this reaction. In contrast to the optimized geometry of an isolated CH_5^+ ion, the hydrocarbon fragment of the optimized TS structure had the character of a slightly negatively charged (-0.29e) CH_3 group bound to a slightly positively charged (+0.52e) H_2 unit. As a result, in the TS the hydrocarbon fragment had a much smaller total Mulliken charge (+0.23e compared to +1.0e). In addition, the two H atoms involved in the transfer were significantly further from the C atom than in the isolated CH_5^+. This analysis indicated that the TS should not be regarded as a fully ionic complex, but rather that there is a significant covalent interaction between the hydrocarbon and the zeolite cluster. Kazansky et al. used the same cluster model but the more approximate HF/3-21G method to calculate this TS.[28] The geometry of the complex was similar to that of Evleth, but the Mulliken charges were quite different. The CH_3 group had a net charge of –0.57e and the H_2 unit had a charge of +1.12e, giving a total hydrocarbon charge of +0.55e.

Blaszkowski et al. re-examined the structural nature and charge character of the TS for the CH_4-assisted H/D exchange reaction using a 3T zeolite cluster and the BP/DZPV nonlocal DFT method for geometry optimization.[44] Their resulting activation energy of 29.9 kcal/mol was significantly lower than in the earlier studies, which may be related to the known tendency of the BP/DZPV method to underestimate barriers in this kind of reaction by an average of about 7 kcal/mol.[58] The proton-oxygen distances in the TS complex were found to be somewhat higher, at 1.33Å, than the values of 1.16Å and 1.20Å found in Ref. 56 and 57, respectively. And the Mulliken charge analysis differed somewhat from that of Evleth et al., yielding a charge of –0.08e on the CH_3 group and an overall charge on the hydrocarbon fragment of +0.51e. This charge is quite a bit lower than for the transition states for ethane cracking and dehydrogenation (see sections 3.1 and 3.2). While confirming the fact that the binding of the TS complex is partially covalent, these results indicate that differences in cluster size, basis set, and computational methodology (HF, MP2, DFT) can cause significant variation in calculated atomic distances and charges in these complexes. In particular, it is clear that Mulliken charges are quite method-dependent and should be interpreted cautiously.

Kramer and van Santen subsequently expanded their earlier study of H/D exchange between CH_4 and a 3T zeolite cluster.[59] First, they considered several different possible reaction pathways for the reaction and determined that the lowest energy pathway was indeed a concerted mechanism of exchange involving different framework O atoms, which had been assumed in the earlier computational studies. Secondly, the authors calculated quantum-mechanical tunneling corrections to the reaction rates found by classical transition state theory for this reaction and found them to be very small. Using deprotonation energies for the faujasite and ZSM-5 structures obtained from force field methods, the ratio of reaction rates in these two zeolites was computed. The higher reaction rate in the ZSM-5 structure was attributed to the greater *difference* in deprotonation energies between O atoms involved in the exchange process than in faujasite.

Blaszkowski et al. extended their earlier work by optimizing the TS structure for H/D exchange between C_2H_6 and a 3T zeolite cluster.[31] The apparent activation energy was obtained using the BP nonlocal functional on structures obtained from the local VWN/DZPV method. The value of 28.2 kcal/mol was only slightly smaller than the result of 29.9 kcal/mol found for the analogous reaction with CH_4, and the TS complexes had nearly identical C-H and O-H distances in both cases. The total Mulliken charge on the hydrocarbon fragment was found to be +0.55e, just slightly higher than found for the exchange reaction with CH_4.

Esteves et al. published an ambitious systematic study of the H/D exchange reaction between a 3T zeolite cluster and CH_4, C_2H_6, C_3H_8, and i-C_4H_{10}.[60] Results for the apparent activation enthalpies were obtained at both the B3LYP/6-31G(d,p) and MP2/6-31G(d,p)//HF/6-31G(d,p) levels of theory. The B3LYP results, corrected for a reaction temperature of 298 K, indicate that when the exchange involves a proton bound to a primary C atom, the activation enthalpy of 32.2-32.3 kcal/mol is essentially the same for all four alkanes studied. When the exchange involves a proton bound to the secondary C of C_3H_8, the enthalpy rises slightly to 33.3 kcal/mol. This small increase is consistent with an experimental ^{1}H MAS NMR study of exchange between deuterated C_3H_8 and H-ZSM-5, in which the activation energy was found to be 26 kcal/mol for exchange at the primary C and 28 kcal/mol for the secondary C.[61] The calculated activation energy for H/D exchange at the tertiary C of i-C_4H_{10} has the even higher value of 36.2 kcal/mol. This higher value is consistent with experimental studies of H/D exchange, which indicate that tertiary alkanes do not readily undergo H/D exchange at the tertiary C.[62] However, the fact that the lowest activation energy calculated for i-C_4H_{10} is the same as those calculated for the linear alkanes does not agree with experimental studies, which show that tertiary alkanes undergo *faster* exchange rates than linear alkanes. The authors believed that this discrepancy indicates that branched alkanes undergo H/D exchange via a different mechanism altogether. However, they cautioned that since the effect of the electrostatic field in the zeolite cage was not included in their calculations, their conclusions were necessarily tentative.

Shortly after the work of Esteves et al., a major computational study of H exchange in zeolite ZSM-5 was undertaken by Ryder, Chakraborty, and Bell.[63] A 5T cluster model with peripheral O atoms frozen at the experimental lattice positions of silicalite (the all-Si analog of ZSM-5) was used to study the exchange process between the zeolite and hydrocarbon molecules CH_4, C_2H_6, C_3H_8, and C_6H_6. While the long-range influence of the more distant zeolite lattice was not included in this work, it was the first published attempt to study this reaction using a cluster model representative of a specific zeolite. The kind of fully optimized 3T or 5T cluster used in previous studies is entirely generic, and its geometry is not found in any bulk zeolite framework.

The BH&HLYP/6-31++G(d,p) DFT method was used to find TS complexes for the H/D exchange reaction for each hydrocarbon species. Results for the apparent activation energies for CH_4 and C_2H_6 were 38.4 and 38.3 kcal/mol, respectively. For exchange at the primary C in C_3H_8, the activation energy was found to be 39.2 kcal/mol, while for the same reaction at the secondary C the value fell slightly to 38.5 kcal/mol. These two values are extremely close but do not agree with the trend found in the previously mentioned experimental ^{1}H MAS NMR study of exchange between deuterated C_3H_8 and H-ZSM-5.[61] For exchange with C_6H_6, the computed activation energy fell to 27.0 kcal/mol, but this is still much higher than the experimental value of 14.4 kcal/mol.[64]

The first attempt to include the long-range effects of the nanoporous zeolite framework on calculated activation energies was carried out by de Vries and co-workers using a combined quantum-mechanical-molecular mechanics (QM/MM) approach.[65] They calculated activation energies for the H/D exchange reaction with CH_4 for faujasite, ZSM-5, beta, erionite, and chabazite structures, and tried to correlate these energies with other energetic and structural properties of the Brønsted acid site in the different frameworks. The quantum chemical region in these calculations was defined to be a 5T cluster centered on a substitutional Al atom, within which geometry optimizations were carried out with the HF/3-21G method. Single point energies were subsequently calculated with the HF/6-31G(d) method. The true activation energy for H/D exchange, with respect to the adsorbed methane structure, varied from 58.8–69.6 kcal/mol for the different frameworks studied. Contrary to the results of Kramer and van Santen,[56,59] de Vries et al. did *not* find a simple correlation between activation energy and the difference in deprotonation energy between the two O atoms involved in the reaction. However, their results suggest a quadratic relation between the activation energy and the framework O-Al-O angle, with the activation energy at a minimum value near an angle of 90 degrees. The calculated activation energies in this study are substantially higher than experimental values for typical zeolites, but the authors found that the use of MP2/6-31G(d,p) energies reduced the calculated values by approximately 20 kcal/mol, bringing them into much closer agreement with experiment. This cautions that structure-reactivity correlations based on energies obtained without including the effects of electron correlation (provided in different ways by MP2 and DFT methods) may not be reliable.

The most recent attempt to compute the effect of the zeolite framework on H/D exchange activation energies is due to Vollmer and Truong.[66] They used an ab initio embedded cluster method that reproduces the Madelung field of the zeolite crystal and includes its effect on the optimized geometry and energy of the structures based on a 3T cluster model of the active site. For zeolite Y they considered two possible exchange pathways in which the acidic proton was initially bound to two different O atoms adjacent to the substitutional Al atom. The geometry of the TS was optimized with the BH&HLYP/6-31G(d,p) DFT method, and final single point energies were calculated with the CCSDT/6-31G(d,p) method. Tunneling corrections were also calculated and found to be quite small (< 1 kcal/mol), while the counterpoise correction for basis set superposition error was found to be more significant. Of particular interest in this study was the authors' comparison of the bare cluster and embedded cluster results, which allowed an assessment of the importance of the long-range influence of the zeolite lattice on the results.

The final embedded cluster results, including all correction terms, for the apparent activation energies were 29.6 and 32.7 kcal/mol for the two pathways considered. The embedding condition lowered these barriers by 7.0 and 2.6 kcal/mol, respectively, relative to the bare cluster results, indicating that the Madelung field preferentially stabilized the TS complex compared to the reactants. The influence of the Madelung potential on the structure and charge of the TS was likewise reported to be significant. The O-H distances were lengthened by approximately 0.10Å and the C-H distances were shortened by roughly 0.05Å due to the embedding condition. The total Mulliken charge on the protonated CH_4 moeity rose from +0.21e to +0.67e when the Madelung field was included. Thus the long-range electrostatic effects of the zeolite increased the degree of ionicity in the TS and made the structure more carbonium-like than bare cluster results predicted.

3.4 ALKENE CHEMISORPTION AND HYDRIDE TRANSFER

Nearly all of the reactions considered thus far have been monomolecular, involving the interaction of just one reactant molecule with the zeolite surface. Such reactions are most easily amenable to computational modeling, and when the experimental conditions are chosen so these reactions predominate, their activation energies can be measured in a straightforward manner. Thus, monomolecular reactions are ideal for comparing computational and experimental results in order to help interpret experimental results qualitatively and test computational methods quantitatively. However, the large majority of industrially significant catalytic transformations of hydrocarbons are believed to occur through bimolecular reactions. These are considerably more difficult to model computationally, and the abundance of side reactions that take place in parallel makes the experimental determination of

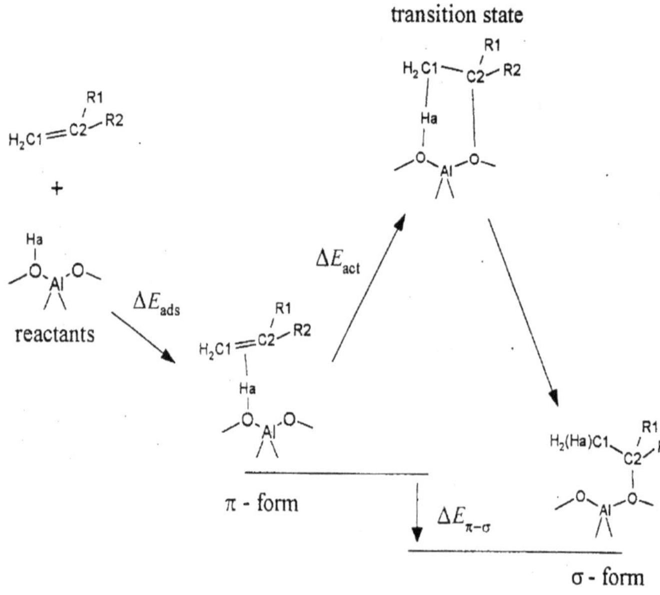

Figure 10. Schematic reaction pathway for alkene chemisorption at a zeolite acid site (O-Ha). The physisorbed species is referred to as the π-form while the chemisorbed alkene or surface-bound alkoxide is designated as the σ-form. Reprinted with permission from Ref. 74.

activation energies more complex and less certain.

Many of the bimolecular hydrocarbon transformation mechanisms proposed in the literature involve the initial step of alkene chemisorption; i.e., the protonation of an alkene by a zeolitic acid site to form a surface-bound alkoxide. A schematic illustration of this reaction is given in Figure 10.

Many subsequent reactions can follow, including hydride transfer from another alkane, skeletal isomerization, bimolecular cracking (already discussed in section 3.1), and MTO/MTG reactions. In this section we will first review a number of studies of alkene chemisorption carried out for both finite clusters and periodic crystalline systems. Afterwards a brief survey will be made of computational studies of the bimolecular hydride transfer reaction. In the language of organic chemistry, alkene chemisorption represents a chain initiation step, while hydride transfer represents a chain propagation step.

3.4.1 Alkene Chemisorption

A quantum-chemical study of alkene chemisorption was first carried out in early work by Kazansky and Senchenya.[67,68] These papers, presenting results obtained using small basis sets (STO-3G and 3-21G), a crude 1T cluster model of a zeolitic acid site, and a planar geometry constraint for this cluster, were extremely influential in persuading chemists that zeolite catalysis was not merely a solid-state analog of liquid superacid chemistry. The results of these studies showed that transfer of a proton from the cluster model to an adsorbed ethene molecule produced *not* a stable carbocationic intermediate, as is the case in liquid superacid media, *but rather* an unstable TS complex. However, the $C_2H_5^+$ geometry in the TS bore a close resemblance to that of a free ethyl carbenium ion. In Kazansky's view, the surface-bound ethoxide $ZOCH_2CH_3$ that results can react with other hydrocarbons in (for example) hydride transfer, isomerization, and oligomerization reactions. Although the qualitative insight provided in this early study was quite valuable, the HF/3-21G activation energy of 8.5 kcal/mol (with respect to the gas phase reactants) was unusually low, perhaps due to the lack of complete geometry optimization in the calculation. A subsequent fully optimized calculation by Sierra et al. using the same basis set and zeolite cluster model gave a much higher activation energy of 16 kcal/mol.[69] A contemporaneous HF/3-21G study of propene and i-butene chemisorption using a 3T cluster model (again with planar symmetry constraints) gave activation energies of 29.8 and 26.2 kcal/mol, respectively.[70] The authors pointed out that the lower activation energy for i-butene occurs due to the greater stability of the tertiary "carbenium-like" ion in the TS compared to the secondary ion formed for propene. However, correlated MP2/3-21G energies calculated for these HF/3-21G geometries reduced the energies to 19.9 and 19.1 kcal/mol, obscuring the trend. This is perhaps due to the unsuitability of a correlated calculation with such a small basis set.

Evleth et al. later extended their work in Ref. 69 to ethene and propene chemisorption with the 3T cluster model, a larger 6-31G(d) basis set, and correlated electronic structure methods (MP2 and BLYP).[71] Activation energies obtained using MP2/6-31G(d) energies calculated for HF/6-31G(d) optimized geometries were 21.5 and 18.4 kcal/mol for ethene and propene, respectively. Using the BLYP/6-31G(d) DFT method, results of 14.8 and 12.9 kcal/mol were obtained. The latter results confirmed the trend of the MP2 results but were significantly lower, consistent with the tendency for DFT to underestimate activation energies. A structural study of the transition states indicated that the degree of proton transfer from zeolite to alkene, in terms of the O-H distance, increased when going from the 1T to 3T cluster and also when going from ethene to propene. This, along with the trend in activation energy for the two alkenes, is consistent with the higher proton affinity for propene (180 kcal/mol) compared to ethene (163 kcal/mol). This difference is a reflection of the greater stability for the secondary ion TS structure of propene compared to the primary ion formed by ethene protonation.

Kazansky et al. used the minimal 1T cluster model, HF/6-31G(d) full geometry optimization, and MP2/6-31++G(d,p) single point energies to calculate the activation energy for i-butene chemisorption.[72] Corrections for harmonic zero-point vibrational energies were also included. An activation energy of 15.3 kcal/mol was obtained, which is somewhat lower than the MP2/3-21G//HF/3-21G result reported above for the 3T cluster model. At least a portion of the discrepancy is likely due to the inability of the 1T cluster to accurately represent a true aluminosilicate acid site. Rigby and Frash subsequently pointed out that the activation energy and overall reaction energy for alkene chemisorption are in principle dependent on the carbon site (primary, secondary, or tertiary) that is bound to the zeolite framework in the alkoxide product.[73] They carried out a comparative study for ethene, propene (s), propene (p), i-butene (t), and i-butene (p). The 3T cluster was used to perform a HF/3-21G geometry optimization with planar geometry constraints, followed by MP2/3-21G single point energies. Relative to the gas phase reactants, the activation energy for ethene chemisorption was 23.2 kcal/mol, in fair agreement with the results of Evleth et al.[71] The results for propene chemisorbed in the (s) and (p) positions were 18.8 and 25.3 kcal/mol, respectively, reflecting the greater stability of the secondary ion TS structure. The same effect is demonstrated by the activation energies of 15.2 and 27.3 kcal/mol for i-butene chemisorbed in the (t) and (p) positions. The propene result is in excellent agreement with Ref. 71, while the i-butene result is somewhat lower than that of Ref. 70. Ribgy and Frash noted that the overall reaction energies for alkene chemisorption at a different site followed the stability trend primary > secondary > tertiary, but that these energy differences were much smaller than the corresponding differences in activation energies, which follow the opposite trend. They therefore concluded that the distribution of chemisorption products for a given alkene is *not* due to the different stabilities of the tertiary, secondary, and primary product alkoxides, *but rather* to the different activation energies for their formation.

Sinclair et al. broke new ground with a comparative study of ethene, propene, and i-butene chemisorption using both finite and embedded-cluster models for the zeolite framework.[74] Their finite cluster model was 4T, for which B3LYP/6-31G(d) as well as MP2/6-31G(d)//HF/6-31G(d) calculations were performed. The 4T model was constrained to mimic the local geometry of the 8-ring window in chabazite by freezing the terminating H atoms at their original positions along the bond directions found in the chabazite lattice. A fully relaxed 4T cluster calculation was performed only for i-butene to facilitate comparison. The embedded cluster study for chabazite was carried out using the QM/MM method implemented in the ChemShell program.[75] The 4T cluster was the fully quantum-mechanical region, surrounded by a classical region of approximately 1700 atoms that interact with the quantum region through a classical force field and point charges located on each atom in the classical region. Due to the complexity of the calculations, especially the optimization of TS geometries, the quantum region was treated using the MP2/6-31G(d)//HF/3-21G method.

The chemisorption activation energies for the most favorable pathways in the 4T finite cluster were found to be 19.1, 12.4, and 9.3 kcal/mol for ethene, propene, and i-butene, respectively (the B3LYP/6-31G(d) result for i-butene was nearly identical). For comparison, the result for i-butene using the *fully relaxed* 4T cluster model was 10.3 kcal/mol. These 4T constrained cluster results are about 5 kcal/mol lower than those of 23.2, 18.8, 15.2 kcal/mol obtained by Rigby and Frash using a planar 3T cluster. The QM/MM embedding scheme produced activation energies of 20.8, 10.3, and 8.4 kcal/mol for the three alkenes, which indicates that the surrounding chabazite lattice has very little net effect on the barrier for this reaction. However, the authors also compared the reaction energies for the finite cluster and QM/MM methods and made an interesting observation. While the chemisorbed product (also called the σ-complex) is in all cases more stable than the physisorbed reactant (also called the π–complex), and while the stability difference (which is identical to the reaction energy) decreases as the alkene chain length increases, two features stand out. First, the reaction energy change as alkene chain length increases is much larger for constrained cluster models than for fully-optimized models. Secondly, even among the constrained models, the magnitude of the reaction energy is much smaller (by 10-15 kcal/mol) when the embedded cluster model is used. Apparently the constraint that forces the zeolite cluster to mimic a crystalline structure produces significant steric repulsion between the longer alkenes and the zeolite framework, reducing the stability of the chemisorbed product compared to a fully-optimized model. Additionally the authors showed that in the embedded cluster, the polarization of the quantum region by the point charges in the classical region causes a structural perturbation in the quantum region that greatly influences the reaction energy. This makes the important point that long-range effects of the lattice may influence details of a reaction pathway even if they do not strongly affect the activation energy.

In the course of their study of double-bond and skeletal isomerization of linear butenes, Boronat et al. also calculated the activation energy for i-butene chemisorption using a 3T cluster with planar constraints and the B3P86/6-31G(d) DFT method.[76] However, these authors considered only the pathway leading to the primary product alkoxy species. Their activation energy of 31.4 kcal.mol was reduced slightly to 29.9 kcal/mol by including zero point vibrational energies. These results are in reasonable agreement with those of Rigby and Frash discussed above.[73] However, Boronat et al. also obtained a reaction energy of only -10.6 kcal/mol (-7.5 kcal/mol including zero point energies), while Rigby and Frash found a value of –19.3 kcal/mol. This illustrates once again how sensitive the stability of surface alkoxide species is to differences in theoretical method, reflected here in the tendency for DFT methods to reduce the stability of the σ-complex with respect to the π–complex. This also agrees with DFT results for i-butene presented by Sinclair et al.[74]

In a very recent series of papers, Boronat and collaborators studied the chemisorption of ethene, propene, but-1-ene, and i-butene using both larger cluster models and a periodic unit cell approach for theta-1 zeolite.[77,78,79] In order to better understand the limitations of the cluster approach to modeling the reactivity of bulk solids, the authors made a detailed study of ethene chemisorption using successively larger 3T, 5T, and 11T cluster models for the local geometry of theta-1 zeolite.[77] Both HF/6-31G(d) and B3LYP/6-31G(d) optimizations were performed. In all clusters larger than the minimal 3T, chemisorption was studied at four different acid sites for completeness. The geometries optimized in the 11T cluster were then embedded in a larger 28T cluster model, which contained two complete 10-rings of the unidimensional channel in theta-1, and single-point energies were calculated. Finally, similar single-point energies were also calculated in the periodic unit cell of theta-1. Planar constraints were employed for the 3T cluster, while in the geometry optimizations carried out for the 5T and 11T models, the positions of the terminal H atoms were fixed to conform the clusters to the theta-1 framework geometry. Single point energies in the largest cluster and periodic models utilized a mixed basis set optimized for periodic calculations. This basis set was also used for single-point calculations on the optimized geometries of the smaller clusters to allow a consistent comparison.

Both activation and reaction energies were calculated for ethene chemisorption, and in reasonable agreement with previously described results, the reaction energy in the generic 3T cluster was found to be –11.7 and –13.8 kcal/mol using the HF/6-31G(d) and B3LYP/6-31G(d) optimizations, respectively. Most strikingly, the energies were quite different for any of the larger clusters representing a specific acid site in the theta-1 framework. For the four acid sites considered in these clusters, the HF calculated energies ranged from –1.5 to +6.3 kcal/mol, and the B3LYP values spanned a range from –5.5 and +2.2 kcal/mol. Here, a positive reaction energy means that the ethoxide σ–complex is actually less stable than the adsorbed π–complex. The periodic reaction energies are all positive and lie between 3.9 and 11.7 kcal/mol. The authors concluded, in agreement with Sinclair et al.,[74] that the structural changes needed to accommodate the ethoxide species are more difficult for a cluster representing a real zeolite framework, causing a destabilizing of the σ–complex relative to the adsorbed π–complex. From a chemical point of view, this is significant because it suggests that surface alkoxide intermediates are less stable and thus more reactive than previous calculations had indicated.

On the other hand, the calculated activation energies vary less dramatically with an increase in cluster size from the generic 3T model. In agreement with previous studies of hydrocarbon reactivity, the zeolite lattice causes an electrostatic stabilization of the somewhat ionic TS complex. Comparing the 3T cluster to the 28T and periodic results, the activation energies are reduced by about 6-7 kcal/mol and 2-3 kcal/mol for HF and B3LYP optimizations, respectively. The size of these effects is similar to those found for the H/D exchange reaction for CH_4, which occurs through a similar concerted mechanism.[66] The TS described by the B3LYP method is less ionic than with HF, so the stabilization is understandably less significant. The results of a parallel DFT study of but-1-ene chemisorption and skeletal isomerization by the same authors confirm the trends in reaction energies and activation energies found for ethene.[78] A more detailed description of these effects was supplied by these authors in a comparative study of the chemisorption of ethene, propene, but-1-ene and i-butene on the same cluster models of theta-1.[79]

However, a very recent fully-optimized periodic study of propene chemisorption in chabazite presents a somewhat conflicting account of the relative stability of the σ– and π–complexes.[80] In this work, the PZW91 DFT method was applied using the VASP periodic plane-wave program with pseudopotentials. The authors specifically studied the effect of zeolite lattice relaxation on the chemisorption reaction and activation energies. In agreement with previous studies, they found that the formation of the secondary alkoxide species was energetically more favorable than the primary species. Including full lattice relaxation for each stationary point along the reaction pathway, the authors found a reaction energy of –6.5 kcal/mol and an activation energy of 13.4 kcal/mol. When, however, they repeated these calculations using zeolite framework constraints, such as fixing the positions of some atoms in the unit cell and fixing the unit cell size and volume, they found that these energies both increased significantly. In some cases, the reaction energy became positive, increasing to as much as +5.5 kcal/mol, just as described in Refs. 77-79. The authors concluded that it is critically important to optimize the positions of a large number of zeolite oxygen atoms in the neighborhood of the active site, and that the energy of the alkoxy species is especially sensitive to the degree of geometry relaxation. Thus, it is possible that the results of Refs. 77-79 may need to be re-examined in light of the substantial use of geometry constraints and single-point energies in those calculations.

In any case, the emergence of such "second-generation" reactivity studies utilizing periodic systems or large cluster models constrained to match the structure of specific zeolites calls for caution to be used in interpreting the results of fully-optimized minimal cluster models, since the energies they produce may not be reliable. At the same time, such ambitious studies indicate that computational modeling is reaching the stage where it may soon be possible to elucidate differences in reactivity and selectivity among different catalyst frameworks.

3.4.2 Hydride Transfer

Once the alkene chemisorption or "chain initiation" step is complete, it is easy to see how subsequent hydrocarbon transformations, which are generally accepted to be chain reactions, can take place via hydride transfer from an alkane (R_2-H) to the surface alkoxide (R_1), leading to the desorption of an R_1 alkene and formation of a new R_2 surface alkoxide. Hydride transfer is self-evidently a crucial elementary step in the reaction chain.

An early computational study of the symmetrical i-butane + surface tert-butoxide reaction was reported by Kazansky et al., using a 1T zeolite cluster and HF/6-31G(d) geometry optimization, followed by single point MP2/6-31++G(d,p) energy calculations.[33] The reaction pathway involved hydride transfer from the tertiary carbon in i-butane to the tertiary carbon of the surface butoxide, along with a concomitant stretching of the butoxide C-O bond. In the TS the hydride is shared nearly equally between the two hydrocarbon units, each of which is stabilized by two hydrogen bonds to the framework oxygens of the zeolite cluster. The activation energy, including zero-point vibrational energies, was found to be 48.4 kcal/mol with respect to the initial i-butene + surface tert-butoxide system. This is significantly higher than the value of 30 kcal/mol obtained from kinetic modeling studies.[41] A subsequent computational study by the same authors traced the pathways for hydride transfer for the methane + surface methoxide, ethane + surface ethoxide, sec-propane + surface sec-propoxide, and sec-propane + surface tert-butoxide reactions.[81] This work indicated that the complexes involved in the methane and ethane reactions were actually local energy minima rather than transition states; however, due to the minimal zeolite cluster size and lack of correlation in the geometry optimization, this result should be regarded as tentative. This is especially true since TS complexes, and *not* local minima, were found for all cases involving hydride transfer between the larger (C3 and C4) hydrocarbons.

The calculated activation energy is largest for the methane-methoxide (66.5 kcal/mol) and ethane-ethoxide (56.4 kcal/mol) transfer reactions. For the other three, the activation energies are all in the range 47-48 kcal/mole. These values confirm the expectation that hydride transfer from secondary and tertiary carbons should be easier than from primary carbons. However, the results also indicate that substitution of the secondary alkane by the tertiary one does not stabilize the TS for hydride transfer. From the previously mentioned kinetic modelling experiments, the activation energy for hydride transfer from propane to the surface tert-butoxide was determined to be 34.9 kcal/mol.[41] If this value is accurate, the quantum-chemical result significantly overestimates the barrier energy, although the experimental result is probably the apparent activation energy measured relative to the gas phase reactants (acid site + propane + i-butene), and Kazansky did not attempt to correct for this. However, in view of the small cluster model used in this study, the discrepancy is not surprising. The R_1-R_2 hydrocarbon fragment in the TS complex has a net Mulliken charge of +0.9e, indicating that it will be significantly stabilized by the electrostatic influence of the zeolite structure not included in the computational study.

Boronat et al., in the same spirit as their work on alkene chemisorption discussed earlier, also carried out a cluster and periodic study of the ethane + surface ethoxide hydride transfer reaction in chabazite, assuming two different Si/Al ratios.[82] The terminal H atoms in the 5T cluster model were fixed at appropriate positions in order to represent the chabazite framework, and constrained BPW91 and hybrid B3PW91 optimizations were performed using the same mixed basis set described in their alkene chemisorption work.[77] In agreement with the 1T results of Kazansky, the structure containing the $(C_2H_5\text{-}H\text{-}C_2H_5)^+$ fragment was found to be a local minimum rather than a TS, although the authors pointed out that when the HF method was used in place of B3PW91, a TS was found instead. The energy of this intermediate structure with respect to the separated reactants (acid site + ethane + ethene) was found to be in a range of 38-48 kcal/mol for a variety of different pure-DFT and hybrid HF-DFT methods. When the optimized geometries were inserted into the periodic chabazite unit cell, this energy was reduced by roughly 3-7 kcal/mol, consistent with the expected effect of the zeolite lattice.

3.5 Skeletal Isomerization and Alkylation

Skeletal isomerization reactions of hydrocarbons are of significant practical importance, because they influence the product distributions of many commercial reactions, such as cracking and alkylation. The well-known alkylation reaction of toluene with methanol in zeolite ZSM-5 to form p-xylene with nearly 100% selectivity is a classic example of a shape selective catalytic transformation. First attempts to model such processes focused on the conceptually simple methyl shift reaction along a surface-bound alkoxide formed at a zeolite acid site through alkene chemisorption.[38,73,83] A proposed reaction pathway for a symmetrical methyl shift reaction for a surface propoxide species is illustrated in Figure 11. The proposed TS complex contains a fragment that resembles a cyclic propenium ion, stabilized in this case by ionic interactions between carbon and framework oxygen atoms. Sherwood et al. [83] studied this reaction using both a 3T cluster and an embedded cluster QM/MM approach to represent the zeolite Y structure. The authors pointed out that while in a 3T cluster the two oxygen atoms bridging adjacent Al and Si atoms are equivalent, this symmetry is broken in a crystalline zeolite, so that activation energies can be computed for both forward

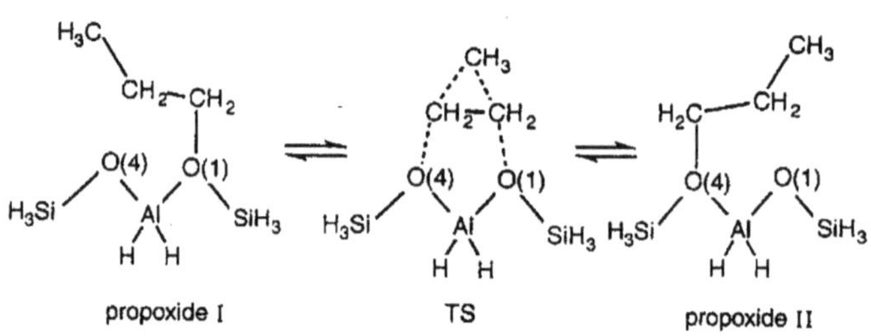

Figure 11. Schematic reaction pathway for methyl shift reaction along a surface-bound propoxide. The TS structure is analogous to a cyclic propenium ion. Reprinted with permission from Ref. 83.

and reverse directions. The bare cluster HF/3-21G result gave an activation energy of 75.5 kcal/mol, but this was significantly reduced by the embedding conditions. Three different embedding conditions were applied, in each case yielding forward and reverse methyl shift activation energies of about 60 and 45 kcal/mol, respectively.

Rigby and Frash used only a 3T cluster model and HF/3-21G geometry optimizations, but performed a comparative study for methyl shift reactions involving surface alkoxides of propane, n-butane, n-pentane, i-pentane, and 2,3-dimethylbutane. All of the TS complexes for these reactions contained the three carbon atom ring structure shown in Figure 11, but Rigby and Frash found two possible conformations of each TS. In one, two carbon atoms were closest to the oxygen atoms of the zeolite, while in the other, two hydrogen atoms on the hydrocarbon fragment played this role. The latter TS structure was found to have the lowest energy for each reaction considered. Interesting trends were observed in this study; namely, the methyl shift activation energies involving the linear alkanes were essentially the same (71.5-73.5 kcal/mol). This, however, contrasts with experimental evidence of a significant difference in isomerization rate for n-butane and n-pentane. The discrepancy may indicate that the isomerization reaction proceeds via a different mechanism for one of these molecules. Both methyl and ethyl shift reactions were considered for the transformation of n-pentane to i-pentane, and their activation energies differed by only an insignificant 1.4 kcal/mol. However, the methyl shift activation energy was found to be strongly dependent on the degree of branching of the reactant. For the reactions of i-pentane and 2,3,-dimethylbutane, the activation energies were 63.6 and 58.8 kcal/mol, respectively. The authors concluded that branch formation in linear alkanes is more difficult than shifting of an existing branch, which agrees with experimental expectations.

These same authors considered a different isomerization pathway in a subsequent and more sophisticated computational study.[84] Starting from a surface alkoxy species, this ring-closing mechanism involves the formation of a substituted cyclopropane and donation of a proton back to the zeolite. Protonation then re-opens the ring, breaking a different bond and thus leading to isomerization. Full B3LYP/6-31G(d) geometry optimizations were performed with a 3T zeolite cluster model in order to model the formation of cyclopropane from a surface-bound prop-1-oxy species. The TS for the ring-closing step resembles a protonated cyclopropane ring stabilized by interaction with the anionic zeolite framework. Two conformations of the TS structure were found, with the cyclopropane ring either coplanar with or nearly perpendicular to the central O-Al-O unit of the 3T cluster. The total Mulliken charge of the hydrocarbon fragment in the TS was approximately +0.8e, indicating that the TS complex is somewhat less ionic than that of other monomolecular hydrocarbon reactions, i.e., cracking and dehydrogenation. Including energy correction for zero-point vibrational energies, the apparent activation energy via the out-of-plane TS was calculated to be 54.9 kcal/mol, far below the value of approximately 77 kcal/mol obtained by Rigby et al. using a 3T cluster with planar symmetry constraints and the MP2/6-31G(d)//HF/3-21G level of theory.[38] This illustrates the significant influence of symmetry constraints on computed activation energies, although it is not clear whether full or constrained optimization provides the most faithful model of the active site in a crystalline zeolite. The fully-optimized result is closer to an unpublished experimental activation energy of about 30 kcal/mol for n-hexane isomerization in Pt/H-ZSM-5 and Pt/H-MOR, but is still far too high. This is due at least in part to the neglect of electrostatic stabilization of the TS complex by the zeolite lattice.

A related study of both skeletal and double-bond isomerization of but-1-ene was performed by the group of Boronat et al.[76] A weakly adsorbed but-1-ene undergoes double-bond isomerization when the acid site protonates the double bond and a basic framework oxygen abstracts a proton from the alkene, forming adsorbed but-2-ene. The skeletal isomerization reaction is slightly more complicated. First the adsorbed but-1-ene is protonated to form a surface-bound secondary alkoxy intermediate. Next, a methyl group is transferred via a cyclic TS to form a branched primary alkoxy species, afterr which a framework oxygen abstracts a proton to give an adsorbed i-butene. These calculations were based on a 3T cluster with planar constraints and B3P86/6-31G(d) geometry optimization. Double-bond isomerization proceeds via a TS whose C_4H_9 hydrocarbon fragment has a net Mulliken charge of only +0.62e, and the apparent activation energy was found to be only 14.0 kcal/mol including zero-point energies. This is in good agreement with typical experimental values of 15-20 kcal/mol cited by the authors, although DFT methods are known to often underestimate reaction barriers. For the skeletal isomerization process the methyl shift TS has the highest energy along the reaction pathway, which leads to a calculated apparent activation energy of 34.1 kcal/mol. The Mulliken charge on the cyclic C_4H_9 fragment is +0.82e, and the authors again point to reasonable agreement with experimental values of about 30 kcal/mol for isomerization of n-butenes. The level of agreement with experiment for these two reactions is surprising in view of the planar constraints of the 3T cluster and the neglect of electrostatic effects, which should be even larger for the skeletal isomerization activation energy. It is possible and even likely (see below) that the B3P86 method of DFT overbinds the TS complexes by enough to overcompensate for the neglected factors.

The same authors further explored the skeletal isomerization of but-1-ene to i-butene using a series of 3T, 5T, 11T, and 28T zeolite cluster models for theta-1 zeolite, which are described in the previous section on alkene chemisorption.[78] Notably, the authors employed a different and more widely-used DFT method, B3PW91, and their 3T cluster gave an apparent activation energy of 43.7 kcal/mol, a full 10 kcal/mol greater than obtained in their previous study. As expected, the activation energy decreased as the cluster model size increased, reaching a value of 32.7 kcal/mol for the 26T cluster, which contains a complete double 10-ring unit found in theta-1. These studies, by the same authors, illustrate that different DFT methods do not give equivalent results, and that good agreement between theory and experiment can be obtained fortuitously and should be carefully evaluated.

We will close this section by discussing a very recent series of computational studies of alkylation and isomerization reactions of toluene and xylene in H-MOR zeolite.[85,86,87,88] These papers describe methods of calculation that represent the "state of the art" in computational materials chemistry, applied to reactions of aromatic molecules that are of significant industrial interest. They present finite cluster and periodic DFT calculations of (1) toluene alkylation with methanol to form xylene; (2) intramolecular isomerization of adsorbed toluene and xylene; (3) transalkylation between toluene and benzene; and (4) disproportionation of toluene and benzene, all catalyzed by the acidic zeolite H-MOR.

The 4T cluster and the MPWPW91/6-31G(d) method of DFT were used in the fully-optimized finite cluster calculations, while the periodic approach implemented within VASP used plane waves and ultrasoft pseudopotentials with the PZPW91 DFT method. Reaction pathways and approximate TS structures were calculated in the periodic calculations with the nudged elastic band (NEB) method.[25] In both cases, energies were corrected for zero-point vibrations. An important synergy between cluster and periodic calculations was achieved by using optimized TS structures from the cluster approach in order to construct the initial reaction path for the NEB method in the periodic case. This undoubtedly saved a great deal of CPU time, and points to the usefulness of both models in studies of chemical reactivity.

The study of toluene alkylation with methanol examined pathways for the formation of *ortho*-, *meta*-, and *para*-xylene. In each case the acidic proton attacks the methanol OH group, and the TS structures show a cleavage of the C-O bond in methanol and concurrent formation of a C-C bond with toluene, with water formed as a reaction product. The finite cluster activation energies calculated with respect to the adsorbed reactants follow the trend expected from the preferred positions for electrophilic substitution, with E_{act} lowest for *o*-xylene (39.3 kcal/mol) and highest for *m*-xylene (41.4 kcal/mol). However, in the medium-pore zeolite H-ZSM-5 this reaction is known to be nearly 100% selective for *p*-xylene, which shows the limitations of a finite cluster model for reactivity phenomena which depend on zeolite framework structure.

The periodic study of the same reactions showed several important differences. First, the reaction pathway was significantly altered from that found using the 4T cluster. In the finite cluster the proton attack on methanol, methylation of toluene to form protonated xylene, and back-donation of a proton from xylene to the zeolite occur in a single step, while in the periodic model this is sterically unfavorable, and the protonated xylene molecule is found to be a quasi-stable reaction intermediate. A low-energy rotation of this molecule is then required to allow back-donation of a proton to the zeolite. Secondly, the orientation of the TS structure for alkylation was modified. In the 4T model, the hydrogen atom bound to the carbon atom undergoing alkylation interacted directly with a basic oxygen atom adjacent to the acid site. In the periodic case it was more energetically favorable for the toluene fragment to orient itself to maximize hydrogen bonding with framework oxygen atoms more distant from the acid site. Of course, these atoms are *not* present in the 4T cluster model. Finally, the activation energies followed a different trend in the periodic study. When corrections for van der Waals interactions of the reactants with the zeolite framework were made, E_{act} was found to be 30.3, 33.9, and 36.1 kcal/mol for *p*-, *o*-, and *m*-xylene, respectively. While these computed energy differences alone are not large enough to explain the high degree of selectivity for *p*-xylene found experimentally, the correct trend was reproduced. This was the first demonstration of transition-state selectivity in a computational study of zeolite catalysis.

In addition to transition state selectivity, differences in product diffusivity are thought to contribute to the selectivity for *p*-xylene in the alkylation of toluene by methanol. The proposed scheme involves the interconversion of different xylene isomers and preferential diffusion of *p*-xylene out of the zeolite micropores. The next paper in the series we are describing studied different reaction pathways for toluene and xylene isomerization in the H-MOR framework. For both toluene and xylene, three different pathways for toluene isomerization were studied, two involving direct methyl shift mechanisms, and the other involving disproportionation and formation of a surface methoxy intermediate. For the case of toluene, the activation energies of all three pathways with respect to adsorbed toluene were found to be about 43 kcal/mol, making it impossible to distinguish between them on this basis. Each isomer of xylene undergoes all three isomerization pathways, and in each case p-xylene has the highest activation energy, although the energy differences are only a few kcal/mol and thus cannot be considered definitive. However, the authors determined that one of the three pathways could not be reproduced with finite cluster models, since it relied on stabilization of the cationic TS structure by framework oxygen atoms far from the Brønsted acid site. This established the principle that a combination of structural constraints and electrostatic effects that are present in a crystalline zeolite and are represented in a periodic model can affect the reaction pathway of intramolcular isomerization reactions.

The final paper in the series considered intermolecular isomerization, or bimolecular transalkylation, reactions between toluene and benzene as a model for understanding the same reaction between toluene molecules. The two possible pathways involve (1) proton attack on a hydrogen atom of the methyl group of toluene, leading to a disproportionation reaction with formation of a benzyloxy intermediate structure; and (2) proton attack on the toluene carbon atom in the α position to the methyl group, causing a direct methyl shift toward the benzene. The latter reaction is very similar to the alkylation of toluene with methanol, which was described earlier. Finite cluster results indicated that the disproportionation reaction was energetically favored by about 6 kcal/mol over the direct transalkylation reaction, while in H-MOR the periodic calculations favored the direct transalkylation route by about 3 kcal/mol. The authors concluded that in larger pore zeolites with less severe steric constraints, the disproportionation mechanism would predominate, but in medium and small pore materials like H-MOR the directly transalkylation reaction would be favored. This work provides yet another example of the influence of the zeolite framework on molecular reactivity and the limitation of finite cluster models for computing reaction pathways. On the other hand, these examples illustrate the impressive computational abilities of periodic DFT methods to investigate reactions of aromatic systems in framework specific environments. They suggest it may soon be possible to rationalize differences in reactivity and selectivity for such reactions in different zeolites, which would truly be a major step forward.

3.6 Methanol-to-Gasoline (MTG) Chemistry

One of the most important industrial applications of nanoporous catalysts is the conversion of methanol to hydrocarbons, in which the catalyst facilitates the formation of C-C bonds. The final products depend on the specific catalyst and the working conditions of the reaction. Specific examples include synthetic processes for methanol-to-olefins and methanol-to-gasoline using zeolite H-ZSM-5. In the study of the mechanism for these reactions, the major unresolved issue is the pathway for the formation of the first C-C bond. The pioneering work of Chang and Silvestri[89] demonstrated that the reaction proceeds in two distinct steps. First, methanol is dehydrated to form dimethyl ether (DME), and secondly, DME reacts with methanol to form a variety of olefins. In the MTG process a number of polymerization and isomerization reactions follow. The exact details of the formation of the first C-C bond have been the subject of much speculation since the 1970s. Clearly this is a question to which computational materials chemistry can in principle be applied. In this section we will briefly survey the results of some recent studies, focusing primarily on the mechanisms proposed and investigated for the C-C bond formation step.

The earliest quantum-chemical studies examined the formation of DME from methanol in the first reaction step. In two separate publications Blaszkowski and van Santen considered DME formation via three distinct pathways using both 3T and 4T zeolite cluster models.[90,91] These studies utilized the BP86/DZVP theoretical method, and corrections were made for zero point energies. The authors reached the conclusion that the most energetically favorable route involved the coadsorption of two methanol molecules at the acid site followed by the elimination of a water molecule and the formation of DME via an associative mechanism. The reaction pathway is depicted in Figure 12, from which it is clear that the calculated reaction barrier is only about 15 kJ/mol or 4 kcal/mol with respect to the reactants. This result was interesting because the two other pathways considered each proceeded via a first step of direct dehydration of either one or two methanol molecules adsorbed at the acid site to form a surface methoxy species. When formed from a *single* methanol molecule the surface methoxy group requires a very high (33.5 kcal/mol) activation energy, while this value is reduced to about 9.3 kcal/mol when two methanol molecules are coadsorbed at the acid site. Apparently the surface methoxy group is not necessary for the formation of DME, although its importance in the C-C bond formation reaction is obvious in what follows.

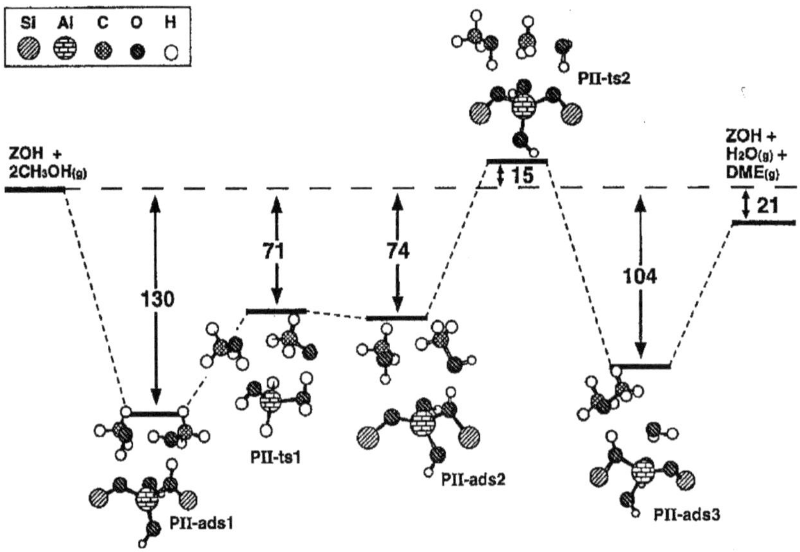

Figure 12. Schematic reaction pathway for dimethyl ether (DME) formation from two adsorbed methanol molecules at a zeolite acid site (ZOH). Relative energies are in kJ/mol. Reprinted with permission from Ref. 90.

The more challenging task of modeling the process of C-C bond formation was taken up shortly afterwards by Blaszkowski and van Santen, who used the same theoretical methods as in their previous studies, but for ease of computation adopted the very small 1T zeolite cluster model.[92] For this reason their energetic results should be viewed with considerable caution. This study computed reaction pathways for the formation of (i) ethanol and (ii) ethyl methyl ether as two candidates for the first species with a C-C bond, and several different mechanisms were considered for each case in order to evaluate which was more energetically favorable. Results showed that a surface methoxy group (presumably formed by initial dehydration of methanol as described above) could combine with methanol to produce ethanol.

The lowest barrier for this reaction, 43.7 kcal/mol with respect to the reactants, occurred via a water-assisted mechanism, compared to a value of 67.9 kcal/mol without water present. In a remarkably similar way, the authors found that DME could combine with the surface methoxy group to form ethyl methyl ether via a water-assisted mechanism with the same activation barrier of 43.7 kcal /mol. An experimental barrier of 46.6 kcal/mol has been deduced with respect to adsorbed methanol.[93] Taking into account the methanol adsorption energy in H-ZSM-5 of roughly 15 kcal/mol, this implies an apparent barrier of about 32 kcal/mol with respect to the gas phase reactants. Despite the inaccuracy of the calculated energies arising from the use of the 1T cluster, the usefulness of water in lowering the barrier for C-C

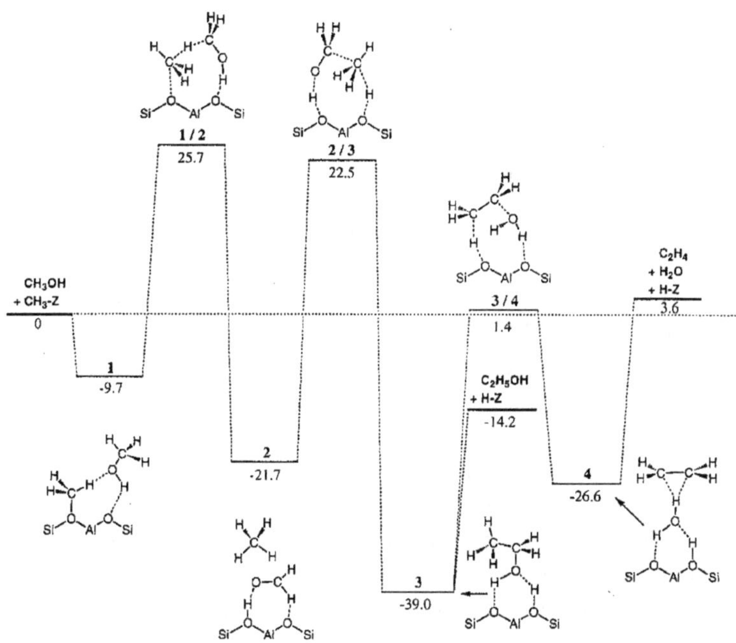

Figure 13. Schematic reaction pathway for formation of first C-C bond in the MTG reaction. Energies are in kcal/mol. Reprinted with permission from Ref. 94.

bond formation reactions was an important result of this study.

An even more realistic computational study was undertaken by Tajima et al.[94] in order to evaluate five different proposed reaction mechanisms for C-C bond formation. The H-ZSM-5 structure was modeled by a 3T cluster in which the terminal Si and hydroxyl O atoms were frozen at experimental positions. Geometry optimizations were carried out at the HF/3-21G level of theory, and final energies were obtained from BLYP/DZVP single-point calculations that also used effective core potentials for the Al and Si atoms.

The most energetically favorable pathway, shown in Figure 13, was for a new mechanism in which methanol initially reacts with the surface methoxy group to form methane and adsorbed formaldehyde. Subsequently, a methane molecule reacts with formaldehyde to regenerate the zeolite acid site and produce ethanol. The activation energy with respect to the initial surface methoxy + methanol system was found to be 25.7 kcal/mol. Blaszkowski and van Santen actually calculated the energy barrier for the formation of methane and formaldehyde to be 32.5 kcal/mol, although they did not map out the subsequent step of ethanol formation. These results leave open the possibility that the rate-limiting step for the reaction may actually be the initial formation of the surface methoxy group, which as noted above has a calculated activation energy of 33.5 kcal/mol from a single methanol.

Two very recent publications present the first attempt to compute reaction pathways for the MTG process using a periodic model of the zeolite ferrierite. Govind et al.[95],[96] used the DMol program with the PBE nonlocal density functional and an all-electron DNP numerical basis set to model the formation of the surface methoxy species and its reaction with methanol to form ethanol. The periodic calculations gave a barrier of 54 (44) kcal/mol for methoxy formation from one (two) methanol molecules, while for the formation of ethanol via a water-assisted mechanism similar to that of Blaszkowski, the barrier was only 25.4 kcal/mol. This indicated that the rate-limiting step for the MTG reaction was indeed the formation of the surface methoxyl species. By comparing the periodic calculations to the finite cluster results of Blaszkowski, we can see a significant influence of the zeolite lattice on the reactivity of the Brønsted acid site. The lattice first causes an *increase* in the activation energy for surface methoxy formation. This is presumably due to steric effects that inhibit its formation in the lattice but are less severe in a finite cluster in which the atomic coordinates have been fully (but unphysically) optimized. However, the lattice actually *reduces* the activation barrier for the formation of ethanol from methanol and the surface methoxy, from 43.7 to 25.4 kcal/mol. This appears to be yet another example of the stabilizing influence of the long-range electrostatic effect of a periodic lattice on an ionic TS structure.

3.7 NO_x Reduction and Decomposition

The potential use of nanoporous catalysts for the reduction or decomposition of nitrogen oxide species formed during fuel efficient "lean-burn" operation of engines has been studied experimentally in great detail since the early 1990s. Zeolites containing the transition metal ions Cu, Co, and Fe have been extensively investigated for this purpose. Progress in this field up to 1995 was nicely summarized by Shelef.[97] Computational quantum chemistry has been used to study these catalysts, but most early efforts focused on determining the coordination and siting of the transition metal ions that are assumed to comprise the active catalytic sites. Only recently have DFT methods been used to investigate complete reaction pathways for NO_x reduction or decomposition at the transition metal sites, and it is these results to which we now turn.

Schneider et al. used a 1T cluster model in which a Cu^+ ion is doubly coordinated to a $Al(OH)4^-$ fragment to represent the active site in a Cu-exchanged zeolite, and calculated a reaction path for NO decomposition to N_2 and O_2 at this site.[98] Geometry optimizations were performed with the LSDA/DZVP method, after which single-point energies were calculated using the BP86/DZVP gradient-corrected DFT method. The pathway, depicted in Figure 14, involves initial adsorption of NO (with the O atom down) at the Cu site (ZCu), followed by reaction with another gas-phase NO molecule to form N_2O and leave behind an O atom bound to the Cu site (ZCuO). Subsequently a gas-phase N_2O recombines with ZCuO to produce N_2 and leave O_2 bound to the Cu site ($ZCuO_2$). The catalytic cycle is completed by desorption of O_2 from the Cu site. The authors found the rate-

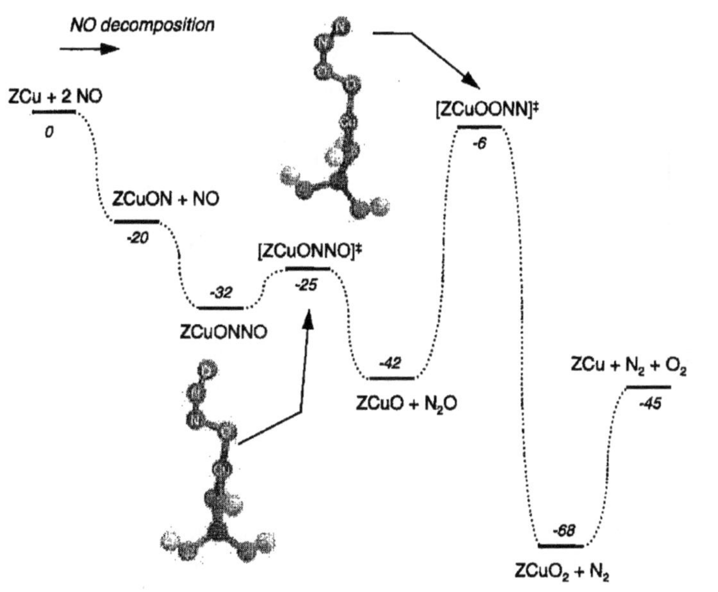

Figure 14. Schematic reaction pathway for catalytic NO decomposition over Cu-exchanged acidic zeolite. Relative energies are in kcal/mol. Reprinted with permission from Ref. 98.

limiting step to be the recombination of N_2O with ZCuO, with an activation energy of 36 kcal/mol, although the TS for this step is still 6 kcal/mol more stable than the initial reactants ZCu + 2NO. A later study by the same authors[99] modeled several additional catalytic reactions of NO_x species in a Cu-exchanged zeolite cluster, including N_2O decomposition and NO oxidation to NO_2.

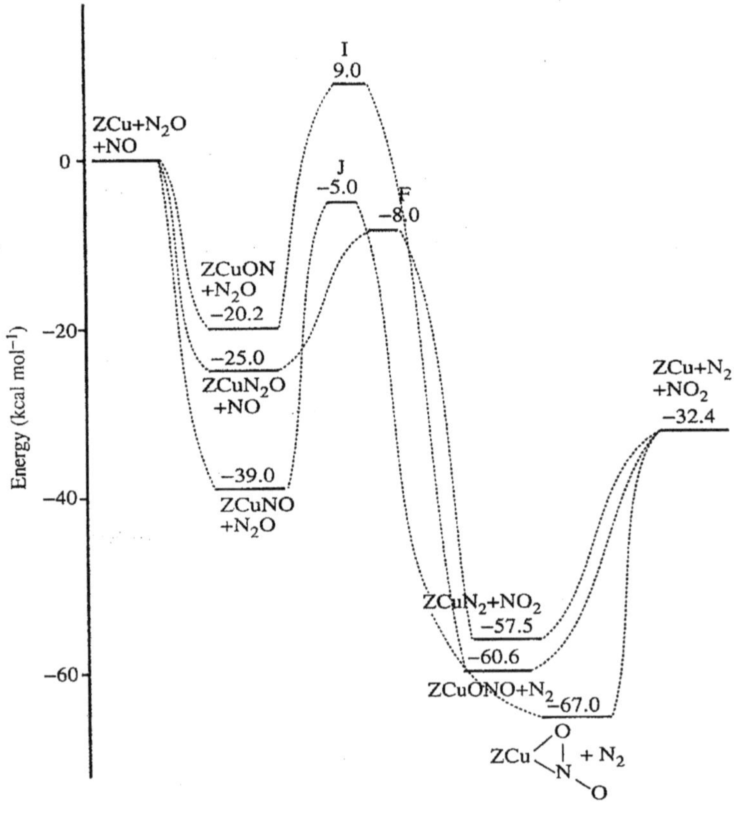

Figure 15. Possible reaction pathways for catalytic N_2O decomposition over Cu-exchanged acidic zeolite. Relative energies are in kcal/mol. Note the most energetically favorable pathway, proceeding through the transition state "F". Reprinted with permission from Ref. [100].

Additional work by this research group was published by Sengupta et al., who explored other possible pathways for the decomposition of N_2O to N_2. in the hope of finding a lower energy reaction channel than that described above.[100] Three new mechanisms were discovered for the ZCu + N_2O + NO reaction, and are illustrated in Figure 15. The three differ from the previous mechanism because in each case once N_2O is formed, its decomposition requires only the reduced ZCu site and not the oxidized site ZCuO. In the reaction channel with the lowest energy TS (F), N_2O initially adsorbs on the ZCu site, after which it reacts with NO to form NO_2, leaving N_2 adsorbed on the ZCu site, from which it subsequently desorbs to close the catalytic cycle.

This process has an activation energy of only 17 kcal/mol with respect to the initially adsorbed N_2O. In both of the other reaction paths, NO is initially adsorbed on the ZCu site, and it is clear that the N-down configuration is much more stable. Each adsorbate then reacts with N_2O to form N_2 and leave NO_2 adsorbed on the ZCu site, with a slightly different structure in each case. The activation energies are much higher for these paths, 29.2 and 34.0 kcal/mol respectively from Figure 15. These energies are comparable to the 36 kcal/mol activation energy initially obtained in Ref. 99. Thus, this series of computational studies has identified a promising candidate for the predominant mechnism for NO decomposition in Cu-exchanged zeolites.

We mention one additional computational treatment of N_2O decomposition in a somewhat different zeolite catalyst, Fe-exchanged ZSM-5. Yakovlev et al. modeled the active site as the oxygen-bridged binuclear Fe complex OH-Fe-O-Fe-OH, in which the Fe atoms were each closely coordinated to two or three framework oxygen atoms in the zeolite cluster model, which was composed of two fused five-membered rings of T atoms.[101] Geometry optimizations were carried out with the VWN/TZVP DFT method, after which energies were calculated with the PW91 nonlocal functional. The positions of terminal H atoms were first optimized in the bare zeolite cluster, and then fixed at those positions for all subsequent calculations of reaction steps. Though this study revealed three potential pathways for N_2O decomposition, only *stable* equilibrium structures were calculated along each reaction channel. Since no transition states connecting these equilibrium structures were determined, definitive conclusions about the preferred pathway could not be drawn. However, this study lays an important foundation for future detailed consideration of this important reaction and how it may differ in Fe-exchanged zeolites compared to the Cu-exchanged form.

3.8 Hydrodesulfurization (HDS) Reactions

Modern environmental regulations have increasingly limited the concentration of sulfur permitted in fuels. Therefore, the development and optimization of desulfurization reactions has emerged as an important field of research in petroleum chemistry and reaction engineering. Nanoporous zeolites have been extensively studied as potential hydrodesulfurization catalysts; in particular, thiophene and dibenzothiophene (DBT) conversion studies have been carried out for ZSM-5 and Y zeolites. However, the details of these acid-catalyzed reactions are poorly understood, and for this reason a few theoretical studies of desulfurization reactions have been undertaken in recent years. These studies are quite ambitious when compared to the modeling of much simpler monomolecular reactions (see sections 3.1-3.3), and well illustrate the level of sophistication reached by modern computational materials chemistry.

The first of these computational studies described calculated reaction pathways for thiophene desulfurization catalyzed by a 3T cluster model of a zeolitic acid site.[102] Thiophene was used as a model compound for this study because it is the smallest sulfur-containing aromatic molecule. Geometry optimizations were performed at the B3LYP/DZV level with no polarization functions in the basis set. Single-point energies including polarizatin functions were calculated in a subsequent step, and corrections for zero-point energies were also included. Two different pathways were considered for the thiophene ring opening and subsequent removal of S, one without H_2 (leading to the formation of butadiyne) and the other with H_2 present (leading to the formation of butadiene). In the absence of H_2 the multistep reaction is characterized by activation barriers of between 50 and 70 kcal/mol and has a very high reaction energy of +72 kcal/mol. These results are consistent with the experimental observation that desulfurization in the absence of H_2 is energetically unfavorable.

The presence of H_2 facilitates the reaction both in the ring opening and the desulfurization steps. A schematic potential energy diagram is shown in Figure 16 and reveals the complexity of this model reaction. Computation of this pathway, which has two minor variants in the step leading to the formation of the butadienethiol intermediate, required the optimization of five major TS structures. The reaction is characterized by reaction barriers of 53 kcal/mol or less and an overall reaction energy of –12 kcal/mol, which makes it thermodynamically much more favorable than when H_2 is absent. In essence the hydrogenation causes a significant stabilization of both the reaction intermediates and the product, butadiene, which is energetically downhill from H_2 + butadiyne.

A more wide-ranging study computational study of thiophene desulfurization was published by Rozanska et al.,[103] who used a nearly identical theoretical method but a larger 4T zeolite cluster. This study examined more carefully the role of the Brønsted acid site in this reaction by computing the reaction pathway for the thiophene ring opening step using not only an acidic ZOH cluster but also a deprotonated ZO^- cluster, a lithium-exchanged ZO^-Li^+ cluster, and a methylated $ZOCH_3$ cluster. This was motivated by the observation that the acidic proton is not involved in the animation of the TS for the this reaction step. The activation energies for this reaction step were found to be 54.0, 47.1, 49.0, and 57.4 kcal/mol for the zeolite models listed above, respectively. These results lead to the conclusion that the ring opening reaction is catalyzed not by a Brønsted acid site but a Lewis base site in the zeolite. This conclusion is consistent with the fact that no protonated TS complexes were identified as part of the reaction pathway.

The authors went on to explore the effect of coadsorbed H_2O and H_2S molecules on the activation energy of the desulfurization reaction. These "partner" molecules facilitated the reaction by minimizing the distortion of

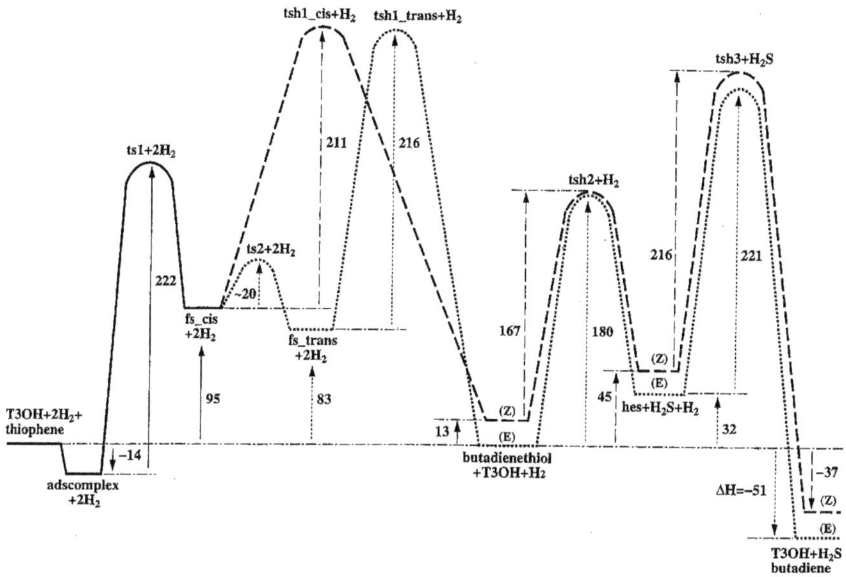

Figure 16. Calculated reaction pathway for desulfurization of thiophene in the presence of H_2. Note that two H_2 molecules are consumed in the reaction. Relative energies are in kJ/mol. Reprinted with permission from Ref. 102.

thiophene in the ring opening reaction and by promoting protonation of the thiophene sulfur. The activation energy was lowered by about 6 and 2 kcal/mol by the presence of H_2O and H_2S, respectively. Finally, the influence of hydrogenation of the thiophene prior to the ring opening reaction step was investigated by calculating the reaction pathways for dihydrothiophene as well as tetrahydrothiophene. In the former case, the activation energy is about 3 kcal/mol *less* than for thiophene, while in the latter case the activation energy is *higher* than for thiophene by a smiliar amount. These differences alone cannot explain the experimentally observed enhancement of desulfurization upon prehydrogenation treatment, but they reveal that the degree of hydrogenation has a significant influence on the magnitude of the enhancement.

The most recent computational treatment of HDS reactions was carried out by Rozanska et al. for the hydrogenation and desulfurization of dibenzothiophene (DBT).[104] Again a 4T cluster model was used, but now the B3LYP/6-31G(d) level of theory included the influence of polarization functions on the non-hydrogen atoms in the system. The direct desulfurization reaction *without* a prehydrogenation step was found to be energetically unfavorable, with a relatively high activation barrier of 68.8 kcal/mol. By contrast, the desulfurization reaction is greatly enhanced by the prehydrogenation of DBT to form hexahydro-DBT, from which sulfur can be removed just as from DBT. All intermediates and TS structures along the pathway are lower in energy for hydrogenated DBT, leading to an overall activation energy of only 56.2 kcal/mol. This is comparable to the activation energy required for the prehydrogenation step, and demonstrates the effectiveness of such a process.

All of the HDS studies described above relied on a finite cluster model of the zeolite framework and thus are unable to distinguish between the reactivity of different framework types. We have seen the promise of periodic DFT-based methods for determining electronic structure in the studies presented in sections 3.4-3.6, and such an approach was recently applied to the study of isomerization reactions among methylated DBT derivatives in acidic mordenite zeolite.[105] Such reactions are regarded as being helpful in facilitating HDS catalysis among these species, since it is assumed that there is a shape-selectivity in the HDS process that favors certain isomers over others. The authors used the VASP plane-wave pseudopotential package to consider isomerization reactions for 4,6-dimethyl DBT and 4-methyl DBT, derivatives that are known to be resistant to the HDS reaction. Using the NEB method for reaction path following, they found in each case that isomerization to a product which is more amenable to HDS could occur with a relatively small activation energy (32.7 kcal/mol for 4,6-dimethyl DBT and 36.1 kcal/mol for 4-methyl DBT), and that the barriers for the reverse reactions were higher by 3-4 kcal/mol. These results are consistent with the idea that the HDS process is shape-selective and that an acidic zeolite can promote reactions to isomers that more readily undergo desulfurization, since the methyl group(s) no longer hinder the thiophenic sulfur atom from accessing catalytic sites of the HDS catalyst. This result again demonstrates the ability of second-generation computational methods to calculate reaction pathways in a realistic model that goes beyond the finite cluster model by including the influence of the entire zeolite framework.

3.9 Reactions in Carbon Nanotubes

This chapter has dealt almost exclusively with computational investigations of molecular reactivity in one particular family of nanoporous solids; namely, zeolites. This is primarily due to the industrial importance of zeolite catalysis in the petroleum and fine chemicals industries and the extensive research this has produced. With the discovery and emergence of carbon nanotubes as a promising template for materials synthesis (quantum wires, for example), a potentially unique surface for designing catalytic sites, and a possible nanoscale sensor for chemical and biological detection, a need has arisen to understand the structure and chemical reactivity of these new materials. Two very recent quantum chemical investigations of specific reaction paths in the carbon nanotube environment are the harbingers of many more such studies to come.

In the first study the ONIOM embedded cluster method was used to investigate the amidation of terminal carboxylic groups at the tips of single-walled carbon nanotubes (SWNTs).[106] The quantum regime (atoms near the tip) was treated at the B3LYP/6-31G(d) level of theory, while the remaining SWNT atoms were treated with universal force field (UFF) molecular mechanics. Two different nanotube geometries were considered, the so-called zig-zag (10,0) and armchair (5,5) SWNTs. Activation energies relative to the adsorbed reactants were found to be 54.0 kcal/mol (armchair) compared to 62.2 kcal/mol (zigzag). These results indicated that direct amidation is significantly more advantageous on the armchair tips than on the zigzag tips, and thus the reactivities of different types of nanoporous structures could be differentiated.

Halls and Schlegel[107] recently undertook a similar study to investigate the influence of the SWNT environment on the reaction energetics of the simplest Menshutkin S_N2 reaction, in which an amine is alkylated by an alkyl halide:

$$NH_3 + CClH_3 \rightarrow NH_3CH_3^+ + Cl^-$$

First the reaction path was calculated in the gas phase at the B3PW91/6-31+G(d) level of theory. Next, the optimized geometries of the reactants, TS structures, and intermediates were placed at the center of a SWNT (whose geometry was optimized at the B3PW91/3-21G level of theory) and single-point energies were calculated using the basis sets and functional used for the original optimization of each portion of the system. Inside the two different SWNTs considered ((8,0) and (9,0)), the reaction endothermicity decreased by 27 and 23 kcal/mol, respectively. In addition, the activation energy decreased by 13 and 12 kcal/mol, respectively. The SWNTs are known to have a large electronic polarizability, and since the reaction involves a large degree of charge separation along the reaction coordinate, the SWNT environment tends to stabilize both TS and product structures compared to the reactants. This is closely analogous to the effect zeolites have on the activation energies of some hydrocarbon reactions, as discussed in sections 3.1-3.3.

4. CONCLUSIONS

We have reviewed the literature describing ab initio and DFT-based computational studies of molecular reactivity in nanoporous materials, focusing primarily on catalytic reactions in zeolite frameworks. Improvements in computing power and algorithmic design have enabled computational chemists and materials scientists to calculate stationary points along reaction pathways for a wide variety of reactions from simple monomolecular hydrocarbon transformations to more complex bimolecular reactions such as the MTG process. Models of the nanoporous solid have progressed from finite cluster descriptions of a generic active site to embedded cluster techniques and most recently to periodic methods that completely specify a particular crystalline environment. The latter models have made it possible to compare the reactivities of structurally distinct nanoporous architectures, and several examples of such studies were presented. In what follows we will state some conclusions suggested by this literature review that will be of interest to future practitioners of computational reactivity studies and to those who read their work.

1. Optimizing the molecular geometry of stable equilibrium structures is a fairly routine computational task, but the determination of TS (i.e., unstable equilibrium) structures along a particular reaction pathway has been difficult, except for reactions with a very simple reaction coordinate involving the motion of only one or two atoms. However, the development of new algorithms such as the NEB method and their use in combination with traditional electronic structure packages (GAUSSIAN, ADF, VASP, etc.) have made it much easier to locate approximate TS structures and indeed to map out a complete minimum energy reaction pathway. We believe that such methods will become the "state-of-the-art" in computational materials chemistry in the near future. They have the promise of allowing the routine exploration of reaction pathways for even very complex reactions, and thus will facilitate the calculation of activation energies that can be used to compare the energetics of different pathways, as well as the intrinsic reactivities of different nanoporous environments.

2. Many studies reviewed in this chapter have revealed the significant energetic impact of long-range interactions between the nanoporous solid lattice and reactant molecules on reaction activation energies and relative stabilities of intermediates and product species. This effect is most significant when the reaction involves large degrees of charge transfer and proceeds through TS complexes with strong ionic character. It is encouraging that embedded cluster and periodic treatments of some reactions have greatly improved the agreement between computed and experimental activation energies. This highlights the need to use embedded cluster or periodic models if one wishes to study these types of reactions.

3. In performing geometry optimizations for finite cluster structural models of a nanoporous material, one must choose whether to do a full optimization or a calculation using some constraints intended to represent the structural features of the material being studied. In the former case, it is easier to carry out the optimization of TS structures with only one imaginary frequency, but the model will of necessity be generic in nature and will not represent a unique crystal structure. In the latter, the local structure of the crystal in the neighborhood of the active site will be preserved, but the reaction path search and normal mode analysis will be complicated by the presence of many imaginary frequency modes resulting from the absence of full optimization to an equilibrium structure. Such are the limits imposed by the finite cluster model. Fortunately these problems are resolved by a periodic model in which a full geometry optimization is performed and the lattice structure is faithfully modeled. Of course the periodic model requires a judicious choice of unit cell size in order to model a particular degree of active site coverage, but this is not a serious difficulty. However, nanoporous solids with large numbers of atoms in the unit cell are very computationally demanding to study in a periodic approach, and their routine study must await future increases in computer speed.

4. Finite cluster models of the nanoporous solid are still valuable and can be usefully combined with embedded cluster or periodic models in the process of reaction path searching. For example, in order to use the NEB method for determining the minimum energy reaction pathway connecting specific reactant and product structures, a series of "image" structures must be defined at points along the hypothetical reaction coordinate. For computational efficiency of the NEB method, which must execute the electronic structure code many hundreds of times during the search process, it is especially helpful to have good initial guesses for these structures. Several of the studies reviewed here used finite cluster models to generate a TS structure that was then used as the initial guess for the TS image in the NEB search. Thus there can be a valuable synergy between finite and periodic models, although the final results of the periodic models are the most accurate.

ACKNOWLEDGEMENTS

This work is supported in part by the U.S. Department of Energy, BES-Materials Sciences, under Contract W-31-109-ENG-38

REFERENCES

1 J. Sauer, P. Ugliengo, E. Garrone, and V. R. Saunders, *Chem. Rev.* **94**, 2095 (1994).
2 G. J. Kramer and R. A. van Santen, *Chem. Rev.* **95**, 637 (1995).
3 S. P. Bates and R. A. van Santen, *Adv. Catal.* **42**, 1 (1998).
4 R. A. van Santen, *Catal. Today* 38, 377 (1997).
5 V. B. Kazansky, *Catal. Today* 51, 419 (1999).
6 E. V. Stefanovich and T. N. Truong, *J. Phys. Chem B*, **102**, 3018 (1998).
7 S. A. Zygmunt, L. A. Curtiss, P. Zapol, and L. E. Iton, *J. Phys. Chem. B* **104** 1944 (2000).
8 J. Shoemaker, L. W. Burggarf and M. S. Gordon, *J. Chem. Phys.* **112**, 2994 (2000).
9 See for example, N. Govind, J. Andzelm, K. Reindel, and G. Fitzgerald, *Int. J. Mol. Sci.* **3**, 423 (2002).
10 G. Kresse and J. Furthmuller, Phys. B **54**, 11169 (1996).
11 F. Jenson, *Introduction to Computational Chemistry*, John Wiley, New York, 1999.
12 W. J. Hehre, L. Radom, J. A. Pople, and P. v. R. Schleyer, *Ab Initio Molecular Orbital Theory*; John Wiley, New York, 1987.
13 W. Kohn, A. D. Becke, and R. G. Parr, *J. Phys. Chem.*, **100**, 12974 (1996).
14 J. C. Slater, *The Self-Consistent Field for Molecules and Solids: Quantum Theory of Molecules and Solids*, Vol. 4, McGraw-Hill, New York, 1974.
15 S. H. Vosko, L. Wilk, and M. Nusair, *Can. J. Phys.*, **58**, 1200 (**1980**).
16 A. D. Becke, *Phys. Rev.*, **A38**, 3098 (1988).
17 J. P. Perdew and Y. Wang, *Phys. Rev.*, **B45**, 13244 (1992).
18 C. Lee, C. Yang, and R. G. Parr, *Phys. Rev.*, **B37**, 785 (1988).
19 L. A. Curtiss, K. Raghavachari, G. W. Trucks, and J. A. Pople, *J. Chem. Phys.*, **94**, 7221 (1991).
20 A. D. Becke, *J. Chem. Phys.*, **98**, 5648 (1993).
21 P. J. Stephens, F. J. Devlin, C. F. Chabalowski, and M. J. Frisch, *J. Phys. Chem.*, **98**, 11623 (1994).
22 B. G. Johnson, P. M. W. Gill, and J. A. Pople, *J. Chem. Phys.*, **98**, 5612 (1993).
23 C.W. Bauschlicher, *Chem. Phys. Lett.*, **246**, 40 (1995).
24 L. A. Curtiss, K. Raghavachari, P. C. Redfern, and J. A. Pople, *J. Chem. Phys.*, **106**, 1063 (1997).
25 G. Mills and H. Jonsson, *Phys. Rev. Lett.* **72**, 1124 (1994).
26 G. Henkelman and H. Jonsson, *J. Chem. Phys.* **113**, 9978 (2000).
27 V. B. Kazansky, I. N. Senchenya, M. Frash, and R. A. van Santen, *Catal. Lett.* **27**, 345 (1994).
28 V. B. Kazansky, M. V. Frash, and R. A. van Santen, *Catal. Lett.* **28**, 211 (1994).
29 S. J. Collins and P. J. O'Malley, *Chem. Phys. Lett.* **246**, 555 (1995).
30 H. Kranilla, W. O. Haag, and B. C. Gates, *J. Catal.* **135**, 115 (1992).
31 S. R. Blaszkowski, M. A. C. Nascimento, and R. A. van Santen, *J. Phys. Chem* **100**, 3463 (1996).
32 T. F. Narbeshuber, H. Vinek, and J. A. Lercher, *J. Catal.* **157**, 388 (1995).
33 V. B. Kazansky, M. V. Frash, and R. A. van Santen, *Appl. Catal. A* **146**, c225 (1996).
34 T. F. Narbeshuber, A. Brait, K. Seshan, and J. A. Lercher, *J. Catal.* **172**, 127 (1997).
35 Y. W. Bizreh and B. C. Gates, *J. Catal.* **88**, 240 (1984).
36 S. Yanping and T. C. Brown, *J. Catal.* **194**, 301 (2000).
37 C. Stefanadis, B. C. Gates, and W. O. Haag, *J. Mol. Catal.* **67**, 363 (1991).
38 A. M. Rigby, G. J. Kramer, and R. A. van Santen, *J. Catal.* **170**, 1 (1997).

39 S. A. Zygmunt, L. A. Curtiss, P. Zapol, and L. E. Iton, *J. Phys. Chem. B* **104** 1944 (2000).
40 M. V. Frash, V. B. Kazansky, A. M. Rigby, and R. A. van Santen, *J. Phys. Chem. B* **102**, 2232 (1998).
41 G. Yaluris, J. E. Rekoske, L. M. Aparicio, R. J. Madon, and J. A. Dumesic, *J. Catal.* **153**, 54 (1995).
42 A. Corma and B. W. Wojciechowski, *Catal. Rev. Sci. Eng.* **27**, 29 (1985).
43 P. J. Hay, A. Redondo, and Y. Guo, *Catal. Today* **50**, 517 (1999).
44 S. R. Blaszkowski, A. P. J. Jansen, M. A. C. Nascimento, and R. A. van Santen, *J. Phys. Chem.* **98**, 12938 (1994).
45 I. Milas and M. A. C. Nascimento, *Chem. Phys. Lett.* **338**, 67 (2001).
46 E. A. Furtado, I. Milas, J. O. M. A. Lins, and M. A. C. Nascimento, *Phys. Stat. Solidi (a)* **187**, 275 (2001).
47 A. Corma, P. J. Miguel, and A. V. Orchiles, *J. Catal.* **145**, 171 (1994).
48 J. R. Mowry, R. F. Anderson, and J. A. Johnson, *Oil Gas J.* **83**, 128 (1985).
49 M. V. Frash and R. A. van Santen, *J. Phys. Chem. A* **104**, 2468 (2000).
50 R. C. Binning and L. A. Curtiss, *J. Comput. Chem.* **11**, 1206 (1990).
51 J. Bandiera and Y. B. Taarit, *Appl. Catal. A* **152**, 43 (1997).
52 M. V. Frash and R. A. van Santen, *Phys. Chem. Chem. Phys.* **2**, 1085 (2000).
53 (a) A. J. H. Wachters, *J. Chem. Phys.* **52**, 1033 (1970); (b) P. J. Hay, *J. Chem. Phys.* **66**, 4379 (1977); (c) K. Raghavachari and G. W. Trucks, *J. Chem. Phys.* **91**, 1062 (1989).
54 A. A. Shubin, G. M. Zhidomirov, A. L. Yakovlev, and R. A. van Santen, *J. Phys. Chem. B* **105**, 4928 (2001).
55 A. L. Yakovlev, A. A. Shubin, G. M. Zhidomirov, and R. A. van Santen, *Catal. Lett.* **70**, 175 (2000).
56 G. J. Kramer, R. A. van Santen, C. A. Emeis, and A. K. Nowak, *Nature* **363**, 529 (1993).
57 E. M. Evleth, E. Kassab, and L. R. Sierra, *J. Phys. Chem.* **98**, 1421 (1994).
58 Q. Zhang, R. Bell, and T. N. Truong, *J. Phys. Chem.* **99**, 592 (1995).
59 G. J. Kramer and R. A. van Santen, *J. Am. Chem. Soc.*, **117**, 1766 (1995).
60 P. M. Esteves, M. A. C. Nascimento, and C. J. A. Mota, *J. Phys. Chem. B* **103**, 10417 (1999).
61 A. G. Stepanov, H. Ernst, and D. Freude, *Catal. Lett.* **54**, 1 (1998).
62 J. Sommer, M. Hachoumy, F. Garin, and D. Barthomeuf, *J. Am. Chem. Soc.* **116**, 5491 (1994).
63 J. A. Ryder, A. K. Chakraborty, and A. T. Bell, *J. Phys. Chem.* B **104**, 6998 (2000).
64 L. W. Beck, T. Xu, J. B. Nicholas, J. F. Haw, *J. Am. Chem. Soc.* **117**, 11594 (1995).
65 A. H. deVries, P. Sherwood, S. J. Collins, A. M. Rigby, M. Riguto, and G. J. Kramer, *J. Phys. Chem. B* **103**, 6133 (1999).
66 J. M. Vollmer and T. N. Truong, *J. Phys. Chem. B* **104**, 6308 (2000).
67 V. B. Kazansky and I. N. Senchenya, *J. Catal.* **119**, 108 (1989).
68 I. N. Senchenya and V. B. Kazansky, *Catal. Lett.* **8**, 317 (1991).
69 L. R. Sierra, E. Kassab, and E. M. Evleth, *J. Phys. Chem.* **97**, 641 (1993) (see ref. 16 of this paper).
70 P. Viruela-Martin, C. M. Zicovich-Wilson, and A. Corma, *J Phys. Chem.* **97**, 13713 (1993).

71 E. M. Evleth, E. Kassab, H. Jessri, M. Allavena, L. Montero, and L. R. Sierra, *J. Phys. Chem.* **100**, 11368 (1996).
72 V. B.Kazansky, M. V. Frash, and R. A. van Santen, *Appl. Catal. A* **146**, 225 (1996).
73 A. M. Rigby and M. V. Frash, *J. Mol. Catal. A* **126**, 61 (1997).
74 P. E. Sinclair, A. de Vries, P. Sherwood, C. R. A. Catlow, and R. A. van Santen, *J. Chem. Soc. Faraday Trans.* **94**, 3401 (1998).
75 ChemShell is a package of molecular simulation routines written by P. Sherwood and A. de Vries, copyright The Council for the Central Laboratory of the Research Council, Daresbury Laboratories, UK.
76 M. Boronat, P. Viruela, and A. Corma, *J. Phys. Chem. A* **102**, 982 (1998).
77 M. Boronat, C. M. Zicovich-Wilson, P. Viruela, and A. Corma, *Chem. Eur. J.* **7**, 1295 (2001).
78 M. Boronat, P. Viruela, and A. Corma, *Phys. Chem. Chem. Phys.* **3**, 3235 (2001).
79 M. Boronat, C. M. Zicovich-Wilson, P. Viruela, and A. Corma, *J. Phys. Chem. B* **105**, 11169 (2001).
80 X. Rozanska, T. Demuth, F. Hutschka, J. Hafner, and R. A. van Santen, *J. Phys. Chem. B* **106**, 3248 (2002).
81 V. B. Kazansky, M. V. Frash, and R. A. van Santen, *Catal. Lett.* **48**, 61 (1997).
82 M. Boronat, C. M. Zicovich-Wilson, A. Corma, and P. Viruela, *Phys. Chem. Chem. Phys.* **1**, 537 (1999).
83 P. Sherwood, A. H. de Vries, S. J. Collins, S. P. Greatbanks, N. A. Burton, M. A. Vincent, and I. H. Hillier, *Faraday Discuss.* **106**, 79 (1997).
84 M. V. Frash, V. B. Kazansky, A. M. Rigby, and R. A. van Santen, *J. Phys. Chem. B* **101**, 5346 (1997).
85 X. Rozanska, X. Saintigny, R. A. van Santen, F. Hutschka, *J. Catal.* **202**, 141 (2001).
86 A. M. Vos, X. Rozanska, R. A. Schoonheydt, R. A. van Santen, F. Hutschka, and J. Hafner, *J. Am. Chem. Soc.* **123**, 2799 (2001).
87 X. Rozanska, R. A. van Santen, F. Hutschka, and J. Hafner, *J. Am. Chem. Soc.* **123**, 7655 (2001).
88 X. Rozanska, X., R. A. van Santen, F. Hutschka, *J. Phys. Chem. B* **106**, 4652 (2002).
89 C. D. Chang, and A. J. Silvestri, *J. Catal.* **47**, 249 (1977).
90 S. R. Blaszkowski, and R. A. van Santen, *J. Am. Chem. Soc.* **118**, 5152 (1996).
91 S. R. Blaszkowski, and R. A. van Santen, *J. Phys. Chem. B* **101**, 2292 (1997).
92 S. R. Blaszkowski, and R. A. van Santen, *J. Am. Chem. Soc.* **119**, 5020 (1997).
93 M. Jayamurthy and S. Vasuvedan, *Catal. Lett.***36**, 111 (1996).
94 N. Tajima, T. Tsuneda, F. Toyama, and K. Hirao, *J. Am. Chem. Soc.* **120**, 8222 (1998).
95 N. Govind, J. Andzelm, K. Reindel, and G. Fitzgerald, *Int. J. Mol. Sci.* **3**, 423 (2002).
96 J. Andzelm, N. Govind, G. Fitzgerald, and A. Maiti, *Int. J. Quant. Chem.* **91**, 467 (2003)
97 M. Shelef, *Chem. Rev.* **95**, 209 (1995).
98 W. F. Schneider, K. C. Hass, R. Ramprasad, and J. B. Adams, *J. Phys. Chem. B* **101**, 4353 (1997).

99 W. F. Schneider, K. C. Hass, R. Ramprasad, and J. B. Adams, *J. Phys. Chem. B* **102**, 3692 (1998).
100 D. Sengupta, J. B. Adams, W. F. Schneider, and K. C. Hass, *Catal. Lett.* **74**, 192 (2001).
101 A. L. Yakovlev, G. M. Zhidomirov, and R. A. van Santen, *J. Phys. Chem. B* **105**, 12297 (2001).
102 X. Saintigny, R. A. van Santen, S. Clemendot, and F. Hutschka, *J. Catal.* **183**, 107 (1999).
103 X. Rozanska, R. A. van Santen, and F. Hutschka, *J. Catal.* **200**, 79 (2001).
104 X. Rozanska, X. Saintigny, R. A. van Santen, S. Clemendot, and F. Hutschka, J. Catal. **208**, 89 (2002).
105 X. Rozanska, R. A. van Santen, F. Hutschka, and J. Hafner, *J. Catal.* **205**, 388 (2002).
106 V. A. Basiuk, E. V. Basiuk, and J. Saniger-Blesa, *Nano Lett.* **1**, 657, (2001).
107 M. D. Halls and H. B. Schlegel, *J. Phys. Chem. B* **106**, 1921 (2002).

Chapter 6

THEORETICAL METHODS FOR MODELING CHEMICAL PROCESSES ON SEMICONDUCTOR SURFACES

J. A. Steckel[a] and K. D. Jordan[b]

[a]*National Energy Technology Laboratory, United States Department of Energy, P. O. Box 10940, Pittsburgh, PA, 15236-0940*

[b]*Department of Chemistry and Center for Molecular and Materials Simulations, University of Pittsburgh, Pittsburgh, PA 15260*

1. INTRODUCTION

Electronic structure calculations are playing an increasingly important role in characterizing the chemistry of semiconductor surfaces. This is a consequence of the rapidly growing power of computers as well as of major advances on the algorithmic front. In this article we will review the major approaches commonly used in modeling semiconductor surfaces, taking care to delineate both the strengths and weaknesses of various approaches. The examples will be drawn from studies of the Si(001) surface. However, the issues considered and the strategies discussed are applicable to semiconductor surfaces in general.

Two different approaches – one using slab models with periodic boundary conditions and the other using cluster models – have evolved for modeling surfaces. Historically, the former has been used predominantly by the physics community and the latter has been more prevalent in the theoretical chemistry community. However, over the past few years this distinction has begun to blur with each community of researchers increasingly adopting the geometrical model that was originally the purview of the other.

L.A. Curtiss and M.S. Gordon (eds.),
Computational Materials Chemistry: Methods and Applications, 246–265.

Although most electronic structure calculations using cluster models are carried out with Gaussian or other atom-centered localized basis functions, both localized-orbital and plane-wave basis sets are used with slab models.

Depending on the nature of the surface and on the issues being addressed, the choice of an appropriate geometrical model can prove to be crucial. Use of too small a cluster or slab model may preclude the treatment of some reaction pathways or introduce large errors in the calculated quantities of interest. It is therefore imperative to carefully assess the strengths and weaknesses of the geometrical model as well as the electronic structure method used in modeling surface processes.

2. GEOMETRICAL MODELS

Both cluster models and slab-models have been employed to study chemical processes on semiconductor surfaces. Each approach has advantages and disadvantages. To illustrate these we consider the bare Si(001) surface, which is the most important semiconductor surface in terms of device applications. This surface adopts a (2 x 1) reconstruction with rows of SiSi dimers as shown in Fig. 1. Until recently, it was widely accepted that

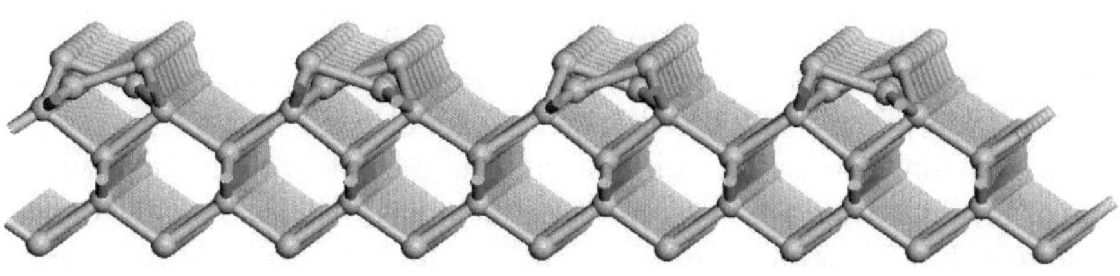

Figure 1. Si(001)-(2 x 1) surface, depicting the rows of surface dimers. The dimers are shown with a buckled arrangement. Five layers of Si atoms are shown.

the surface is further stabilized by buckling of the dimers.[1-3] Different buckling patterns are possible, with the most important being shown in Fig. 2. Results from scanning tunneling microscopy (STM) studies have been interpreted in terms of buckled dimers, which rapidly switch orientation at room temperature leading to an averaged symmetric appearance in STM images, but at lower temperatures become locked into an antisymmetric, buckled configuration.[3-7] There are two reports that, at very low temperatures (*e.g.*, T~5 K), the STM images associated with the dimers appear to revert back to a symmetric structure,[8-9] although still more recent results from STM,[10,11] photoelectron[12] and AFM[13] studies at similar temperatures have been interpreted in terms of buckled dimers.

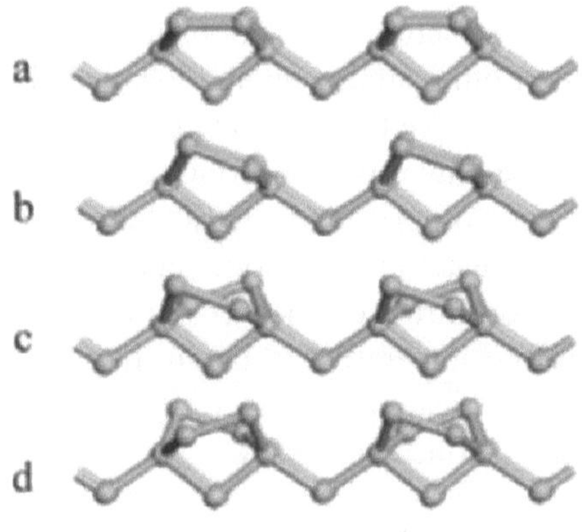

Figure 2. Si(001) surface with various possible buckling arrangements. a) unbuckled, b) (2 x 1) buckling arrangement, c) p(2 x 2) buckling arrangement, and d) c(4 x 2) buckling arrangement.

Density functional calculations using both cluster and slab models give buckled surface dimers, with the energy lowering associated with buckling being ~0.1 eV/dimer.[14-19] Diffusion Monte Carlo calculations using cluster models also give a buckled surface, but with the stabilization associated with buckling being less than that predicted by DFT calculations.[20] In contrast, multiconfiguration self-consistent-field (MCSCF) calculations on cluster models of the Si(001) surface do not give buckled dimers.[21-24]

Each Si atom of a surface dimer forms three bonds: one to the other Si atom of the dimer and two to second-layer Si atoms. This leaves each dimer Si atom with a "dangling bond" with an unpaired electron. Alternatively, one can view the dimers as having weak π bonds, with small π/π^* energy gaps.[21] These small gaps lead to considerable diradical character, and appreciable configuration mixing in MCSCF calculations. This has led some researchers to question the suitability of density functional methods for describing the surface dimers and for addressing the question of surface buckling.[22-24] The history of this problem is discussed by Gordon in another chapter in this volume. The question as to whether the ideal Si(001) surface is buckled at low temperatures will need further experimental and theoretical work to resolve. Here we note that even if the ideal reconstructed surface is not buckled, there is compelling evidence that various defects can induce buckling over extended regions.[3,4,25] In the following discussion, intended primarily to illustrate factors that must be considered in designing cluster models for surfaces, it will be assumed that the surface is buckled.

The simplest "reasonable" cluster model for describing the Si(001)-(2 x 1) surface is the Si_9H_{12} cluster depicted in Fig 3a. Cluster models similar to this have been used in many early theoretical studies of the Si(001) surface.[15,19,21,26-31] This model has one surface dimer, four Si atoms in the second layer, two in the third, and one in the fourth and employs H atoms to terminate the dangling bonds of the subsurface Si atoms. Several problems with the use of such a small cluster model are immediately obvious. These include: (1) unrealistic charge distributions resulting from the use of the terminating H atoms in place of the neighboring Si atoms of the true surface, (2) ambiguities as to how to choose the geometry, and (3) the inappropriateness of small cluster methods for describing long-range interactions, *e.g.*, intradimer interactions, which would, for example, be crucial in establishing the preferred buckling arrangement.

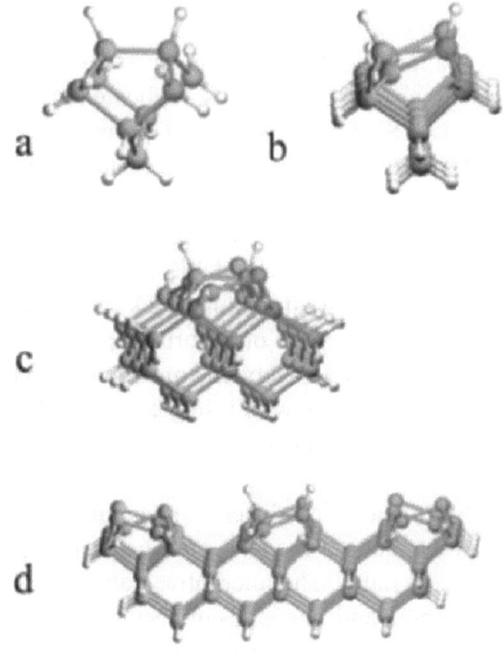

Figure 3. Cluster models for studying H_2 desorption from a dimer on the Si(001) –(2 x 1) surface. a) Si_9H_{14} model with one surface dimer, b) $Si_{23}H_{22}$ model with three surface dimers along a row, c) $Si_{41}H_{38}$ model with three surface dimers along a row, d) $Si_{89}H_{62}$ cluster model with nine surface dimers, three in each of three adjacent rows.

Full, unconstrained geometry optimizations using small cluster models can result in surface structures appreciably distorted from those of the actual surfaces of interest. This has led some researchers using cluster models to impose constraints in the geometry optimizations, for example, fixing the lower layer Si atoms in the cluster at the ideal crystallographic positions for bulk silicon or allowing relaxation to occur only along selected axes.[19,26,27,30] Unfortunately, it is not clear how to choose the constraints so as to produce a cluster model that best mimics the actual surface. In spite of its obvious shortcomings, the Si_9H_{12} cluster model has been employed for examining a wide range of processes on the Si(001) surface, *e.g.*, it has been used by several groups to study H_2 desorption from a single dimer site on the surface.[27,29,30,32]

The problems raised above can be addressed by monitoring the convergence of the quantities of interest with increasingly large cluster models such as those depicted in Figures 3b-3d.[18,33] Cluster models have an advantage over slab models in that they may be used with any electronic structure method, although highly computationally demanding methods such as CCSD(T)[34-37] are prohibitive for large cluster models. Cluster models are also well suited for use with QM/MM methods such as the SIMOMM method of Shoemaker *et al.*[38]

An alternative approach to modeling surfaces is to use slab models, which make use of a supercell that is replicated by the use of periodic boundary conditions. Slab models are not prone to edge effects than are cluster models. However, the choice of too small a supercell can result in significant errors in geometries and energetics and may preclude description of some processes. For example, a supercell containing only a single SiSi dimer in the surface layer is not suitable for addressing the problem of whether the surface adopts an alternating buckling pattern. Similar considerations arise in describing defects or adsorbed species. Consider, for example, modeling the binding of an acetylene molecule atop a SiSi dimer.[39] The supercell, depicted in Figure 4a, with a single surface dimer and an adsorbed acetylene molecule would be most appropriate for describing the surface at full coverage (*i.e.*, with one C_2H_2 molecule per dimer). To the extent that the adsorbed C_2H_2 molecules on different sites interact either directly or indirectly *via* the lattice, such a model would be inappropriate for describing the surface at low coverage. The alternative supercell, shown in Figure 4b, with four surface dimers and one C_2H_2 molecule, corresponds to the 25 ML coverage limit, and is a more appropriate model for the medium-to-low-coverage limit.[39] However, to be sure that such a supercell is appropriate for describing low coverages, it is necessary to calculate the properties of interest using still larger supercells to establish convergence with respect to supercell size.

A related issue in constructing supercells for modeling surface processes is the number of layers of atoms to employ. Shifts in the positions of Si atoms (with respect to their bulk positions) have been detected for atoms down to the sixth layer for the Si(001)-(2 x 1) surface,[40,41] and introduction of a defect or adsorption of an atom or molecule can induce sizable geometrical distortions through several layers of a semiconductor surface. A few theoretical studies employing slab models to study reactions on semiconductor surfaces have allowed the reaction to proceed on both the top and bottom surfaces.[42,43] However, it has been more common to have the process of interest occur on only one of the two surfaces.[18,33,44-46] In the latter case, the dangling bonds on the bottom semiconductor layer are generally terminated with H atoms. Many slab-model calculations on semiconductor

a

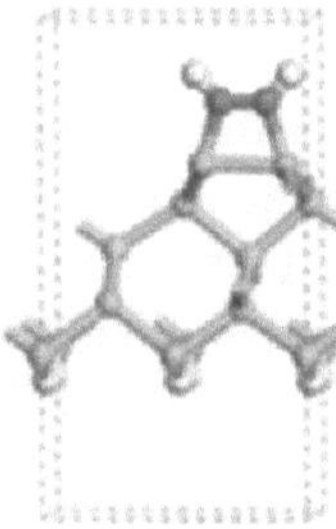

b

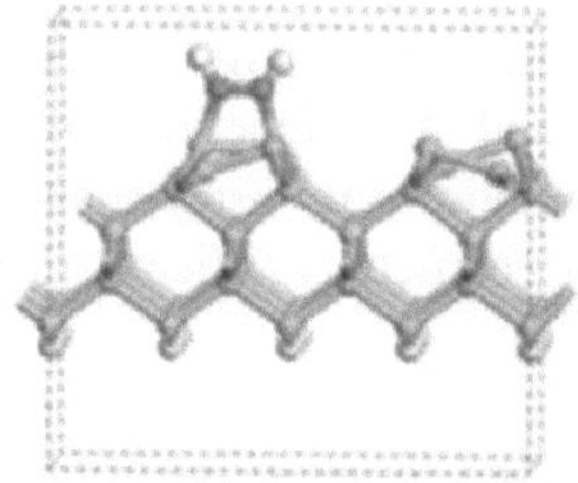

Figure 4. Two supercells for modeling acetylene adsorption on the Si(001)-(2 x 1) surface. a) model with a single surface dimer, corresponding to saturation coverage, and b) model with 25% surface coverage.

surfaces have employed five or six layers of semiconductor atoms with the bottommost one or two layers of atoms being frozen in their bulk positions. In cases such as the Si(001) surface, where there is a significant surface reconstruction down to the sixth layer, such slab models cannot fully recover the effects of geometrical relaxation. Still, depending on the process of interest, the error introduced into the reaction and activation energies due to the adoption of such models may be acceptable. Again, it is important to estimate the error introduced by using a slab model with a given number of layers by comparison with the results of test calculations with models employing additional layers.

Another issue concerning the use of slab models is that, when used with plane-wave basis sets, it is necessary (at least in commonly employed codes such as VASP,[47-49] Dacapo,[50] and CASTEP[51]), to replicate the system in the direction perpendicular to the surface. This requires introducing a vacuum layer in the supercell, (see Fig. 4). The vacuum layer has to be thick enough so that the atoms on the top surface in one cell do not interact appreciably with the atoms on the bottom surface of the replicated slab located above it. Generally, the thickness of the vacuum layer is chosen to be on the order of 8 Å. The introduction of the vacuum layer is accompanied by an increase in the size of the plane-wave basis set, thereby increasing the cost of such calculations.

3. BASIS SETS

With few exceptions, cluster-model calculations are carried out with localized, generally atom-centered, Gaussian type orbital (GTO) or numerical basis sets. On the other hand, both plane-wave and localized function basis sets are commonly used with slab models and periodic boundary conditions. Even with the adoption of localized basis functions, slab-model calculations are, in fact, carried out with periodic linear combinations of the basis functions.[52]

The diffusion Monte Carlo (DMC) method,[53] which avoids the basis set problem, can also be used in conjunction with either cluster or slab models. Due to the high computational cost associated with DMC calculations and the lack of analytical geometry gradients, they have not received widespread use in tackling problems dealing with the chemistry of surfaces. However, we note that recently DMC calculations have been combined with cluster models to address the issue of buckling of the SiSi dimers on the Si(001) surface[20] and to estimate the barrier for H_2 desorption from the monohydride.[54]

At present, calculations using plane-wave basis sets are limited to "pure" density functional theory (DFT) methods. Calculations using wave function-based methods such as Hartree-Fock or MP2, or hybrid functionals, such as Becke3LYP,[55-57] containing a contribution from the exact exchange, are presently not feasible with plane-wave basis sets because of the large number of exchange integrals that would need to be evaluated. (However, efforts are underway to adapt plane-wave codes for use with hybrid functionals.[58])

Many periodic codes, *e.g.*, Dmol,[3,59] ADF,[60] NWChem,[61] SIESTA,[62,63] and deMon,[64,65] using localized atomic basis functions, are restricted to "pure" density functional methods. However, others, *e.g.*, Crystal,[66] Mondo[67] and Gaussian 03,[68] can perform periodic calculations using the Hartree-Fock method or DFT functionals including exact exchange. At least two groups have developed codes that can perform periodic MP2 calculations,[52,69] however, analytical gradients for optimizing geometries have not yet been implemented for periodic MP2 calculations which greatly restricts the range of problems that can be addressed with this method.

In concluding this section, we note that although plane-wave basis sets are generally used for periodic systems, they can also be used to study isolated molecules or cluster models of surfaces. This can be useful, for example, for estimating errors due to the use of pseudopotentials in the plane-wave calculations.[70] When using a code employing periodic boundary conditions, it is necessary to choose a supercell large enough so as to minimize interactions between molecules or clusters in adjacent cells. Although plane-wave codes have been developed that can treat isolated molecules or clusters without the use of periodic boundary conditions,[70] this capability is not implemented in any of the widely used plane-wave codes.

As for calculations with localized basis functions, the adoption of pseudopotentials results in a sizable savings in CPU time because of the reduction both in the number of electrons treated explicitly and in the size of the basis set required to achieve convergence. The savings for using pseudopotentials is much greater with plane-wave basis sets because high energy cutoffs (translating to very large basis set expansions) are required to describe tightly bound core electrons and the inner parts of the valence orbitals. For this reason, essentially all plane-wave calculations employ pseudopotentials (although all-electron calculations are feasible with augmented plane-wave approaches). The pseudopotentials used with plane-wave codes generally have been determined using the results of DFT calculations on atoms, whereas most pseudopotentials used with localized orbital codes have been parameterized using the results of atomic Hartree-Fock calculations.

4. SYSTEMATIC CONSTRUCTION OF CLUSTER MODELS FROM SLAB-MODEL GEOMETRIES

A strategy employed in several recent studies of reactions on semiconductor surfaces is to optimize the geometries of the relevant species (minima and transition states) using slab models with period boundary conditions and then to use these optimized geometries to construct cluster models for subsequent single-point calculations using theoretical methods, *e.g.*, diffusion Monte Carlo, DFT with hybrid density functionals, or MP2, not accessible through readily available periodic codes.[33,54] Such an approach was recently used by the authors to examine H_2 desorption from an SiSi dimer on the Si(001)-(2 x 1) surface.[33] In that work, the geometries of the reactant, transition state, and products, *i.e.*, the bare Si(001)-(2 x 1) surface plus a free H_2 molecule, were optimized using plane-wave DFT in conjunction with the PW91 exchange-correlation functional[71] and a supercell (depicted in Fig. 5) containing eight surface dimers, five layers, and a total of 80 Si atoms. The resulting slab-model geometries were used to construct Si_9H_{14}, $Si_{21}H_{22}$, $Si_{41}H_{38}$, and $Si_{89}H_{62}$, cluster models, containing one, three, three, and nine surface dimers, respectively (depicted in Fig. 3).

In generating these cluster models, the positions of the Si atoms and of the two chemically relevant H atoms were taken directly from the slab-model structures. The subsurface terminating H atoms were located such that the HSiSi angles and HSiSiSi dihedral angles were the same as the corresponding SiSiSi and SiSiSiSi dihedral angles in the slab model structures.

The resulting cluster models were used to perform single-point calculations using the PW91,[71] BLYP,[56,72] and Becke3LYP[55-57] exchange-correlation functionals as well as MP2 calculations using flexible Gaussian basis sets. Comparison of the cluster-model and slab-model results, using the same (PW91) functional allows one to assess the reliability of cluster models of various sizes to describe the H_2 desorption process. The activation energies are summarized in Fig. 6, from which it is seen that the Si_9H_{16} cluster model, not surprisingly, is quite poor for describing the H_2 desorption process. However, the $Si_{41}H_{38}$ cluster model with three surface dimers yields an activation energy within 1 kcal/mol of the slab-model result, and the nine-dimer cluster model gives essentially the same activation energy as obtained from the slab-model calculations, when both approaches employ the same functional.

a

b

c

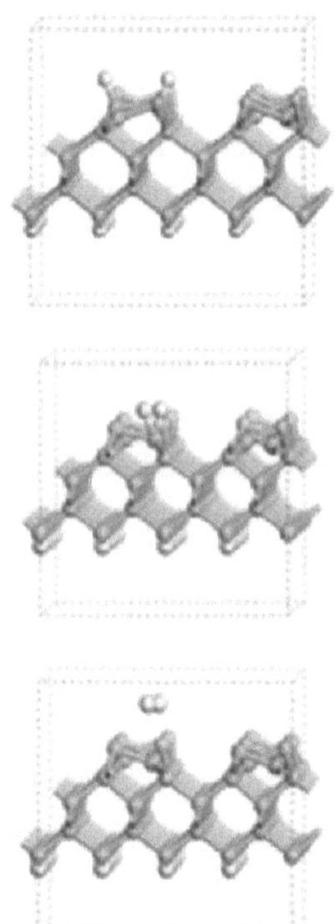

Figure 5. Supercell used to study H_2 desorption from an SiSi dimer on the Si(001)-(2 x 1) surface. a) HSiSiH monohydride species, b) transition state, and c) bare surface plus free H_2 molecule.

Nachtigall *et al.*[73] previously concluded, based on calculations on small silanes and silenes as well as on a Si_9H_{16} cluster model of H_2 interacting with the Si(001) surface that the PW91 and Becke-Perdew[72,74] (BP) functionals (both of which have been used in numerous studies of chemical processes on semiconductor surfaces) consistently underestimate, as compared to experiment and high-level QCISD(T)[75] or CCSD(T) calculations, the activation energy for H_2 loss.[73] These authors found that the DFT calculations with the Becke3LYP functional more closely reproduced the activation energies obtained from the high level wave function-based calculations. However, there remained the possibility that the deficiencies in the PW91 and BP functionals for calculating the activation energy would not be as great when using more realistic models of the surface. From Fig. 6, it is seen that for each of the four cluster models discussed above, the activation energy for H_2 desorption calculated using the Becke3LYP functional is energetically about 10 kcal/mol higher than that calculated using the PW91 functional. These results show conclusively that the PW91 functional considerably underestimates the activation energy for H_2 loss from a surface dimer *via* a 1,2-shift-

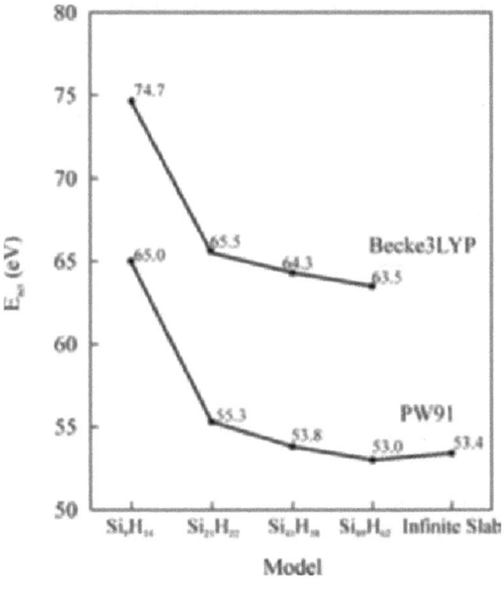

Figure. 6. Calculated activation energies for H_2 desorption from a dimer site on the Si(001)-(2 x 1) surface. Results are reported for the four cluster models depicted in Figure 3 and at both the PW91(lower curve) and Becke3LYP (upper curve) levels of theory. The activation energy calculated using the PW91 functional and the slab model shown in Figure 5 is also reported. Adapted with permission from Ref. 33.

type transition state even when realistic models of the surface are employed.

Several mechanisms for H_2 desorption from the Si(001)-(2 x 1) surface have been proposed in the literature.[76-80] Although early work on this problem tended to focus on 1,2-shift or "preparing" mechanism, (in which H_2 desorbs from a single SiSi dimer *via* the asymmetric transition state pictured in Fig. 5b), there is now considerable experimental and theoretical evidence that the adsorption/desorption process occurs *via* an interdimer mechanism.[81-84]

The strategy of using slab-model geometries optimized using DFT methods to generate cluster models for subsequent calculations with hybrid density functionals or with wavefunction-based methods should be applicable to a wide range of surface problems. However, we caution that this approach will not be appropriate for those cases that DFT methods give a poor description of the electronic structure, and hence, the geometry.

5. HOW RELIABLE ARE PLANE-WAVE DFT METHODS?

There are two major questions concerning the reliability of plane-wave DFT calculations: (1) is the density functional method adopted suitable for describing the process of interest, and (2) are sizable errors introduced in the DFT results due to the adoption of a plane-wave rather than a localized orbital basis set? The first question concerning the reliability of density functional theory is relevant regardless of the basis set or whether slab-model or cluster models are employed. There are two reasons for the latter concern. First, the plane-wave expansion may not be large enough to accurately describe localized chemical bonds, and second, plane-wave calculations are necessarily carried out using pseudopotentials, which could introduce errors.

Ascertaining convergence of energetics in calculations using plane-wave basis sets is challenging since the flexibility of a plane-wave basis set depends on both the energy cutoff and the size of the supercell. An increase in the size of the supercell, for a given energy cutoff, is accompanied by an increase in the flexibility of the basis set. The convergence of the energy also depends on the number of *k* points used for the sampling, with the importance of multiple *k* points also depending inversely on the size of the supercell. As a result, changes in the energetics for going to a larger supercell could reflect physical effects, *e.g.*, the incorporation of long-range effects not treated with a small supercell or numerical effects, *e.g.*, the results of increased basis set flexibility or less sensitivity to the *k*-point sampling. The errors in the plane-wave calculations resulting from adoption of a particular energy cutoff in the basis set expansion depend also on the choice of pseudopotential and on the elements that are present. Oxygen and fluorine atoms are particularly challenging in this regard, requiring very high-energy cutoffs to achieve convergence. To some extent, this problem can be solved by using ultrasoft pseudopotentials.[85] The presence of H atoms also necessitates the use of rather high-energy cutoffs, prompting many researchers to adopt a pseudopotential on H atoms. In this case, the pseudopotential does not reduce the number of electrons being treated explicitly, but simply cuts off the short-range part of the Coulomb potential. It may come as a surprise that the adoption of such a pseudopotential (if chosen carefully) introduces only small errors in the energies associated with XH bonds.[86]

One of the most detailed investigations of these issues is that mentioned above by Nachtigall *et al.* who compared, for several small silanes and silenes as well as for small cluster models of the Si(001) surface, the activation and reaction energies for H_2 loss as described by plane-wave DFT, as well as DFT and wavefunction methods using GTOs.[74] This study revealed that:

(1) the all-electron DFT calculations are already well converged with the 6-311G(d,p) basis set.[87,88] Specifically the DFT calculations with this basis set gave reaction and activation energies within 1.5 kcal/mol of those obtained using much larger basis sets. Somewhat larger errors were found with the 6-31G(d) or 6-31G(d,p) basis sets[89,90] that have been used in some studies of chemical processes on the Si(001) surface.[17,23]

(2) Basis sets much larger than 6-311G(d,p) are required to converge to ±1.5 kcal/mol the energy differences calculated using the MP2 or higher-order wave function based methods.

(3) Overall, both the Becke3LYP and MP2 methods give reaction and activation energies in fairly good agreement with the results of high-level QCISD(T) calculations, providing sufficiently flexible basis sets are used. The largest discrepancies between the reaction and activation energies obtained from the Becke3LYP and MP2 methods and those obtained at the QCISD(T) level are only about 3 kcal/mol. The agreement among these three sets of results is somewhat surprising given the partial diradical character in some of the product species.

(4) Significantly larger errors in the energy differences, particularly for the activation energies, are found for the pure density functional methods. In particular, as already noted, errors as large as 10.5 kcal/mol are found in the activation energies calculated using the BP and PW91 functionals. (Here errors are defined as deviations from the QCISD(T) results.)

(5) The adoption of the LANL pseudopotential[91] on the Si atoms introduced errors as large as 8.2 kcal/mol in the energy differences obtained from the Gaussian orbital calculations.

(6) The energy differences from the plane-wave calculations using a cutoff of 472 eV in conjunction with a cubic box of 25 a.u. sides, a single *k* point, and pseudopotentials on both the Si and H atoms, were converged to within 1 kcal/mol. (Here convergence was established by comparison with results obtained using still larger energy cutoffs.)

(7) In the local density approximation, the energy differences from the plane-wave calculations and the large basis set all-electron calculations agree to within 1 kcal/mol. However, with other functionals

(*e.g.*, BLYP), the plane-wave and all-electron GTO calculations give energy differences that differ by as much as 5.6 kcal/mol from the QCISD(T) or CCSD(T) results. This presumably reflects the inadequacy of using pseudopotentials generated in the LSD approximation in calculations using gradient-corrected functionals.

In summary, the study of Nachtigall *et al.* revealed that excellent agreement can be obtained between the predictions of plane-wave and Gaussian orbital DFT calculations, but also indicated that sizable errors in reaction and activation energies can be introduced upon the adoption of pseudopotentials. We believe that these errors would be minimized by use of pseudopotentials specifically designed for use with the exchange-correlation method employed. Finally, this study confirms the conclusions reached by other researchers that pure exchange correlation functionals may perform quite poorly for the calculation of activation energies.

6. TRANSITION STATES AND REACTION PATHS

For understanding chemical reactivity, knowledge of barrier heights and reaction pathways is essential. Until recently, most periodic electronic structure codes lacked algorithms for automatic location of transition states or for mapping out reaction pathways. This gave a clear-cut advantage to cluster models for such applications. However, with the development of robust methods such as the nudged elastic band (NEB) algorithm of Jónsson and coworkers[92,93] for optimizing minimum energy paths using energies and gradients alone, it is now fairly routine to locate transition states using slab models. The NEB algorithm has been implemented into the VASP and DACAPO plane-wave codes and has been extensively used for mapping out reaction pathways for chemical processes occurring on silicon and other semiconductor surfaces.[33,39,45,94] Our group has developed an external driving routine for carrying out NEB optimizations with Gaussian-orbital based codes.[95] We have used the NEB procedure as implemented in VASP for locating the transition state for the H_2 desorption process described above,[33] for mapping out pathways for (3 x 1) $\rightarrow$ (2 x 1) rearrangements of H-covered silicon surfaces following STM-induced desorption of H atoms,[45] as well as for characterizing the pathways for interconversion between various binding sites for an acetylene molecule adsorbed on the Si(001) surface.[39] A representative pathway for the Si(001)/acetylene system as described by a slab model with four surface dimers and using the PW91 functional is shown in Fig. 7. These NEB optimizations were carried out using up to 256 CPUs, typically employing, 8 to 32 images, with 8 to 16 CPUs being used per image.

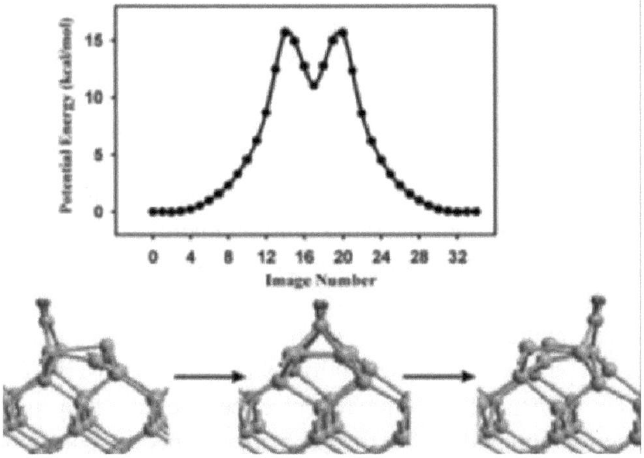

Figure 7. Minimum energy pathway for migration of an adsorbed acetylene molecule from a structure bridging two surface dimers at one end to the other, showing a tetracoordinated intermediate. Results obtained using density functional calculations with the PW91 functional, a slab model, and the NEB procedure for optimizing the pathway. Reproduced with permission from Ref. 39.

7. ELECTRONIC STRUCTURE OF CHEMICALLY REACTED SEMICONDUCTOR SURFACES

Scanning tunneling microscopy (STM) has proven to be an exceedingly valuable tool for imaging and manipulating semiconductor surfaces.[96] However, the interpretation of STM images is often far from straightforward, and theoretical calculations of the tunneling currents can prove especially valuable for assigning the observed images. Several groups have developed codes, generally based on the Tersoff-Hamann algorithm,[97] for calculating STM images.[98,99] Such calculations necessarily involve approximations, *e.g.*, modeling the tip by a single atom or by a small cluster of atoms. Most calculations of STM images of semiconductor surfaces (either pristine or with adsorbed species) have also ignored field effects, which can be quite important at the biases commonly used in experiments.

Stokbro has implemented a Greens function code for calculating STM tunneling currents allowing for field effects and has interfaced this to versions of the PWSCF and Siesta DFT codes.[98,100] The former is limited to plane-wave basis sets, whereas the latter supports both plane-wave and numerical basis sets. We have used Stokbro's modified PWSCF code to calculate STM images for various arrangements of acetylene on the Si(001)-(2 x 1) surface.[101] This problem has attracted considerable attention as a result of STM measurements by Mezhenny *et al.* which provided evidence for three different bonding sites for acetylene on the Si(001) surface.[102] Figure 8 reports the calculated image for an arrangement with two acetylene molecules each bridging two adjacent SiSi dimers, giving the so-called (di-bridge)$_2$ species.[101] This calculated image closely resembles that attributed by Mezhenny *et. al.*[102] to a single tetracoordinated C_2H_2 molecule. A combination of the trends in the calculated binding energies as a function of coverage and comparison of the calculated and measured STM images led us to reassign one of the observed binding sites to a species comprised of two C_2H_2 molecules bound. Recently two other groups have also concluded that this di-bridge type species is especially stable and is observed experimentally.[103-105] However, still another recent experimental study using photoelectron diffraction concluded that the (di-bridge)$_2$ species is unimportant.[106]

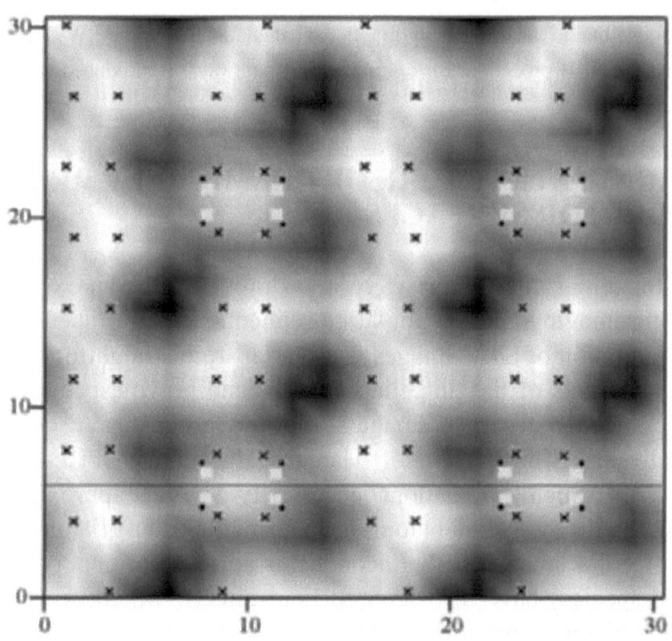

Figure 8. Calculated STM image (30 Å x 30 Å) of the SI(001)-(2 x 1) surface with two acetylene molecules bound in an end-bridging manner across two adjacent SiSi dimers. The X's denote the positions of the surface Si atoms and the shaded boxes and dark dots, the C and H atoms, respectively. Adapted with permission from Ref. 101.

8. SUMMARY

As a result of the advancements in algorithms and the huge increase in speed of computers over the past decade, electronic structure calculations have evolved into a valuable tool for characterizing surface species and for elucidating the pathways for their formation and reactivity. It is also now possible to calculate, including electric field effects, STM images for surface structures. To date the calculation of such images has been dominated by density functional methods, primarily because the computational cost of accurate wave-function based calculations using either realistic cluster or slab models would be prohibitive.

DFT calculations have proven especially valuable for elucidating chemical processes on silicon and other semiconductor surfaces. However, it is also clear that some of the systems to which DFT methods have been applied have large non-dynamical correlation effects, which may not be properly handled by the current generation of Kohn-Sham-based density functionals. For example, our CASSCF calculations on the Si(001)/acetylene system reveal that at some geometries there is extensive configuration mixing.[86] This, in turn, could signal problems for DFT calculations on these systems. Some of these problem systems can be addressed using ONIOM or other "layering" methods, treating the primary region of interest with a CASMP2 or other multireference-based method, and treating the secondary region by a lower level of electronic structure theory or by use of a molecular mechanics method.

ACKNOWLEDGEMENTS

We wish to thank H. Jónsson, C. Sosa, D. Sorescu, P. Nachtigall, and T.-C. Shen with whom we have collaborated in theoretical studies of chemically modified silicon surfaces and Prof. K. Stokbro for use of his code for calculating tunneling currents. We are also grateful to the National Science Foundation for support of this work and to NCSA and PSC for computational resources used in our studies.

REFERENCES

1. R. E. Schlier and H. E. Farnsworth, *J. Chem. Phys.* **30**, 917 (1959).
2. D. J. Chadi, *Phys. Rev. Lett.* **43**, 43 (1979).
3. R. J. Hamers, R. M. Tromp, and J. E. Demuth, *Phys. Rev. B* **34**, 5343 (1986).
4. R. A. Wolkow, *Phys. Rev. Lett.* **68**, 2636 (1992).
5. A. R. Smith, F. K. Men, K.-J. Chao, and C. K. Shih, *J. Vac. Sci. Tech. B* **14**, 914 (1995).
6. T. Yokoyama and K. Takayanagi, *Phys. Rev. B* **56**, 10483 (1997).

7. T. Yokoyama and K. Takayanagi, *Phys. Rev. B* **57**, R4226 (1998).
8. Y. Kondo, T. Amakusa, M. Iwatsuki, and H. Tokumoto, *Surf. Sci.* **453**, L318 (2000).
9. T. Yokoyama and K. Takayanagi, *Phys. Rev. B* **61**, 5078 (2000).
10. K. Hata, S. Yoshida, H. Shigekawa, *Phys. Rev. Lett.* **89**, 286104 (2002).
11. K. Sagisaka, M. Kitahara, D. Fujita, *Jpn. J. Appl. Phys.* **242**, 126 (2003).
12. Y. Yamashita, S. Machida, M. Nagao, S. Yamamoto, Y. Kakefuda, K. Mukai, and J. Yoshinobu, *Jpn. J. Appl. Phys.* **241**, L272 (2002).
13. T. Uozumi, Y. Tomiyoshi, N. Suehira, Y. Sugawara, S. Morita, *Appl. Surf. Sci.* **188**, 279 (2002).
14. J. Dabrowski and M. Scheffler, *Appl. Surf. Sci.* **56**, 15 (1992).
15. R. Konečný and D. J. Doren, *J. Chem. Phys.* **106**, 2426 (1997).
16. J. S. Hess and D. J. Doren, *J. Chem. Phys* **113**, 9353 (2000).
17. C. Yang and H. C. Kang, *J. Chem. Phys.* **110**, 11029 (1999).
18. E. Penev, P. Kratzer, and M. Scheffler, *J. Chem. Phys.* **110**, 3986 (1999).
19. C. Yang, S. Y. Lee, and H. C. Kang, *J. Chem. Phys.* **107**, 3295 (1997).
20. S. B. Healy, C. Filippi, P. Kratzer, E. Penev, M. Scheffler, *Phys. Rev. Lett.* **87**, 16105 (2001).
21. A. Redondo and W. A. Goddard, III, *J. Vac. Sci. Tech.* **21**, 344 (1982).
22. B. Paulus, *Surf. Sci.* **48**, 195 (1998).
23. J. R. Shoemaker, L. W. Burggraf, and M. S. Gordon, *J. Chem. Phys.* **112**, 2994 (2000).
24. M. S. Gordon, J. R. Shoemaker, and L. W. Burggraf, *J. Chem. Phys.* **113**, 9355 (2000).
25. E. Buehler and J. J. Boland, *Science* **290**, 506 (2000).
26. Y. J. Chabal and K. Raghavachari, *Phys. Rev. Lett.* **53**, 282 (1984).
27. P. Nachtigall, K. D. Jordan, K. C. Janda, *J. Chem. Phys.* **95**, 8652 (1991).
28. Z. Jing, G. Lucovsky, and J. L. Whitten, *Surf. Sci. Lett.* **296**, L33 (1993).
29. C. J. Wu, I. V. Ionova, and E. A. Carter, *Surf. Sci.* **295**, 64 (1993).
30. P. Nachtigall, C. Sosa, and K. D. Jordan, *J. Chem. Phys.* **101**, 8073 (1994).
31. X. Lu, Q. Zhang, and M. C. Lin, Phys. *Chem. Chem. Phys.* **3**, 2156 (2001).
32. M. R. Radeke and E. A. Carter, *Surf. Sci.* **355**, L289 (1996).
33. J. A. Steckel, T. Phung, K. D. Jordan, and P. Nachtigall, *J. Phys. Chem. B* **105**, 4031 (2001).
34. J. Cizek, *Adv. Chem. Phys.* **14**, 35 (1969).
35. G. D. Purvis and R. J. Bartlett, *J. Chem. Phys.* **76**, 1910 (1982).
36. G. E. Scuseria, C. L. Janssen, and H. F. Schaefer, III, *J. Chem. Phys.* **89**, 7382 (1988).
37. J. A. Pople, R. Krishnan, H. B. Schlegel, and J. S. Binkley, *Int. J. Quant. Chem.* **XIV**, 545 (1978).
38. J. R. Shoemaker, L. W. Burgraff, and M. S. Gordon, *J. Phys. Chem. A* **103**, 3245 (1999).
39. D. C. Sorescu and K. D. Jordan, *J. Phys. Chem. B* **104**, 8259 (2000).

40. H. Over, J. Wasserfall, W. Ranke, C. Ambiatello, R. Sawitzki, D. Wolf, and W. Moritz, *Phys. Rev. B* **55**, 4731 (1997).
41. R. Felici, I. K. Robinson, C. Ottaviani, P. Imperatori, P. Eng, and P. Perfetti, *Surf. Sci.* **375**, 55 (1997).
42. P. Kratzer, B. Hammer, and J. K. Nørskov, *Chem. Phys. Lett.* **229**, 645 (1994).
43. P. Kratzer, B. Hammer, and J. K. Nørskov, *Phys. Rev. B* **51**, 13432 (1995).
44. E. Pehlke and M. Scheffler, *Phys. Rev. Lett.* **74**, 952 (1995).
45. T.-C. Shen, J. A. Steckel, and K. D. Jordan, *Surf. Sci.* **446**, 211 (2000).
46. A. Vittadini and A. Selloni, *Chem. Phys. Lett.* **235**, 334 (1995).
47. G. Kresse, J. Hafner, *Phys. Rev. B* **48**, 13115 (1993).
48. G. Kresse, Furthmüller, *J. Comput. Mater. Sci.* **6**, 15 (1996).
49. G. Kresse, Furthmüller, *J. Phys. Rev. B* **54**, 11169 (1996).
50. The program *Dacapo* is available at http://www.fysik.dtu.dk/CAMPOS.
51. M. C. Payne, M. P. Teter, D. C. Allan, T. A. Arias, and J. D. Joannopoulos, *Rev. Mod. Phys.* **64**, 1045 (1992).
52. P. Y. Ayala, K. N. Kudin, and G. E. Scuseria, *J. Chem. Phys.* **115**, 9698 (2001).
53. W. M. C. Foulkes, L. Mitas, R. J. Needs, and G. Rajagopal, *Rev. Mod. Phys.* 73, *33* (2001).
54. C. Filippi, S. B. Healy, P. Kratzer, E. Pehlke, M. Scheffler, *Phys. Rev. Lett* **89**, 166102 (2002).
55. A. D. Becke, *J. Chem. Phys.* **98**, 5648 (1993).
56. C. Lee, W. Yang, and R. G. Parr, *Phys. Rev. B* **37**, 785 (1988).
57. S. H. Vosko, L. Wilk, and M. Nusir, *Can. J. Phys.* **58**, 1200 (1980).
58. S. Chawla and G. A. Voth, *J. Chem. Phys.* **108**, 4697 (1998).
59. B. Delley, *J. Chem. Phys.* **113**, 7756 (2000).
60. G. te Velde and E. J. Baerends, *J. Comput. Phys.* **99**, 84 (1992).
61. High Performance Computational Chemistry Group, NWChem, A Computational Chemistry Package for Parallel Computers, Version 4.5, Pacific Northwest National Laboratory, Richland, Washington, 99352, 2003.
62. D. Sánchez-Portal, P. Ordejó, E. Artacho, and J. M. Soler, *Int. J. Quant. Chem.* **65**, 453 (1997).
63. E. Artacho, D. Sánchez-Portal, P. Ordejó, A. Garcia, and J. M. Soler, *Phys. Stat. Sol. B* **215**, 809 (1999).
64. D. R. Salahub, M. E. Castro, and E. I. Proynov in *Relativistic and Electron Correlation Effects in Molecules and Solids*, vol. 318 of NATO ASI Series B: Physics, G. L. Malli, ed., Plenum Press, 1994, p. 411.
65. D. R. Salahub, M. E. Castro, R. Fournier, P. Calaminici, N. Godbout, A. Goursot, C. Jamorski, H. Kobayashi, A. Martinez, I. Papai, E. Proynov, N. Russo, S. Sirois, J. Usio, and A. Vela in *Theoretical and Compuational Approaches to Interface Phenomena*, H. Sellers and J. T. Golab, eds., Plenum Press, 1995, p. 187.

66. R. Dovesi, V. R. Sanders, C. Roetti, M. Causà, N. M. Harrison, R. Orlando, and E. Aprà, CRYSTAL95 User Manual (Università di Torino, Torino, Italy, and CCLRC Daresbury Laboratory, Daresbury, UK) 1996.
67. M. Challacombe and E. Schwegler, MONDOSCF, a program suite for massively parallel, linear scaling SCF theory, http://www.t12.lanl.gov/home/mchalla.
68. Gaussian 03, Revision A.1, M. J. Frisch, G. W. Trucks, H. B. Schlegel, G. E. Scuseria, M. A. Robb, J. R. Cheeseman, J. A. Montgomery, Jr., T. Vreven, K. N. Kudin, J. C. Burant, J. M. Millam, S. S. Iyengar, J. Tomasi, V. Barone, B. Mennucci, M. Cossi, G. Scalmani, N. Rega, G. A. Petersson, H. Nakatsuji, M. Hada, M. Ehara, K. Toyota, R. Fukuda, J. Hasegawa, M. Ishida, T. Nakajima, Y. Honda, O. Kitao, H. Nakai, M. Klene, X. Li, J. E. Knox, H. P. Hratchian, J. B. Cross, C. Adamo, J. Jaramillo, R. Gomperts, R. E. Stratmann, O. Yazyev, A. J. Austin, R. Cammi, C. Pomelli, J. W. Ochterski. P. Y. Ayala, K. Morokuma, G. A. Voth, P. Salvador, J. J. Dannenberg, V. G. Zakrzewski, S. Dapprich, A. D. Daniels, M. C. Strain, O. Farkas, D. K. Malick, A. D. Rabuck, K. Raghavachari, J. B. Foresman, J. V. Ortiz, Q. Cui, A. G. Baboul, S. Clifford, J. Cioslowski, B. B. Stefanov, G. Liu, A. Liashenko, P. Piskorz, I. Komaromi, R. L. Martin, D. J. Fox, T. Keith, M. A. Al-Laham, C. Y. Peng, A. Nanayakkara, M. Challacombe, P. M. W. Gill, B. Johnson, W. Chen. M. W. Wong, C. Gonzalez, and J. A. Pople, Gaussian, Inc., Pittsburgh, PA, 2003.
69. A. Abdurahman, M. Albrecht, A. Shukla, and M. Dolg, *J. Chem. Phys.* **110**, 8819 (1999).
70. R. N. Barnett and U. Landman, *Phys. Rev. B* **48**, 2081 (1993).
71. J. D. Perdew, J. A. Chevary, S. H. Vosko, K. A. Jackson, M. R. Pederson, D. J. Singh, and C. Fiolhais, *Phys. Rev. B* **46**, 6671 (1992).
72. A. D. Becke, *Phys. Rev. A* **38**, 3098 (1988).
73. P. Nachtigall, K. D. Jordan, A. Smith, and H. Jónsson, *J. Chem. Phys.* **104**, 148 (1996).
74. J. P. Perdew, *Phys. Rev. B* **33**, 8822 (1986).
75. J. A. Pople, M. Head-Gordon, and K. Raghavachari, *J. Chem. Phys.* **87**, 5968 (1987).
76. K. W. Kolasinski, W. Nessler, A. de Meijere, and E. Hasselbrink, *Phys. Rev. Lett.* **72**, 1356 (1994).
77. K. W. Kolasinski, W. Nessler, K.-H. Bornscheuer, and E. Hasselbrink, *J. Chem. Phys.* **101**, 7082 (1994).
78. K. Sinniah, M. G. Sherman, L. B. Lewis, W. H. Weinberg, J. T. Yates, Jr., and K. C. Janda, *J. Chem. Phys.* **92**, 5700 (1990).
79. K. Sinniah, M. G. Sherman, L. B. Lewis, W. H. Weinberg, J. T. Yates, Jr., and K. C. Janda, *Phys. Rev. Lett.* **62**, 567 (1989).
80. M. L. Wise, B. G. Koehler, P. Gupta, P. A. Coon, and S. M. George, *Surf. Sci.* *258*, 166 (1991).

81. M. Dürr, Z. Hu, A. Biederman, U. Höfer, and T. F. Heinz, *Phys. Rev. Lett.* **88**, 046104 (2002).
82. M. Dürr, Z. Hu, M. B. Raschke, E. Pehlke, and U. Höfer, *Phys. Rev. Lett.* **86**, 123 (2001).
83. M. Dürr, Z. Hu, A. Biederman, U. Höfer, and T. F. Heinz, *Science* **296**, 1838 (2002).
84. M. Dürr and U. Höfer, *Phys. Rev. Lett* **88**, 076107 (2002).
85. D. Vanderbilt, *Phys. Rev. B* **41**, 7892 (1990).
86. J. A. Steckel and K. D. Jordon, unpublished results.
87. R. Krishnan, J. S. Binkley, R. Seeger, and J. A. Pople, *J. Chem. Phys.* **72**, 650 (1980).
88. A. D. McLean and G. S. Chandler, *J. Chem. Phys.* **72**, 5639 (1980).
89. M. S. Gordon, *Chem. Phys. Lett.* **76**, 163 (1980).
90. W. J. Hehre, R. Ditchfield, and J. A. Pople, *J. Chem. Phys.* **56**, 2257 (1972).
91. P. J. Hay and W. R. Wadt, *J. Chem. Phys.* **82**, 284 (1985).
92. G. Mills and H. Jónsson, *Phys. Rev. Lett.* **72**, 1124 (1994).
93. G. Mills, H. Jónsson, and G. K. Schenter, *Surf. Sci.* **324**, 305 (1995).
94. B. Uberuaga, M. Levskovar, A. P. Smith, H. Jónsson, and M. Olmstead, *Phys. Rev. Lett.* **84**, 2441 (2000).
95. D. Alfonso and K. D. Jordan, *J. Comp. Chem.* **24**, 990 (2003).
96. G. Binnig and H. Rohrer, *Surf. Sci.* **126**, 236 (1983).
97. J. Tersoff and D. R. Hamann, *Phys. Rev. B* **31**, 805 (1985).
98. K. Stokbro, U. Quaade, and F. Grey, *Appl. Phys. A* **66**, S907 (1998).
99. W. A. Hofer and J. Redinger, *Surf. Sci.* **447**, 51 (2000).
100. The Transiesta code is described at www.transiesta.com. See also, M. Brandbyge, J.-L. Mozos, P. Ordejon, J. Taylor, and K. Stokbro, Phys. Rev. B. 65, 165401 (2002).
101. F. Wang, D. C. Sorescu, and K. D. Jordan, *J. Chem. Phys. B* **106**, 1316 (2002).
102. S. Mezhenny, I. Lyubinetsky, W. J. Choyke, R. A. Wolkow, J. T. Yates, Jr., *Chem. Phys. Lett.* **344**, 7 (2001).
103. Y. Morikawa, *Phys. Rev. B* **63**, 33405 (2001).
104. W. Kim, H. Kim, G. Lee, J. Chung, S. Y. You, Y. K. Hong, J. Y. Koo, *Surf. Sci.* **514**, 376 (2002).
105. W. Kim, H. Kim, G. Lee, Y. K. Hong, K. Lee, C. Hwang, D. H. Kim, J. Y. Koo, *Phys. Rev. B* **64**, 193313 (2001).
106. R. Terborg, M. Polcik, J. T. Hoeft, M. Kittel, D. I. Sayago, R. L. Toomes, and D. P. Woodruff, *Phys. Rev. B* **66**, 085333 (2002).

Chapter 7

THEORETICAL STUDIES OF GROWTH REACTIONS ON DIAMOND SURFACES

P. Zapol,[a] L. A. Curtiss,[a] H. Tamura,[b] and M. S. Gordon[c]

[a] *Materials Science and Chemistry Divisions, Argonne National Laboratory, Argonne, Illinois 60439 USA*
[b] *Department of Computational Science, Kanazawa University, Kakuma, Kanazawa 920-1192 Japan*
[c] *Chemistry Department, Iowa State University, Ames Iowa 51001 USA*

1. INTRODUCTION

Diamond is of great interest because of its unique properties.[1] It is the hardest substance found in nature due to its very high atom number density and strong covalent bonding. It also has the highest known thermal conductivity and the highest elastic modulus of any material. In addition, it is an insulator (E_g = 5.5 eV) and is transparent over a large range of wavelengths. Thermodynamically, diamond is slightly less stable than graphite and the two materials have very different structures and properties. These and other properties of diamond make it useful in a variety of applications and potential applications such as tool coatings, thermal management, optical windows and optoelectronics. Nanocrystalline diamond has potential for new applications such as seal coatings, electrochemical electrodes, microelectromechanical systems, biomolecular materials, and electron-emitting surfaces for flat-panel displays.

As a result of these many uses, synthesis of diamond has been of considerable interest for many years. The initial synthesis work focused on methods based on high pressures and such methods are now widely used for production of commercial diamonds.[2] There has also been an intensive effort aimed at low-pressure chemical vapor deposition (CVD) of crystalline diamond.[3] Efforts to synthesize crystalline diamond by CVD were initially limited by low growth rates, but in the last 20 years there has been much prog-

L.A. Curtiss and M.S. Gordon (eds.),
Computational Materials Chemistry: Methods and Applications, 266–307.

ress in achieving higher growth rates. The advantages of CVD methods are low cost and flexibility with respect to the form of diamond that is deposited leading to many potential uses.

Several types of CVD methods have been used for growing diamond films including thermal methods such as hot filament and thermal plasmas and nonthermal methods such as microwave plasmas. CVD methods have been the subject of several reviews.[3,4] The most successful CVD growth methods have been from hydrocarbon-hydrogen gas mixtures. In most cases growth is initiated by the dissociation of a gaseous mixture of H_2 and a simple hydrocarbon precursor such as CH_4. There are a variety of species that can be present in the gas dependent on the conditions and method used to activate the precursor gas. These include radicals, such as CH_3, CH_2, C_2H, and CH, acetylene, methane, and atomic and molecular hydrogen. In conventional methods atomic hydrogen present in the plasma is likely to saturate the diamond surfaces as well as etch non-diamond phases and abstract hydrogen from some surface sites. The latter process will provide surface radical sites that can react with the hydrocarbon species. The growth mechanisms are not well understood, but there has been considerable speculation about growth sites and vapour species involved. In nonthermal methods methyl radicals are believed to be the growth species, while in thermal methods acetylene is believed to be the major species involved. Recently, it has been discovered by Gruen *et al.*[5,6] that diamond can be grown in low hydrogen argon plasmas with C_2 as the growth species. The morphology of the diamond films obtained in the CVD processes depends on the C/H ratio in the plasmas. For example, in low hydrogen plasma growth it has been found that the crystallite sizes of diamond are very small (3-10 nm) due to a high renucleation rate.[5] Understanding the growth mechanism is difficult because of the large number of experimental parameters and the difficulty of making accurate measurements under the CVD conditions.

Theoretical studies have played an important role in gaining an understanding of the possible growth mechanisms of diamond in CVD. This work started in the late 1980's[7] and has largely been based on quantum chemical studies of reactions of growth species on diamond surfaces with a focus on methyl radical or acetylene as the growth species. Generally reaction mechanisms are postulated to be initiated by the creation of a surface radical via the abstraction of hydrogen, followed by addition of the growth species to the surface radical site. In addition, some molecular dynamics and kinetics studies have been reported. An excellent review of the early theoretical work on mechanisms for CVD diamond growth has been presented by Spear and Frenklach.[8]

In this chapter we present a review of the theoretical studies of growth reactions on diamond surfaces. The focus of this review is largely on quantum chemical studies of reaction mechanisms of the major growth species methyl radical, methylene, and ethylene. In addition, theoretical studies of mechanisms involving C_2 are reviewed. This review is divided as follows. Section 2 contains a survey of the theoretical methodologies used for diamond surface reaction including two new promising approaches for dealing with surface reactions: density functional-based tight binding and SIMOMM. Section 3 reviews theoretical studies of the surface structure of diamond. Section 4 reviews theoretical studies of reactions involving the hydrogen abstraction and important growth species. In this chapter we also present some of our recent work on growth mechanisms involving C_2 and methylene. Finally, Section 5 reviews other adsorbates on diamond surfaces that play a role in growth mechanisms.

2. THEORETICAL METHODS AND STRUCTURAL MODELS

2.1 Models

Application of computational chemistry to study diamond surface reactions requires that a representative model for the surface be chosen. For a typical surface study there are two types of models: (1) a finite cluster of atoms and (2) a periodic slab with a unit cell that is repeated in all directions. Most of the computational studies reported here use one of these two types of models.

The cluster model includes a finite number of carbon atoms that are

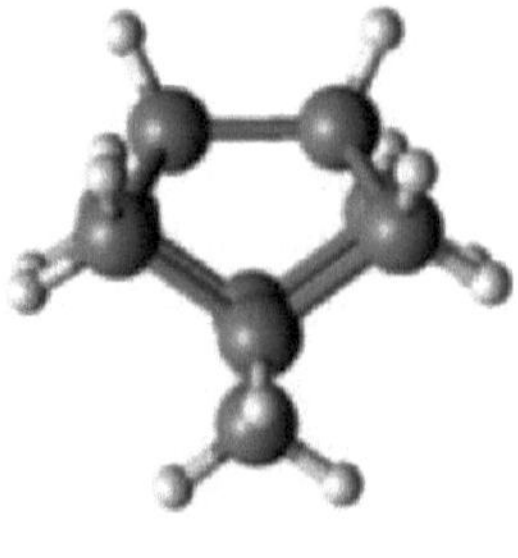

Figure 1. C_9H_{14} cluster used to model the diamond 100 surface.

terminated by hydrogen. The hydrogens replace carbons at the cluster surface, but with normal C-H distances. The number of carbons in the clusters used to model a diamond surface ranges from two to as many as about 60 atoms. The C_9H_{14} cluster shown in Fig. 1 is a typical small cluster used to model the reconstructed diamond (100) surface. An important consideration is the amount of geometry relaxation of the clusters used in calculations.

While for molecular reactions full geometry relaxations are preferable, in the case of surface studies this can sometimes be unrealistic because the surface atoms move differently than if they were constrained by the missing carbons. One approach to approximate the surface reaction with limited geometry optimization is described in Section 2.3.1 below.

The periodic models used for surface studies have translational symmetry in two directions parallel to the surface. In the periodic slab model, translational symmetry is also applied in the third direction, however, the basis vector is chosen to be sufficiently larger than the slab thickness to avoid interactions between repeating slabs. The periodic slab model is a convenient tool for surface calculations, since many codes that have only 3D periodic boundary conditions implemented can be used for surface studies. Typically, separation between slabs more than 10 Å is adequate for nonpolar surfaces, which all surfaces are in homonuclear systems such as diamond. Separations that are too large may cause unnecessary computational costs for calculations with plane wave basis sets, but can typically be tolerated for localised, i.e. Gaussian or Slater, basis sets. In the codes where true two-dimensional periodic boundary conditions are implemented, these problems are non-existent.

Periodic models perform best in adsorption studies under conditions of high surface coverage because they permit the study of periodic adsorbate structures using relatively small surface unit cells. Adsorption of isolated molecules at the surface represents a challenge for periodic models. In this case, a sufficiently large unit cell should be chosen to avoid interactions between the molecules in the neighbouring cells making calculations more expensive. When comparing adsorption energies and structures at different coverages, the same size unit cell should be used to obtain reliable results. Adsorbates can be placed on either one side or both sides of the slab, each approach having its merits. Adsorption on just one side reduces the overall number of atoms in the cell, while adsorption on both sides allows higher symmetry of the system. In the former case, the clean side of the diamond slab can be frozen at the reconstructed surface geometry or saturated with hydrogen atoms. The positions of atoms in the adsorbed molecule and several surface layers need to be optimized to obtain reliable results.

2.2 Quantum Mechanical Methods

A variety of quantum chemical methods have been used in the study of reactions occuring on surfaces. Most of these methods fall into three general categories: *ab initio* molecular orbital theory, density functional theory, and semi-empirical molecular orbital theory. These three methods are briefly summarized in this section. The reader is referred to other reviews of these methods[9,10,11] for more details.

In this section we also describe two specific approaches that have been developed for handling large numbers of atoms in surface reactions on diamond. The first, Surface Integrated Molecular Orbital/Molecular Mechanics (SIMOMM), combines several low and high levels of theory to provide a more realistic cluster model for a surface reaction. The second, self consistent charge density functional based tight binding, is a semi-empirical approach that uses parameters obtained from careful fitting to density functional calculations. This method is related to tight binding methods that have proved very successful in solid state physics for inorganic materials.[12] In addition to these quantum mechanical methods, a number of other non-quantum mechanical methods have been used for investigating diamond surface growth mechanisms including molecular dynamics simulations. Molecular dynamics results depend greatly on the interatomic potentials that are used and how well they represent the interactions.

2.2.1 *Ab initio molecular orbital theory*

The most commonly used *ab initio* molecular orbital methods[10] for diamond surface reactions include Hartree-Fock (HF) theory and second order perturbation theory (MP2) for inclusion of correlation effects The HF level calculations are often used for geometries and vibrational frequencies. They are not very reliable for reaction energies or barriers, where correlation effects are important and need to be included. Very high level calculations are needed to assess the reliability of these relatively low level calculations, especially for bond breaking. Very high level methods that are used for this purpose include Gaussian-n (G2 and G3) theories,[13] multi-reference methods[14] and coupled cluster theory (CCSD(T))[15] with large basis sets.

2.2.2 *Density functional theory*

Density functional theory (DFT) is a cost-effective method for studying properties of molecular systems and has been found to be useful for the study of diamond surface reactions. The most popular of the functionals is B3LYP with a double zeta plus polarization basis set such as 6-31G*[16,17] In several validation studies for molecular geometries and frequencies, DFT has given results of quality similar to that of MP2 theory[18] at lower computational cost. It has also been examined for use in calculation of thermochemical data.[19] While DFT is not as accurate as very high level *ab initio* molecular orbital methods such as G3 theory or CCSD(T), it is more cost effective and the B3LYP functional is more accurate for reaction energies than HF method and often MP2 theory. None of the commonly used single configuration methods, such as DFT or MP2, are reliable if significant diradical character is present.

2.2.3 Semi-empirical molecular orbital theory

Semiempirical molecular orbital methods have been used in some studies of diamond growth, especially the early work. These methods treat only valence electrons and neglect certain interaction terms or replace them with empirical terms. This greatly reduces the computational cost, but make the outcome dependent on having a good set of experimental data to use in the parameterization. Among the methods that have been used are Modified Neglect of Diatomic Overlap (MNDO),[20] Austin Model 1 (AM1),[21] and Parametric Method Number 3 (PM3).[22] All three methods include only valence s- and p-functions, which are taken as Slater-type orbitals. The AM1, MNDO, and PM3 methods work well for ground state properties such as enthalpies of formation and geometries because they have been parameterized for these quantities. Thus, these methods are useful for geometry optimizations of large clusters, but are unreliable for transition state or strained structures in diamond growth reactions studies.

2.2.4 Hybrid quantum mechanics/molecular mechanics

In the hybrid quantum mechanics/molecular mechanics (QM/MM) techniques a large molecular system is partitioned into a small, chemically active part where a reaction will occur, and a larger, chemically inactive piece. The embedded cluster approach called SIMOMM[23] (Surface Integrated Molecular Orbital/Molecular Mechanics) of Shoemaker *et al.* was developed for surface chemistry calculations to minimize time-consuming electronic structure computations while maintaining the effect of the "bulk". This is a modification of the IMOMM method (integrated molecular orbital/molecular mechanics).[24] In the SIMOMM method, the region where the actual chemical reaction occurs is treated quantum mechanically, while the other regions are treated using molecular mechanics. The transition from the QM to the MM region is of critical importance for surface systems. The SIMOMM approach uses hydrogens or model atoms as "place holders" in the link region.

2.2.5 Density Functional-based Tight Binding (DFTB)

The modeling of materials problems such as growth reactions on diamond surfaces requires accuracy under diverse conditions such as stretched bonds, strained bond angles, etc. Also since the reactions occur on surfaces, accurate simulations often will require a large number atoms be included in the calculations. This is not always possible with the traditional quantum chemical methods described in this section. Thus, several methodologies have been developed and implemented that include quantum mechanics in an approximate way so that a sufficiently large number of atoms can been handled. The tight binding methodology has proved very successful in solid state physics for inorganic materials.[12] The tight-binding method is similar to the semi-empirical molecular orbital approach in that they both use parameterization of integrals. Recently the tight binding methodology has been extended to handle a wider variety of systems. In the DFTB approach the energy is written as a band energy term plus a sum over repulsive pair potentials. The band energy term is the sum of occupied orbital energies obtained with a minimal basis set. In this section we describe a self-consistent charge density functional based tight binding method that is parameterized by fitting to density functional results for appropriate systems as developed by Frauenheim and coworkers.[25,26,27] This approach is referred to as Self-Consistent-Charge Density-Functional-based Tight-Binding (SCC-DFTB, denoted DFTB in short). The method is also sufficiently fast that it can be used in molecular dynamics simulations to include electronic structure into the calculations at each step.

The SCC-DFTB method is based on the expansion of density functional $E[\rho]$ to second order over the density fluctuations $\delta\rho$ around a suitable reference electron density, ρ_0. Diagonal matrix elements of the Hamiltonian are taken as the one-electron eigenvalues of the neutral atom and among non-diagonal matrix elements only two-center contributions are evaluated. The diatomic repulsive potential $U_{rep}{}^{A\text{-}B}$ can be determined by the following expression:

$$U_{rep}^{A\text{-}B}(R_{AB}) = [E_{tot}(R_{AB}) - E_{tot}(\infty)] - \sum_i n_i \varepsilon_i^0 (R_{AB})$$
$$- \frac{1}{2}\sum_{AB} \gamma_{AB}(R_{AB}) \Delta q_A \Delta q_B,$$

where the second term, the band energy, is the sum over the eigenstates ε_0 of the reference system with density ρ_0, with n_i being the occupation numbers. The third term is the second-order energy correction expressed by Mulliken charges Δq and γ, which describes Madelung-like distance dependence.

The total energy difference $[E_{tot}(R_{AB}) - E_{tot}(\infty)]$ can be determined for model systems (e.g. for selected molecules) using a DFT method, and the remaining two terms can be calculated using DFTB after the atomic parameters are obtained. The accuracy of the calculation of $U_{rep}{}^{A\text{-}B}(R_{AB})$ is crucial for the overall accuracy of DFTB as the changes in the repulsive potential's shape and magnitude greatly affect the chemistry describing the element in question.

The performance of the SCC-DFTB method has been tested on a series of some typical organic reactions involving small molecules.[25] The method has been tested for a small set of small molecules,[25] and found to give reasonable results, with some exceptions, for geometries, vibrational frequencies, and bond energies. We have recently carried out some test calculations on a set of larger organic compounds for a more comprehensive assessment.[28] The SCC-DFTB method does reasonably well for most systems. The only exceptions are strained ring systems such as bicyclobutane and spiropentane where the parameterization apparently fails. Due to the possible failure of the parameterization it is advisable to assess tight binding methods against high level *ab initio* methods for reliability on the type of reaction being studied.

3. DIAMOND SURFACE STRUCTURE

The three prominent low-index surfaces of diamond are (100), (110), and (111). The unreconstructed (100) surface has two dangling bonds per atom at the bulk-terminated surface. The (110) and (111) surfaces have only one dangling bond per surface atom (the 111 surface can also have three dangling bonds). The presence of dangling bonds means that the surfaces are easily hydrogenated or reconstructions can occur. The (100) surface has been of most interest because of its importance in the CVD of diamond, probably because it is the slowest growing surface and it is the most prominent surface in diamond films. Knowledge of all of three surface structures is important for modeling growth mechanisms. We review theoretical studies of diamond surface structure with particular relevance to growth mechanisms.

3.1 (100) Surface

The saturation of the dangling bonds on the (100) surface occurs by formation of rows of π-bonded double bonds. The saturation can also occur by partial or full hydrogenation of the surface. The partial hydrogenation also involves reconstruction. Possible structures are illustrated in Fig. 2 including the fully hydrogenated 1x1:2H surface, the clean 2x1 surface, the monohydride 2x1:H surface, and a mixed 3x1:1.33H surface.

The structures and energetics of (100) diamond surfaces have been studied both experimentally and theoretically.[29] Experimentally, Hamza *et al.* found 1x1 dihydride and 2x1 monohydride structures on the hydrogenated diamond (100) surfaces using low-energy electron diffraction (LEED).[30] The 2x1 monohydride structure is generally known to be stable on the diamond

(100) surface. The hydrogen is desorbed from the 2x1 monohydride structure above 1300 K and a 2x1 clean dimer row structure appears.[31,32]

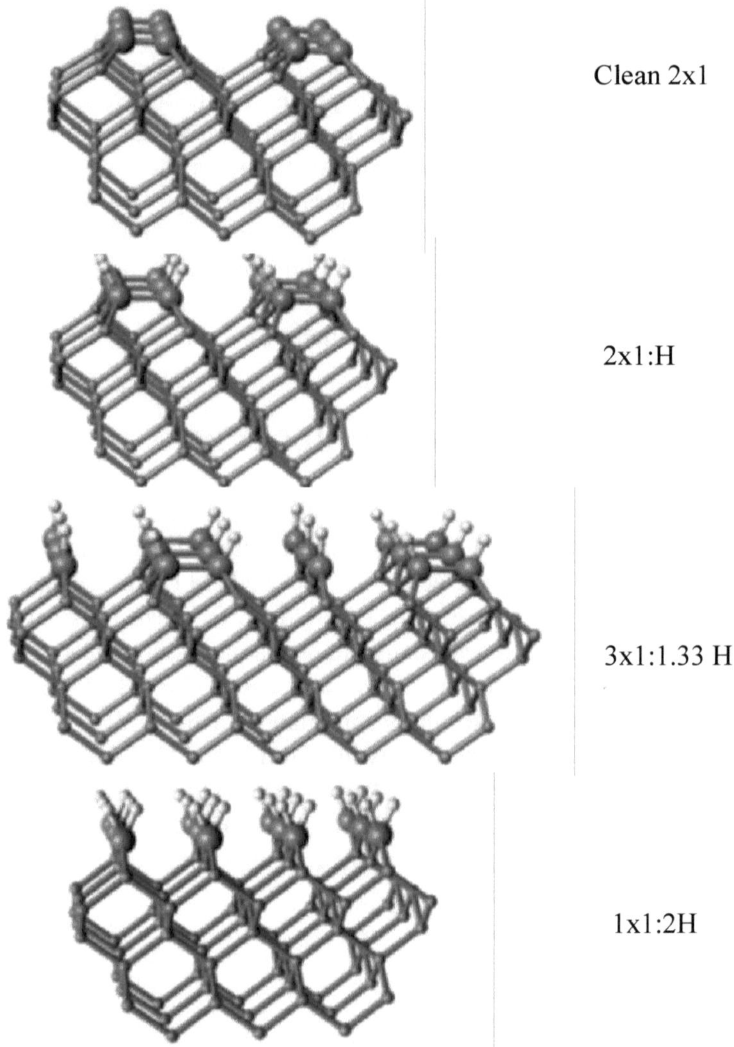

Figure 2. Side view of diamond (100) surface structures.

The structures of the diamond (100) 2x1 clean, 2x1 monohydride, and 1x1 dihydride surfaces have been exhaustively investigated by various theoretical methods. These include empirical potentials,[33,34] semiempirical[35,36,37] and *ab-initio*[38,39] quantum mechanics (QM), semiempirical QM/molecular dynamics (MD),[40] and density functional theory (DFT) using the local density approximation (LDA)[41,42,43,44,45,46,47] or the generalized gra-

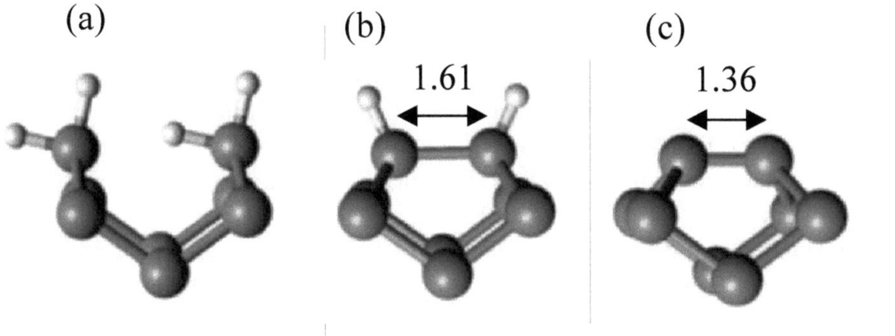

Figure 3. Equilibrium geometries of (a) 1x1 dihydride, (2) 2x1 monohydride dimer, and (c) 2x1 clean dimer.[48]

dient approximation (GGA).[48,49,50] Other methods include the tight-binding (TB) approximation using empirical parameters[51,52] and the TB potential determined by DFT (DFTB).[53]

Fig. 3 shows the equilibrium geometries of the diamond (100) surfaces optimized using periodic DFT calculations with the GGA.[48] The DFT calculations predict that the 1x1 dihydride (Fig. 3a) is canted to decrease steric repulsions between the hydrogen atoms.[44,47,48] Figures 3b and 3c illustrate the equilibrium geometries of the 2x1 monohydride and 2x1 clean dimer, respectively. Table 1 summarizes the bond length of the clean and monohydride dimers predicted by theoretical calculations. The C-C distance in the clean dimer is generally predicted to be smaller than that in the monohydride, since the (partial) π bond in the clean dimer has been broken in the monohydride. These calculations predict a symmetric unbuckled 2x1 clean dimer on the diamond (100) surface. This is in contrast to the analogous DFT predictions for the Si(100) surface, most likely because of the much stronger π bonding in the diamond dimer. The Si(100) dimer has much greater multi-reference character, so that single reference methods are less reliable.

The GGA calculations[48] indicate that reconstruction from the 1x1 dihydride to the 2x1 monohydride with H_2 desorption is exothermic (-61 kcal/mol), while the H_2 desorption from the monohydride is endothermic (84 kcal/mol). These calculations indicate that the 2x1 monohydride is stable on the diamond (100) surface, while the 1x1 dihydride is unstable due to steric repulsions between the hydrogen atoms. For desorption of a single H atom from the monohydride surface, second order perturbation theory (MP2)[39] and GGA[49] calculations predict that the desorption of the second H atom is easier than the first, since a π bond is formed when the second H is removed. This enhances the associative desorption of hydrogen from the monohydride dimer and contributes to the first order kinetics of the H_2 thermal desorption.[29] Another way for the diamond (100) 1x1:2H surface to relieve strain is to form a 3x1:1.33H structure with alternating monohydride and dihydride units (see Fig. 3). This structure has been investigated in several studies studied and found to be stable.[41,40]

The formation energy of (100) surface structures with different hydrogen coverages (2x1:H, 5x1:1.2H, 3x1:1.33H, 4x1:1.5H, and 1x1:2H) was calculated using GGA as a function of the hydrogen chemical potential.[50] This most comprehensive study shows that increase in hydrogen chemical potential leads to change of the most stable phase from the clean surface to monohydride to mixed monohydride and dyhidride phases and finally to the dihydride surface. However, at hydrogen chemical potentials where phases

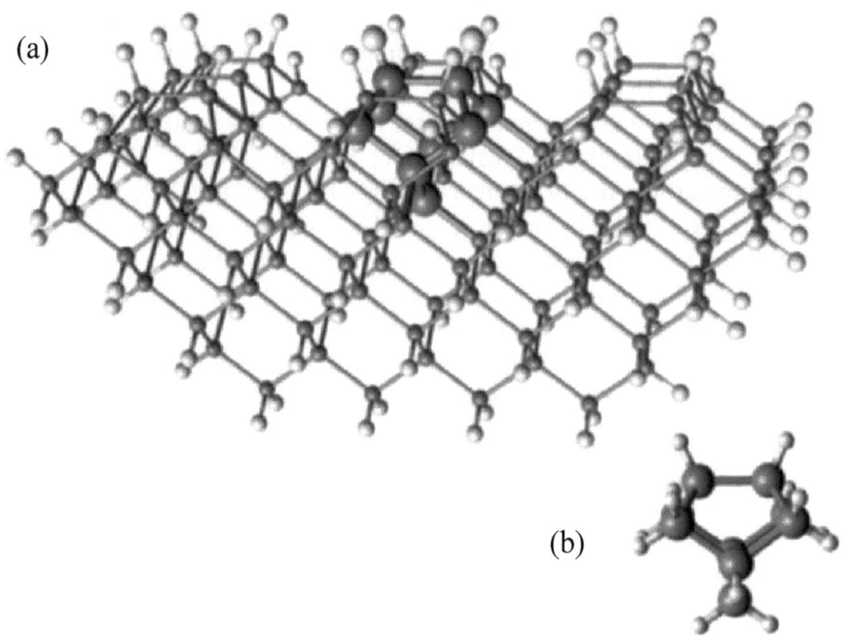

Figure 4. Cluster models for QM/MM calculations, (a) 9-dimer and 9-layer MM cluster and (b) 1-dimer ab-initio cluster (C_9H_{14}). The ab-initio cluster is embedded into the MM cluster (marked with black). The dangling bonds of ab-initio cluster at QM/MM boundary are terminated with hydrogen.

with dihydride units are more stable than monohydride, these phases are not thermodynamically favorable compared to the formation of methane. If such phases still exist on the surface because of the barrier for methane etching, they can be experimentally identified by the H-C-H bending mode around 1480 cm^{-1}.

QM/MM methods are particularly useful for providing accurate *ab-initio* calculations including bulk steric effects at an acceptable computational cost. The SIMOMM (surface integrated molecular orbital molecular mechanics) method[23] was developed to perform QM/MM calculations to describe surface phenomena. The equilibrium geometries of both clean and hydrogenated dimer on diamond (100) have been calculated using the SIMOMM code. Fig. 4 shows the cluster models that were used in the QM/MM calculations of diamond (100) surfaces. A small *ab-initio* cluster (C_9H_{14}) is embedded into a much larger MM cluster (9-dimer and 9-layer) so that the steric strain from the bulk lattice is efficiently taken into account by the empirical (MM) force field. The 6-31G basis set is used with d and p polarization basis functions on heavy atoms and hydrogens, respectively. The DFT results using the B3LYP functional[16] and multi-configurational self-consistent field (MCSCF)[54] calculations using the SIMOMM method are shown in Table 1. The (2,2) active space of the MCSCF calculation con-

Table 1. Calculated C-C length of clean and monohydride 2x1 dimers [Å].

Reference	Clean	Mono-hydride
32 (empirical)	1.38	1.63
33 (empirical)	1.46	1.63
34 (semiempirical)	1.43	1.67
35 (semiempirical)	1.58	1.73
36 (semiempirical)	1.38	1.56
37 (MP2/6-31G(d)) con strained cluster	1.44	1.71
38 (MP2/6-31G(d)) relaxed cluster	1.38	1.58
39 (semiempirical)	1.53	1.66
40 (LDA)	1.40	1.67
41 (LDA)	1.37	
42 (LDA)	1.37	1.61
43 (LDA)	1.37	
44 (LDA)	1.38	1.62
45 (GGA)	1.36	1.61
46 (TB)	1.40	1.62
47 (TB)	1.54	
48 (DFTB)	1.41	1.58
QM/MM: SIMOMM B3LYP/6-31G(d)	1.38	1.65
QM/MM: SIMOMM MCSCF/6-31G(d)	1.42	1.65

sists of two electrons in the dimer π and π^* orbitals on the diamond (100) surface. The bond lengths of the clean and hydrogenated dimers predicted by the DFT/SIMOMM method are close to those obtained from DFT calculations that employ periodic boundary conditions.[41,42,43,45,46,48] The clean dimer MCSCF bond length is slightly longer than the DFT value due to the anti-bonding contribution from the π^* orbital in the MCSCF calculation.

3.2 (110) and (111) Surfaces

The saturation of the dangling bonds on the diamond (111) and (110) surfaces can occur by reconstruction to form π-bonded chains. In the case of (110) there is one dangling bond per carbon atom and in the case of (111) there can be one or three dangling bonds per carbon atom. The saturation of the dangling bonds can also occur by hydrogenation of the surface.

The structure of the (110) diamond surface is the simplest of the diamond surfaces. The hydrogen covered 1x1:H surface is shown in Fig. 5 and the reconstructed surface with dimer chains is shown in Fig. 6. There have

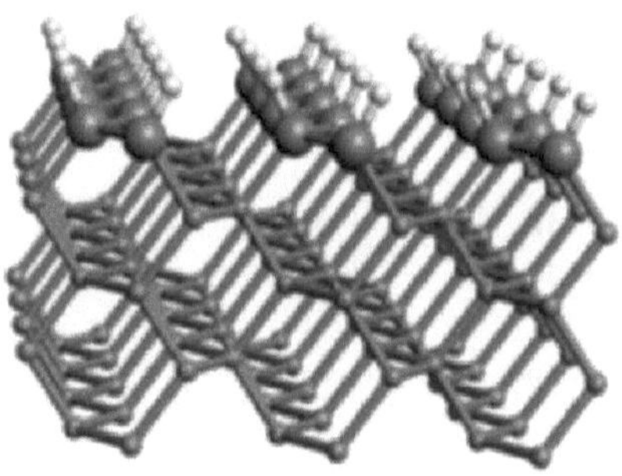

Figure 5. Diamond (110) 1x1:H surface.

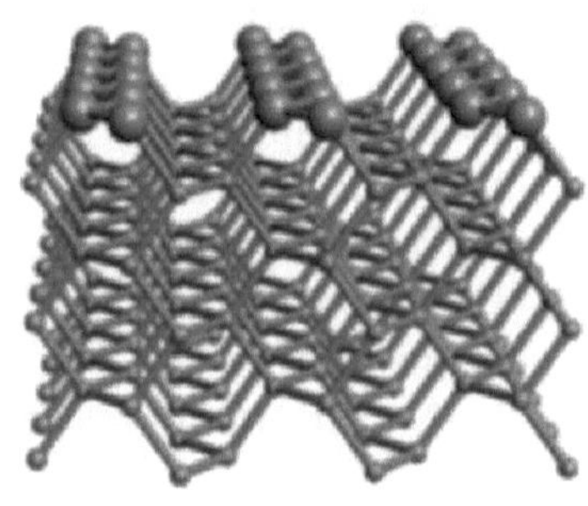

Figure 6. Clean diamond (110) surface.

been several theoretical studies of the clean and hydrogenated (110) surfaces.[55,56,57] The clean surface does not require reconstruction due to surface topology. Addition of hydrogen atoms allows relaxation to their ideal bulk-terminated positions.

The structure of the (111) surface is more complicated than the (110) surface because of the possibility of single and triple-dangling bonds. The hydrogen covered (111) surface with one hydrogen per carbon is shown in Figure 7(a). Theoretical studies[58,59,60,61,62,63,64,65] have focussed on the structures and relative stabilities of the different possible (111) surfaces. Calculations have generally found that at high hydrogen coverage the (111) surface with a single dangling bond has a 1x1 structure with hydrogen termination

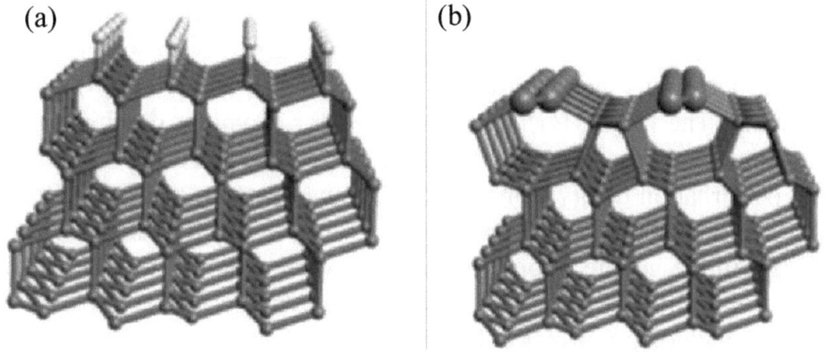

Figure 7. Diamond (111) surfaces: (a) (1x1):H surface, (b) reconstructed (2x1) surface.

as shown in Fig. 7 (a). The most stable clean (111) surface undergoes a 2x1 reconstruction with a Pandey π-bonded chain structure, shown in Fig. 7(b). The π-bonded chains are unbuckled. The (111) surface with three dangling bonds has been found to have a 2x1 reconstruction for both the clean and hydrogenated surfaces.[64]

4. GROWTH REACTION MECHANISMS

4.1 Hydrogen abstraction and migration

One of the key steps in many diamond chemical vapor deposition methods is the interaction of hydrogen atoms with hydrogen atoms on a diamond surface and with C-atom radical sites on the surface. The fraction of radical sites on the surface will depend on two reactions

$$C_d\text{-H} + H^\bullet(g) \rightarrow C_d^\bullet + H_2(g) \qquad (1)$$

$$C_d^\bullet + H\,(g) \rightarrow C_d\text{-H}\ , \qquad (2)$$

where C_d-H represents a hydrogen terminated diamond surface and $C_d^\bullet$ represents radical site on the surface. The potential energy surfaces for these reactions have been studied using various cluster models for the diamond surface and in some cases rates for these reactions have been derived from the calculated barriers.[7,66,67,68,69,70] In most of these studies small clusters have been used to investigate the potential energy surfaces. The largest clusters were used in an *ab initio* molecular orbital study of hydrogen abstraction (Reaction (1)) from the (111) and (100) diamond surfaces by

Brown *et al.*[66] This study represents the highest level theoretical calculation of hydrogen abstraction. The calculations were carried out on clusters of 46 and 35 carbon atoms modeling the (111) and reconstructured (100) surfaces, respectively. The HF/6-31G* level was used for geometry optimization and single point energies were calculated at the level of second order Möller Plesset perturbation theory using the 6-31G* basis set. In addition, effects of higher level correlation were investigated with the coupled cluster method, CCSD(T). In the geometry optimization, the carbon atoms near the abstraction site were allowed to relax. The calculated energy barrier for hydrogen abstraction including correction for higher level theory was 8.3 kcal/mol for the (111) surface and 11.1 kcal/mol for the (100) surface.[66] The higher barrier for (100) was ascribed to the incomplete relaxation of the C(100) surface carbon dimer towards planarity.

The migration of hydrogen on diamond surfaces can play a role in diamond growth mechanisms and has been studied theoretically for the various diamond surfaces.[71,72,73,74,75] In an *ab initio* molecular orbital (MP2) study of hydrogen migration on the (111) surface Larsson and Carlson[71] found a barrier of 59 kcal/mol for migration between radical sites. A barrier of 76 kcal/mol was reported by Chang *et al.*[73] from the semi-empirical Brenner potential. This result agrees with several previous theoretical studies of migration on the (111) surface including Frenklach and Skokov[74], that have investigated hydrogen migration barriers and rates between carbon radical sites on (100) surfaces using quantum chemical methods. They found barriers ranging from 66 to 110 kcal/mol. Thus, these studies have generally concluded that hydrogen migration by itself on diamond surfaces is likely to be slow and not have a significant role in diamond growth processes.

4.2 Acetylene

Hydrocarbon-hydrogen gas mixtures have proven to be the most successful for CVD growth. The dissociation of a gaseous mixture of H_2 and a simple hydrocarbon precursor results in a variety of species that can be dependent on the conditions and method used to activate the precursor gas. In conventional methods the presence of atomic hydrogen is likely to saturate the diamond surfaces with hydrogen as well as abstract hydrogen from the surface as discussed in the previous section. The latter process will provide surface radical sites that can react with the hydrocarbon species. In thermal methods acetylene is believed to be the major species involved. In this section we review theoretical studies of growth reactions involving acetylene.

A variety of computational studies have modeled the growth of diamond surfaces with acetylene as the growth species. These investigations generally postulate reaction mechanisms initiated by the creation of a surface radical via the abstraction of hydrogen (see Section 4.1), followed by addition of the growth species to the surface radical site.

4.2.1 Modeling using empirical thermodynamic data

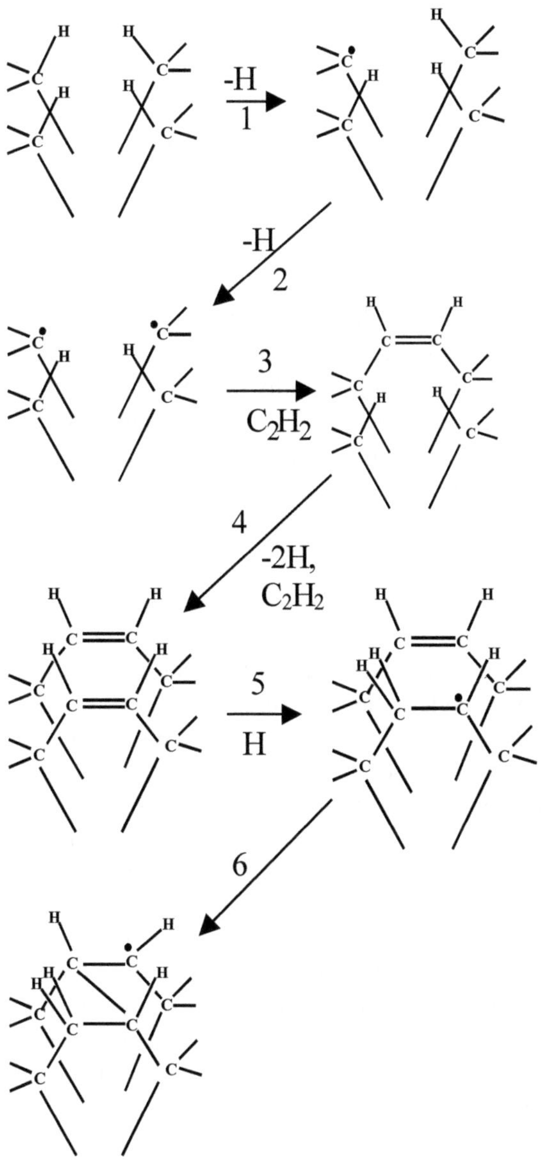

Figure 8. Illustration of the mechanism investigated by Belton and Harris (Ref. 75) for acetylene based growth on the (110) surface.

Harris and co-workers[76,77,78,79] used rate constants obtained from a combination of group-additivity methods and gas-phase reaction analogs to perform kinetic simulations of possible growth mechanisms on all three low-index surfaces, (111), (110), and (100), with C_2H_2 and CH_3 as the growth species. These investigations were not based on any quantum mechanics, but the mechanisms that they proposed proved to be very useful in future quantum chemical studies.

For acetylene-mediated growth on diamond (110) the mechanism proposed by Belton and Harris[77] is illustrated in Fig. 8. It involves hydrogen abstraction (Steps 1,2,4) and acetylene addition across the trough (Steps 3, 4). The completion of a new layer on the diamond surface occurs when two adsorbed ethylene-like groups link up after a hydrogen atom adds to one of the ethylene-like groups (Step 5). This changes the ethylene-like structure into an ethyl-like radical, which then attacks the other ethylene-like group to form a butyl-like radical (Step 6). Subsequent free-radical addition to other adsorbed ethylene-like groups rapidly completes the surface layer. The theoretical growth rate from this model was consistent with experiments.

Harris and Goodwin[79] also examined an abstraction/addition mechanism using C_2H_2 on a (100)-(2x1): H diamond surface. In this case they could not account for the experimental growth rates.

4.2.2 *Quantum chemical studies*

There have been numerous quantum chemical studies of the growth of diamond with acetylene as the growth species.[7,67,68,80,81,82,83,84,85,86,87,88,89] These studies have used cluster models for the diamond surfaces and various quantum chemical methods to calculate potential energy surfaces of the reaction mechanisms and reaction rates. In a series of papers Frenklach and coworkers[7,67,68,80,81,82,83,84] have investigated the energetics of acetylene growth reactions on the (111) and (100) surfaces using semiempirical, density functional, and *ab initio* molecular orbital methods. They have also carried out kinetics analysis of the presumed gas-phase and surface reactions involved. These kinetics simulations were based on data from similar hydrocarbon chemistry and provided insight into possible mechanisms. In their initial studies,[7] they suggested that removal of hydrogen from the surface is required to create vacant carbon radical sites that are reactive towards chemisorption of gaseous precursors. They found that the addition of C_2H_2 to a surface radical site on the (100) surface was a thermodynamically unstable adsorbate with a lifetime too short to allow its incorporation into the diamond lattice. They also found that addition to a biradical site was thermodynamically stable, but kinetically unstable as the adsorbate desorbs after a reaction with a hydrogen atom followed by β-scission of its C-C bond.

However, in calculations on the (100) surface they found that incorporation of acetylene into the diamond lattice is possible at several types of biradical sites as well as at a dimer monoradical if the surface migration of hydrogen atoms from the adsorbate is included in the reaction mechanism.[82] They carried out Projected MP2/6-31G* cluster calculations on a series of different reaction mechanism including hydrogen migration from the adsorbate to surface sites in some of the steps to calculate adsorption energies, barriers, and rates. Skokov *et al.*[82] found that reactive adsorption results in a $>C=CH_2$ adsorbate that can either add to a neighboring site, migrate along a dimer chain, or be etched away by reactions with hydrogen atoms.

There have been a number of other theoretical studies of diamond growth mechanisms based on acetylene. Latham et al.[86] examined the barriers to the addition of acetylene across two dimers on a (100) 2x1:H monohydrogenated surface using a cluster model. They used a local density functional method with pseudopotentials and found that the reaction is exothermic, but with a very high barrier. Besler, Hase, and Hass[85] used the semiempirical PM3 and MNDO methods to examine a growth mechanism in which acetylene adds to the (110) surface in a series of steps to yield an ethylene-like structure similar to that investigated by Belton and Harris.[77] They studied two mechanisms for the completion of a new surface layer via the linking of adjacent ethylene-like groups. Both methods predicted that the reaction of adjacent ethylene-like groups to form a singly bonded, butyl-like diradical is exothermic resulting in formation of a new diamond layer, with activation barriers of less than 20 kcal/mol. Yang et al.[90] used periodic density functional calculations to investigate addition of acetylene to a nonhydrogenated (111)-(2x1) surface. They found that the most favorable configuration is the "on-top" site of the Pandey chain that provides a template for assembly of polyethylene that runs parallel to the chain.

4.3 Methyl radical

The CVD diamond (100) surfaces are generally terminated by a 2x1 monohydride dimer. The 2x1 and 1x2 domains of the dimer arrays coexist on the diamond (100) surfaces. Single layer steps perpendicular and parallel to the dimer bond of the upper terrace are called S_A and S_B, respectively, as shown in Figure 9.[91,92] In typical diamond CVD using a CH_4/H_2 gas mixture, CH_3 radical is presumed to be the precursor for crystal growth. As the CH_4/H_2 ratio in the source gas decreases, the CVD diamond growth rate decreases, and the diamond crystal quality is improved. Flat surfaces are grown along the dimer arrays at a low (e.g., 2%[92]) CH_4/H_2 ratio, producing the so-called step-flow growth.[91,92] Step-flow growth selectively proceeds at the S_B step (Fig. 9(e)). Nucleations on the terrace increase at a higher (e.g., 6%[92]) CH_4/H_2 ratio, and the selectivity of the growth direction decreases.

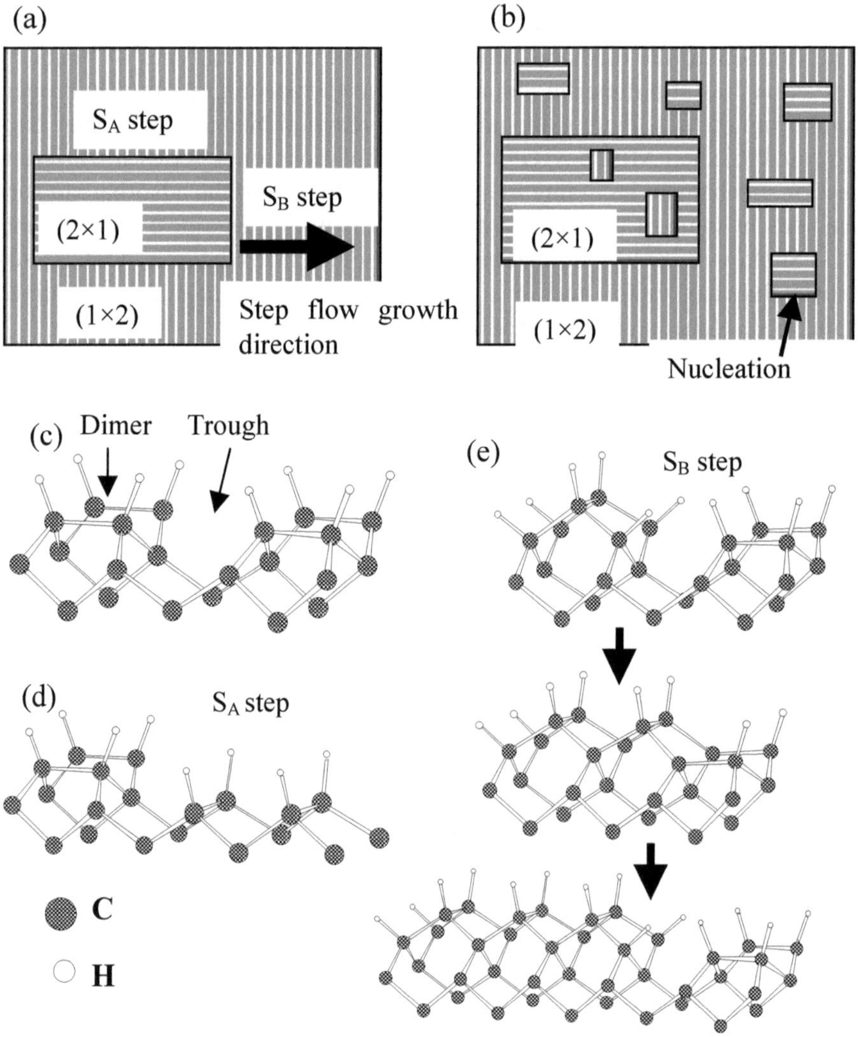

Figure 9. Schematic illustrations of diamond (100) surfaces, (a) step-flow growth at low CH_4/H_2 ratio, (b) nucleations on terrace at high CH_4/H_2 ratio, (c) dimer and trough on flat terrace. (d) S_A step, and (e) step-flow growth at S_B step.

The growth mechanisms of diamond surfaces have been theoretically investigated using empirical potentials,[79] semi-empirical[81,82,84,93] and *ab-initio* quantum mechanics (QM),[94] DFT calculations,[49,95,96,97] Monte Carlo (MC) simulations,[75,97,98] and tight-binding potentials determined by DFT (DFTB).[93,99] Harris and Goodwin investigated the chemical reactions of hydrogenated diamond (100) surfaces with the CH_3 radical and with hydrogen using molecular mechanics (MM) calculations and proposed two kinds of reaction mechanisms, the dimer mechanism and the trough mechanism, for diamond (100) growth.[79] Tamura et al.[49] investigated these mechanisms using GGA and the results are shown in Figs. 10 and 11 for the dimer and trough mechanisms, respectively. The dimer mechanism is initiated by an H abstraction from the monohydride dimer that causes new nucleation on the

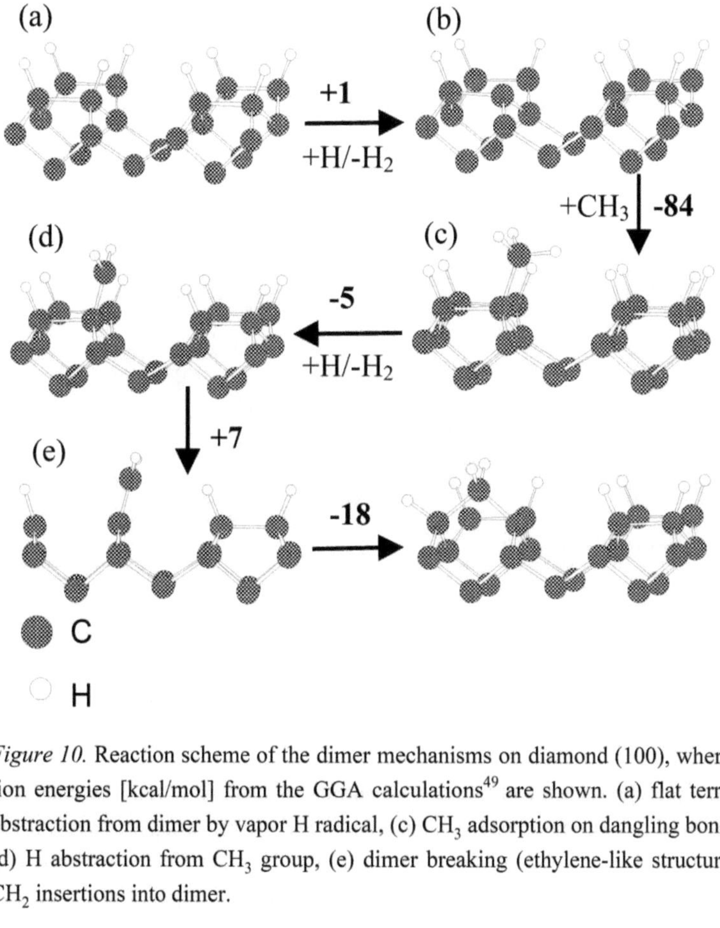

Figure 10. Reaction scheme of the dimer mechanisms on diamond (100), where the reaction energies [kcal/mol] from the GGA calculations[49] are shown. (a) flat terrace, (b) H abstraction from dimer by vapor H radical, (c) CH_3 adsorption on dangling bond at dimer, (d) H abstraction from CH_3 group, (e) dimer breaking (ethylene-like structure), and (f) CH_2 insertions into dimer.

terrace. The trough mechanism is initiated by an H abstraction from the S_B step edge, thereby promoting step-flow growth. The trough mechanism can also proceed at the CH_2-inserted dimer created by the dimer mechanism as well as at the S_B step edge. H abstraction from the S_B step edge is predicted by DFT calculations to be preferred to abstraction from the monohydride on the terrace, due to both steric repulsion between H atoms and the relaxation of the dangling bond at the S_B step edge.[49] The high probability of H abstraction at the S_B step edge can rationalize the selectivity of step-flow growth at low CH_4/H_2 ratio. Semi-empirical (PM3) and DFTB calculations indicate that diffusion of CH_2 groups on the diamond (100) surface plays an important role in the step-flow growth.[93]

Calculations predict substantial activation energies for the CH_2 insertion in the dimer mechanism.[95] Table 2 compares the SIMOMM[23] energy changes that occur during the CH_2 insertion process with previous DFT/B3LYP calculations.[95] The SIMOMM calculations employed several methods as the QM part of the QM/MM calculation: DFT/B3LYP,[16] second order perturbation theory (MP2),[100] MCSCF,[54] and multi-reference MP2 (MRMP2).[101] For the QM/MM calculations, the CH_2 group and reactive dimer are modeled

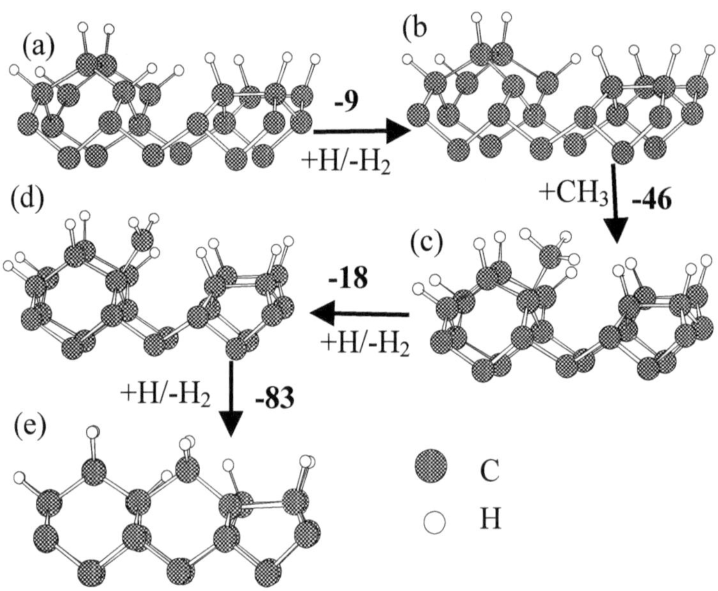

Figure 11. Reaction scheme of trough mechanisms on diamond (100), where the reaction energies [kcal/mol] from the GGA calculations[49] are shown. (a) S_B step edge, (b) H abstraction from S_B step edge by vapor H radical, (c) CH_3 adsorption on dangling bond at S_B step edge, (d) H abstraction from CH_3 group, and (e) CH_2 bridging across trough.

with an *ab-initio* cluster, while the surrounding diamond lattice is modeled with a MM cluster. The small $C_{10}H_{15}$ *ab-initio* cluster is embedded into a larger MM cluster (Fig. 12(a)). In this model, the bond breaking and formation via the CH_2 insertion are treated with *ab-initio* methods, while the "observer" interactions (e.g., steric repulsions from surrounded dimers) and steric stain from the diamond lattice are efficiently calculated with an empirical force field. Transition states are found for the dimer breaking and ring-closing processes. The activation energy for closing the ring is predicted by all methods to be larger than that for breaking the dimer. There

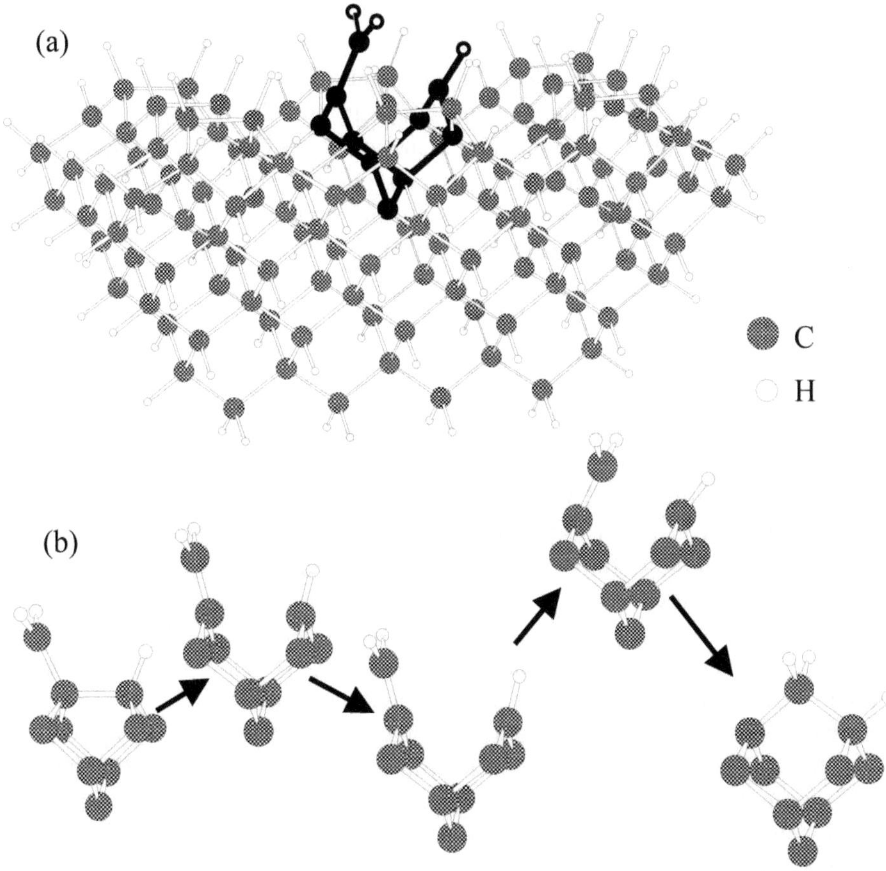

	TS1	dimer-breaking	TS2	ring-closing
MCSCF	10.2	-2.6	16.1	-14.6
MRMP2	2.0	-6.5	7.4	-18.2
MP2	17.4	3.6	25.4	-15.1
B3LYP	9.6	-0.5	13.6	-10.5

Table 2. Energy changes (relative to the initial structure) in the dimer mechanism (kcal/mol) in Fig. 12 The values of MCSCF, MRMP2, and MP2 were calculated using SIMOMM. The B3LYP values are from Ref. 35. The TS1 is the first transition state in Fig. 12b and TS2 is the second transition state in Fig. 12b.

fore, the ring-closing reaction is considered to be the rate-determining step of the dimer mechanism. Note also that single reference methods such as DFT and MP2 are quantitatively unreliable for this mechanism, with their predicted barriers for both steps being substantially too large (see Fig. 12).

The effects of etching on the CVD growth of diamond (100) have also been investigated theoretically.[97] Kinetic Monte Carlo simulations using theoretically calculated activation energies indicate that the etching of surface carbon during CVD growth plays an important role in the slow growth rate and smoothing of the diamond (100) surface.[97]

4.4 Carbon dimer

Diamond films grown by the conventional plasma CVD methods from hydrogen/hydrocarbon mixtures typically have a microstructure with micron-size crystallites.[102] The fraction of atoms in the grain boundaries relative to the bulk in the case of microcrystalline diamond is extremely small. In contrast, diamond films grown from hydrogen poor argon/fullerene or argon/methane microwave plasmas are characterized by a microstructure consisting of crystallites with an average size of 3-10 nm.[5,6,103,104] The latter films are called ultrananocrystalline diamond (UNCD). Use of hydrogen-poor growth conditions systematically produces uniform films with controlled grain size in a few nanometer range and with thicknesses up to several tens of microns. These films have a large percentage of atoms at the grain boundaries that reaches 10% for an experimental grain size of 3 nm. Many physical properties such as electrical conductivity, mechanical, and optical properties of ultrananocrystalline diamond differ from both natural and microcrystalline diamond[5] and open opportunities for specific applications such as tribology, MEMS and electronic devices. The carbon dimer has been proposed as the principal growth species in hydrogen poor plasmas used to grow diamond.[5,104] A number of computational studies have modeled the growth of diamond surfaces with carbon dimer as the growth species. These investigations have generally postulated reaction mechanisms with diamond growth occurring by insertion of C_2 into the C-H bonds of the hydrogen-terminated diamond surface or into π bonded carbon dimers on dehydrogenated diamond surfaces. The growth mechanism for diamond films grown from hydrogen poor argon/fullerene or argon/methane microwave plasmas are significantly different than those in hydrogen rich plasma since the latter depends on hydrogen abstraction by the hydrogen present in the plasma. In this section we review the work on C_2 growth mechanisms that has been done on various diamond surfaces.

4.4.1 C_2 growth reactions on (110) 1x1:H surface

Several quantum chemical studies of C_2 growth reactions on a hydrogen

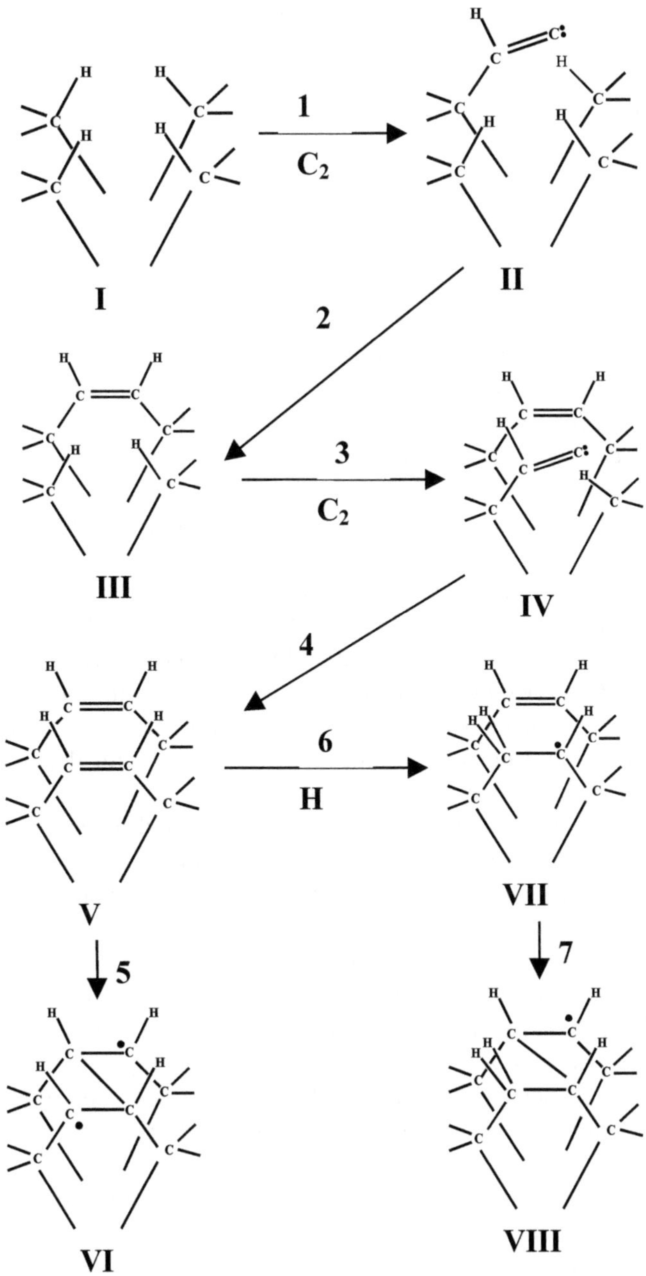

Figure 13. Illustration of the growth mechanism on the (110) surface with C_2 as the growth species.

covered diamond (110) surface have been reported.[105,106] A mechanism similar to that of Harris and Belton[77] for acetylene addition to the (110):1x1:H surface was proposed for C_2 growth. This mechanism is illustrated in Fig. 13. *Ab initio* molecular orbital methods were used to examine the energetics of small-molecule models of the initial steps of this growth mechanism, namely, insertion of C_2 into a C-H bond in methane, insertion of CCH into a C-H bond in methane, and the rearrangement of methylvinylidene to methylacetylene. At the G2 level of theory[13] it was found that C_2 could insert into a C-H bond of methane with no energy barrier, and that insertion of CCH_2 into a C-H bond of methane had a barrier of only 17 kcal/mol.

A subsequent more thorough quantum chemical study of C_2 addition to the diamond (110) surface was carried out using molecular orbital and density functional methods.[106] Hydrogen-terminated diamond-like clusters were used to represent the (110) diamond surface. In the stepwise mechanism in Fig. 13, one C_2 unit adds to the surface by inserting itself into first one C-H bond (step 1) and then a second (step 2), producing an adsorbed ethylene-like structure (**III**). A second C_2 can then insert itself into two other C-H bonds (steps 3 and 4) to produce a surface with two adjacent ethylene-like groups (**V**). Formation of a C-C single bond between adjacent ethylene-like groups can produce a new layer on the diamond surface via step 5 or via steps 6 and 7 (producing structures **VI** and **VIII**, respectively). All of these steps were examined as well as a concerted addition step (replacing steps 1 and 2), in which the C_2 molecule inserts into two C-H bonds simultaneously. Compared to previous mechanisms proposed for diamond growth under CVD conditions, this mechanism was unique in that it was not dependent upon the abstraction of hydrogen atoms from the surface.

In this study the (110) surface of diamond was modeled using 18- and 48-atom carbon clusters, with dangling bonds terminated by hydrogens. The $C_{18}H_{26}$ cluster with C_{2h} symmetry, illustrated in Fig. 14, can accept two added C_2 molecules across the trough of the (110) surface. This structure was fully optimized at the AM1, HF/3-21G, and HF/6-31G* levels. A larger, $C_{48}H_{50}$ model of the diamond (110) surface was also used and is illustrated in Fig. 14. Calculations on these two clusters indicated that addition of a C_2 to the diamond (110) surface is very favorable, with large energy lowerings of about 150-180 kcal/mol per C_2. The barriers for addition of C_2 to the (110) surface are small. The two-step addition of C_2 to the C-H bonds on the surface has a barrier only 5 kcal/mol at the BLYP/6-31G*//AM1 level. A single-step concerted mechanism for C_2 addition has no barrier at the BLYP/6-31G*//AM1 level. Adjacent C_2 moieties on the (110) surface, adsorbed in ethylene-like arrangements, can be connected via a radical mechanism involving initiation by hydrogen atom addition to the double bond of one ethylene-like group. Calculations indicate that there is little or no energy barrier for this reaction, and that the ethyl-like radical intermediate is stable with respect to ß-scission. A path for linking adjacent surface ethylene-like species directly, without the assistance of hydrogen addition, also exists. The BLYP/6-31G*//AM1 barrier for this process is ca. 2 kcal/mol and corresponds to the linking of the first two ethylene-like units to form a singly-bonded diradical structure. The completion of a new surface layer via formation of single bonds between radical structures and ethylene-like groups results in an energy lowering of about 20 kcal/mol.

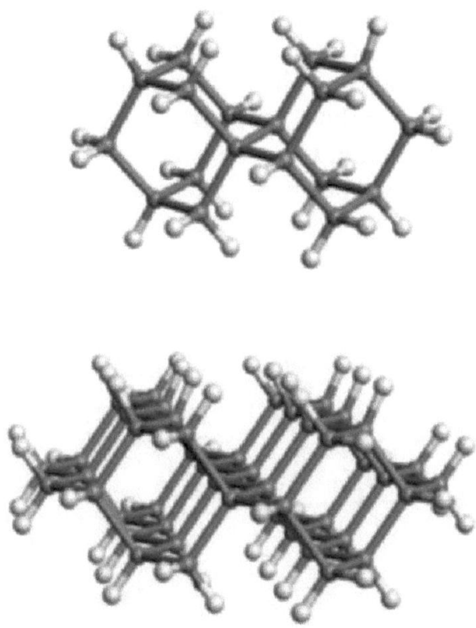

Figure 14. Illustration of $C_{18}H_{26}$ and $C_{48}H_{50}$ clusters used for modeling the (110) surface for investigation of C_2 addition reactions.

4.4.2 C_2 growth reactions on clean (110) surface

A density functional based tight binding study of growth steps involving addition of C_2 on a clean diamond (110) surface (Fig. 6) has been reported by Sternberg *et al.*[107] The surface was simulated using a two-dimensional slab geometry with six to eight carbon monolayers. The calculations indicated that initial C_2 adsorption onto a clean (110) surface proceeds with small barriers (2-4 kcal/mol) into a lattice site. The adsorption energy for one C_2 is about 180 kcal/mol. The reaction pathway and energetics of this insertion are very similar to that on the hydrogenated surface (see above).

Sternberg *et al.*[107] found that the addition of more carbon dimers near the first one, leads to C_{2n} chains along the [110] direction on the surface. The adsorption energies are in the range 160-230 kcal/mol per C_2 at adsorption sites that lead to chain growth and slightly smaller 115-160 kcal/mol for sites that do not result in chain continuation, i.e. defect sites. The barriers for additional C_2 insertion are small (0-12 kcal/mol). The calculations indicate that the C_2 chain addition mechanism eventually leads to coalescing chains with broken backbonds on either side. Relaxation was carried out on a model in which every other trough along the [110] direction was covered with a contiguous chain. This results in a C(110):2x1 reconstruction shown in Figure 15(a) that has multiple bent graphene sheets along the [110] direction. The bending direction is that of a carbon nanotube of the (n,n) type, i.e. the armchair tube. Continued addition of C_2 in the valley between the two arches revealed that it was possible, without barrier, to cause the sp^2-like carbons

(a)

(b)

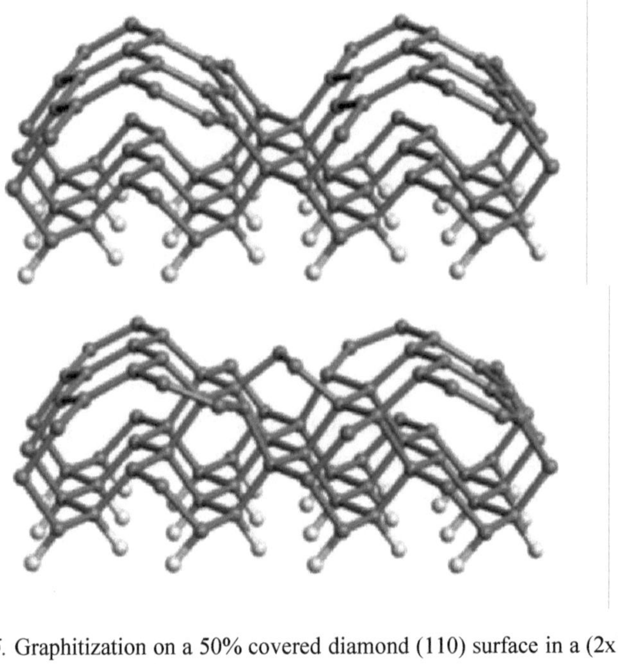

Figure 15. Graphitization on a 50% covered diamond (110) surface in a (2x1) reonstruction (a) and induced rebonding after depostion of additional C_2 (b). Adapted with permission from Ref. 107

near the absorbate to return to an sp^3 configuration and rebond in the diamond structure, as shown in Fig 15(b).

Sternberg *et al.*[107] also investigated diffusion of C_2 on the clean (110) diamond surface. They found that diffusion of C_2 that inserted in tightly bonded sites is highly unlikely. The diffusion barriers are greater than 75 kcal/mol due to the strong covalent bonds that are formed. The only case for which an inter-island diffusion path exists occurs when a C_2 is added on top of a C_{2n} chain. The energy barrier along an adsorbate ridge is on the order of 20 kcal/mol. Such a C_2 will diffuse until it reaches the end of the chain and will then be incorporated there. Thus, increased diffusion rates could increase growth rates.

4.4.3 C_2 growth reactions on (100) 2x1:H surface

The (100) diamond surface is the slowest growing surface and thus plays an important role in the growth mechanism. Under growth conditions used to grow diamond films with the C_2 dimer, it is likely that the (100) surface has a substantial fraction (0.1-1%) of sites devoid of hydrogen.[5] The depositions are carried out with small amounts of hydrogen admixtures (1-2%), while the gas species present in overwhelming quantity is argon, which may act to remove some of the surface hydrogen, particularly at the substrate temperatures of 700-900°C in which the depositions are carried out. The structures of both the unhydrided (clean) and monohydrided diamond (100) surfaces have been investigated for C_2 growth reactions.[108,109,110] In this section we review results for the 1x1:H surface. In the next section growth reactions on the clean surface are reviewed.

Gruen *et al.*[108] have investigated C_2 insertion reactions on the (100) 2x1:H surface using density functional theory. A small cluster (C_9H_{14}, Fig.

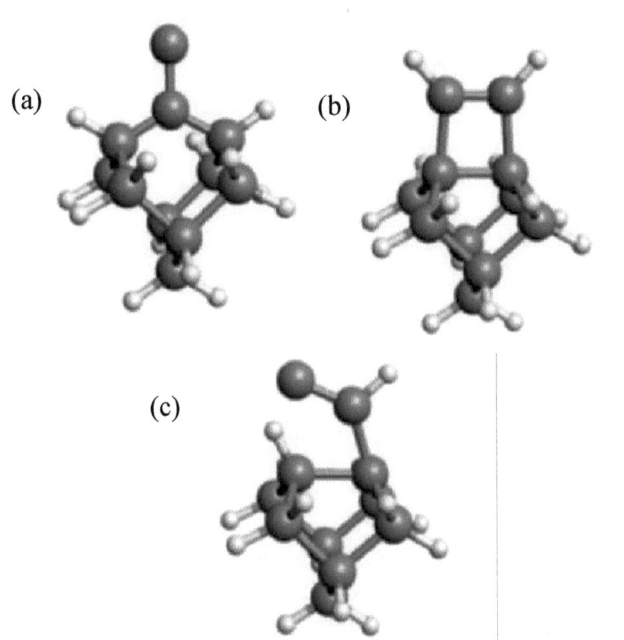

Figure 16. Products from addition of C_2 to C_9C_{14}: (a) carbene-like structure, (b) cyclobutene-like structure, (c) vynilidene intermediate

1) was used to model the initial steps of the reaction of C_2 with the surface. A carbene-like structure shown in Fig. 16a resulting from insertion of one end of C_2 into a C-C single bond on the C_9H_{14} cluster is about 32 kcal/mol higher in energy than the cyclobutene-like structure (Fig. 16b) resulting from insertion of the C_2 into two C-H bonds. For both the carbene-like and cyclobutene-like structures the reaction paths from the C_2 and C_9H_{14} reactants involve the initial insertion of C_2 into a C-H bond to produce a common, monosubstituted vinylidene intermediate (Fig. 16c), but the energy barrier between this intermediate and the carbene-like structure is about 60 kcal/mol higher than the barrier along the path from this intermediate to the cyclobutene-like product. Thus, insertion of C_2 into C-H bonds to form the cyclobutene-like structure (Fig. 16b is favored on the monohydrided surface. Starting from the cyclobutene-like structure, formation of a new layer of the existing monohydrided surface proceeds readily.[82]

4.4.4 *C2 growth reactions on clean (100) 2x1 surface*

The C_2 growth reactions on the clean (100) diamond surface have been modeled in its early stages on cluster models[108] and a more comprehensive investigation was carried out with two-dimensional periodic calculations.[109] In this section we review these two studies and the implications for growth

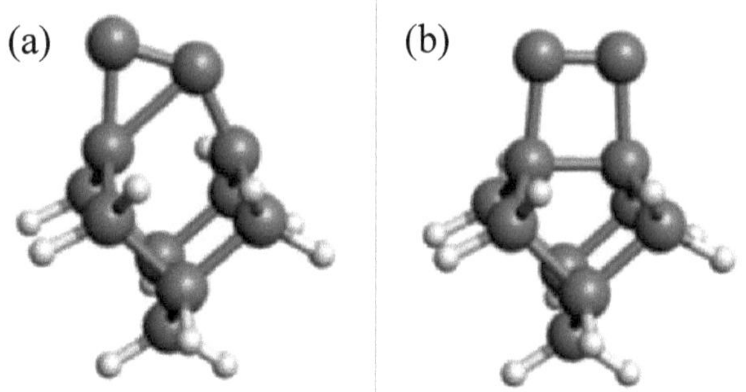

Figure 17. Products from addition of C_2 to C_9C_{17}: (a) carbene-like structure, (b) cyclobutyne-like structure.

on this surface.

Gruen *et al.*[108] have investigated C_2 insertion reactions on the (100) 2x1 surface using density functional theory with a small cluster (C_9H_{12}) to model the initial steps of the reaction of C_2 with the surface. The calculations indicate that insertion of C_2 into a carbon-carbon double bond leads to a large energy lowering of about 120 kcal/mol for the C_9H_{12} cluster. The insertion of a single end of the C_2 into a C=C bond to yield a carbene-like structure shown in Fig. 17a, occurs with no energy barrier and is 33 kcal/mol more exothermic than the double insertion that yields a cyclobutyne-like structure (Fig. 17b). These results suggest that the mechanism of diamond growth on the diamond (100) surface based on C_2 addition is quite dependent on the degree of hydrogenation of the surface. On the clean surface the most energetically favorable addition of C_2 is insertion of C_2 into a C=C bond to produce the carbene-like structure (Fig. 17a), while insertion of C_2 into C-H bonds to form the cyclobutene-like structure (Fig. 16b) is favored

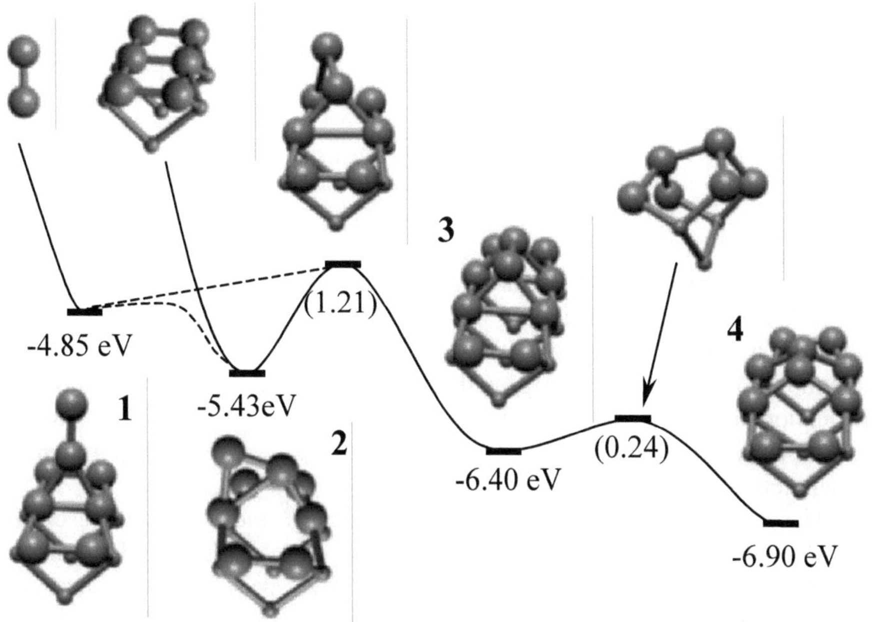

Figure 18. Potential energy surface from density functional based tight binding study of C_2 addition to a clean diamond (100) 2x1 surface. Energies are in eV. Barrier heights are in parentheses. Adapted with permission from Ref. 109.

on the monohydrided surface. Sternberg *et al.*[109] have used a density-functional based tight binding method (DFTB) to carry out a more comprehensive study of diamond growth by C_2 on a clean (100) surface. The clean (100) diamond surface was modeled using a supercell containing 16 atoms per layer and 8 carbon monolayers. The potential energy surface for addition of C_2 to the surface is summarized in Fig. 18. The most stable configuration (labeled structure **4** in Fig. 18) is a bridge structure between two adjacent surface dimers along a dimer row. The surface dimers are opened by the insertion. This insertion product has an adsorption energy of 159 kcal/mol and was not found in the cluster study[108] due to the small cluster size that was used. A barrier of ~28 kcal/mol must be overcome to reach it from a higher energy configuration labeled **2** in Fig. 18. Other adsorption configurations were located[109] in the investigation with adsorption energies differing by up to 60 kcal/mol including the structures found in the cluster study.[108] Thus the results for isolated adsorbates indicate that there are relatively large barriers of >20 kcal/mol that must be overcome to reach growth positions. A some-

Figure 19. Structures resulting from an agglomeration of two adsorbates labeled **2** in Fig. 18.

what lower barrier of 12-18 kcal/mol was found if the structure **2** in Fig. 18 is neighboring a trough-bridging ad-dimer. Conversely, neighboring adsorbates on a dimer row (Fig. 19) were found to have high barriers (~50 kcal/mol) to conversion into growth positions and therefore are likely re-nucleation sites. The barriers to diffusion were found to be very high (40-60 kcal/mol) effectively precluding diffusion on the surface at experimental growth temperatures.

The results of the study by Sternberg *et al.*[109] indicate that the clean (100) surface constitutes sites for both growth and re-nucleation. In contrast C_2 addition on the hydrogenated (100) surface was found to lead to growth.[108] Because only a small fraction of the (100) surface sites are unhydrided, most C_2 insertion events likely lead to growth, with only a few leading to re-nucleation. However, this is likely to be sufficient to explain the small grain size observed in ultra-nanocrystalline diamond.[5]

4.4.5 C_2 growth reactions on clean (111) 2x1 surface

The (111) surface has been found experimentally to dehydrogenate at temperatures above 1000 K and to transform into the Pandey-chain reconstruction.[111] Yang *et al.*[90] have investigated addition of C_2 to the reconstructed (111) surface using periodic density functional theory. They found a growth path for Pandey chains starting along the unreconstructed surface and leading to a graphite ad-layer.

5. ADSORBATES ON DIAMOND SURFACES

5.1 Oxidation reactions

Since oxidation is important for polishing, etching and wear of diamond surfaces, it is useful to employ theory to understand oxidation mechanisms. Oxidation of diamond surfaces has been studied experimentally[112,113,114,115] and theoretically.[44,48,116,117,118,119,120] The chemisorbed species on the diamond surface were investigated using Fourier transform infrared (FTIR) spectroscopy and temperature programmed desorption (TPD) . Various functional groups, such as carbonyl (C=O) and cyclic ether (COC), were found on the diamond surface.[113] Thermal desorption of the C=O groups from the diamond (100) surface was observed at high temperatures, while the C-O-C groups remain on the surface[114] at these temperatures. Oxidation of diamond

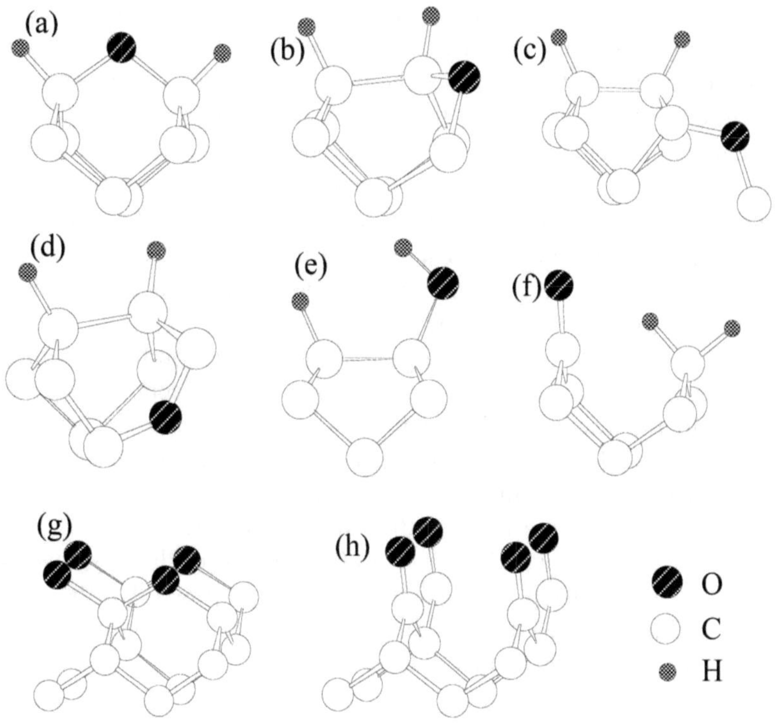

Figure 20. Equilibrium geometries of oxygenated diamond (100) surface. (a) C-O-C (ether) structure, O insertions into (b) second layer and (c),(d) third layer, (e) hydroxyl group, (f) C=O (ketone) group, (g) C-O-C monolayer and (h) C=O monolayer.

is known to proceed only on the top surface. In contrast, oxidation of silicon extends into lower layers. Semi-empirical[120] and GGA[48] calculations indicate that the ether (Fig. 20(a)), hydroxyl (Fig. 20(e)) and ketone (Fig. 20(f)) structures formed on the diamond (100) surface are stable, while oxidation of lower layers (Fig. 20(b), 20(c) and 20(d)) is found to be unfavorable.[48] Skokov et al.[120] performed semi-empirical (PM3) molecular dynamics calculations and interpreted the vibrational spectra of the oxygenated diamond (100) surface. These calculations support the stabilities of the experimentally observed functional groups on the diamond (100) surface and the fact that oxidation of diamond proceeds only on the top surface layer.

Based on LEED patterns, Thomas *et al.* revealed that the completely oxygenated diamond (100) surface exhibits a 1x1 structure.[112] There are two possibilities for the oxygenated diamond (100) 1x1 structure. One of these is the bridge ether type (Fig. 20(g)), and the other is the on-top ketone type (Fig. 20(h)). The GGA calculations predict that the most stable structure of the 1x1 oxygen monolayer depends sensitively on the lattice parameters.[48] The diamond (100) lattice stretches the periodic C-O-C structures on the surface, while steric repulsions are found between C=O groups.[48,120] Therefore, as the cell parameters increase, the bridge ether becomes unstable and the on top ketone becomes more stable.[48] Oxidation reactions on the diamond (100) surface as a function of oxygen coverage have also been extensively investigated using computations.[48,120] The ether structures are found to be formed preferentially at low oxygen coverage, while ketone structures increase in occurrence as oxygen coverage increases.[48]

The diamond band-gap is so large that the conduction band minimum is higher than the vacuum level. Therefore, the excited electron at the surface escapes into the vacuum. This is the so-called negative electron affinity (NEA).[44,121] The NEA of the hydrogenated diamond surface disappears upon oxidation. The changes in the electronic properties of the diamond (100) surface upon oxidation have been studied with DFT calculations.[44,115] Loh *et al.*[115] investigated the local density of states (LDOS) of the oxygenated diamond (100) surface using DFT calculations and proposed that the occupied lone-pair and unoccupied anti-bonding π^* orbital of the C=O group create a smaller band-gap than does bulk diamond.

5.2 Effects of sulfur and oxygen on growth mechanisms

Diamond is potentially useful as a material for electronic devices, due to its extreme properties, e.g., the highest hardness, high thermal conductivity, wide band-gap, and high carrier mobility. These properties are suitable for electronic devices working at high power, high speed, high temperatures, and other extreme environments. Therefore, diamond devices are expected to overcome the limitations of silicon devices.

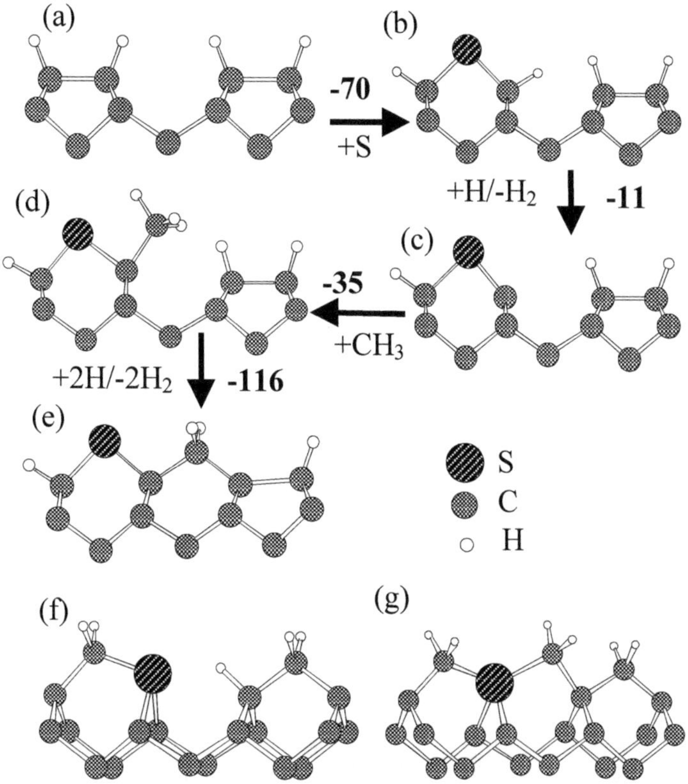

Figure 21. Reaction scheme of S-assisted growth mechanism of diamond (100), where the reaction energies [kcal/mol] from GGA calculations[96] are shown. (a) flat terrace, (b) S insertion into dimer bond, (c) H abstraction from S-inserted dimer, (d) CH_3 adsorption on dangling bond at S-inserted dimer, and (e) CH_2 bridging across trough. (f) 3-coordinated and (g) 4-coordinated S on diamond (100).

Since electronic devices are generally composed of p- and n-type semiconductors, both the p- and n-type semiconducting diamond must be developed to realize diamond electronic devices. P-type semiconducting diamond is easily obtained by boron doping,[122,123] and is expected to realize high speed and high power devices due to its high carrier mobility. However, n-type diamond is still difficult to obtain, even though some impurities (e.g. N, P, S, and O) have been studied as possible donor dopants.[123,124,125,126,127,128] Unlike silicon semiconductors, annealing techniques cannot be used for diamond, because it undergoes a transformation to graphite. Therefore, impurity doping using the CVD growth reaction[124,125,126] and ion-implantations[127,128] are important techniques for the development of diamond electronic devices. Recently, n-type diamond was successfully produced by incorporating phosphorous[124] and sulfur[125] into CVD diamond.

It is known that the crystal quality of CVD diamond can be improved by including S[125] or O[129] in the source gas (e.g. CVD using a $CH_4/H_2/H_2S$ or a $CH_4/H_2/O_2$ gas mixture). While sulfur can be incorporated into the diamond lattice by the CVD process,[125] oxygen incorporation during the CVD growth process has never been observed. The effects of sulfur and oxygen on the growth mechanisms of CVD diamond (100) have been investigated using GGA calculations.[96] Figure 21 shows a reaction scheme for the sulfur-assisted growth of CVD diamond (100). The S insertion into the dimer bond makes the H abstraction easier than for the pure monohydride because of the steric repulsion between the H atoms introduced by the presence of the sulfur. Then, the formation of the CH_2 bridge across the trough proceeds via CH_3 adsorption and H abstraction. Furthermore, GGA calculations predict

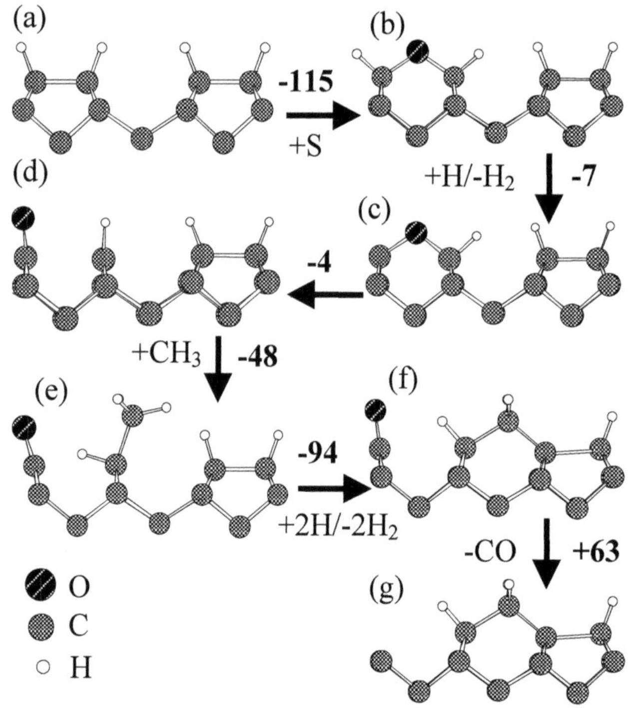

Figure 22. Reaction scheme of O-assisted growth mechanism of diamond (100), where the reaction energies [kcal/mol] from GGA calculations[96] are shown. (a) flat terrace, (b) O insertion into dimer bond, (c) H abstraction from O-inserted dimer, (d) C=O (ketone) formation with C-O breaking, (e) CH_3 adsorption on dangling bond adjacent to C=O, (f) CH_2 bridging across trough, and (f) CO desorption.

that the inserted S decreases the tensile stress of the CH_2 bridged structure.[96]

The stress is calculated as the derivative of total energy (E), dE/dx, with respect to cell strain (x) in periodic boundary calculations.[130] A reaction mechanism for the incorporation of S into the diamond lattice has also been proposed based on GGA calculations.[96] Stable 3-coordinate and 4-coordinate sulfur (Figs. 21(f) and 21(g)) are formed on the diamond (100) surface via CVD growth reactions. Although 3- and 4-coordinated sulfur are often unstable in organic molecules, they may be formed on diamond (100) surfaces due to steric strain from lower layers of the diamond lattice.

A similar reaction scheme might be proposed for the oxygen-assisted CVD growth of diamond (100). Unlike sulfur, in the case of oxygen, the on-top C=O (ketone) structure is also stable on the diamond (100) surface. Therefore, a competing reaction scheme has also been proposed for oxygen-assisted growth shown in Fig. 22[96] In analogy with S-assisted growth, O insertion into the dimer initiates growth reactions. However, the ketone structure can be formed during the growth reactions. This C=O group is expected to be desorbed as CO during the high temperature CVD process. So, oxygen functions as a catalyst for CVD growth. The GGA calculations predict that 3-coordinated oxygen is less stable than 3-coordinated sulfur on the diamond (100) surface, due to the presence of an orbital that is antibonding between the C-O-C and CH_2 groups.[96] The difference between sulfur and oxygen apparently originates form the stronger core interaction and shorter C-O bond length of oxygen. These calculations predict that S incorporation into the diamond lattice during CVD growth is energetically preferred relative to O incorporation due to the relative stabilities of intermediates on the surface, even though in the bulk crystal, S-substituted diamond is less stable than O-substituted diamond.[131]

6. SUMMARY AND OUTLOOK

It is clear that growth reactions on diamond surfaces are very complex and that computational studies have helped to advance our understanding of CVD diamond growth mechanisms. In recent years, density functional theory has emerged as an important theoretical tool to study these reactions, due to the accuracy of hybrid functionals such as B3LYP. However, for reaction mechanisms in which diradicals play a role, any single configuration-based method, including DFT, is unlikely to produce reliable results. Modeling of surface reactions on a diamond surface requires considering large numbers of surface atoms in the calculations of surface reactions. This has been achieved by using less expensive methods. Two promising examples of such approaches are a density functional based tight binding method that involves careful parameterization using density functional results, and SIMOMM, which combines molecular orbital calculations in the chemically active region with a molecular mechanics approach in the surrounding region.

We have reviewed calculations on the three prominent low-index surfaces of diamond (100), (110), and (111). The unreconstructed surfaces have dangling bonds and the presence of dangling bonds means that the surfaces are easily hydrogenated or reconstructions can occur. The (100) surface has been of most interest because of its importance in the CVD of diamond. Understanding all three surface structures is important for modeling growth mechanisms and theoretical studies have provided structural and energetic information that is relevant to understanding the growth mechanisms.

Most quantum chemical studies have investigated mechanisms for diamond growth that involve methyl radical and acetylene as the growth species. One of the key steps in growth involving methyl and acetylene is the abstraction of hydrogen from the surface by interaction of hydrogen atoms and the migration of hydrogen on the surface. Theoretical studies have provided information on the barrier for abstraction and migration. Studies of acetylene addition to the radical sites have found that even though some additions are stable thermodynamically, they are unstable kinetically. Thus, it is important to consider the kinetics of the growth mechanisms. In addition, it is necessary to consider different adsorption sites. It is believed that incorporation of acetylene into the diamond lattice is possible at several types of sites on the surface if the surface migration of hydrogen atoms from the adsorbate is included in the reaction mechanism.

Recently, the carbon dimer has been proposed as the principal growth species in hydrogen poor plasmas used to grow diamond. In this chapter we have reviewed several theoretical studies of growth mechanisms involving C_2. These growth mechanisms are different from the conventional ones in that hydrogen abstraction is not required as a first step. It has been found that C2 can add with little or no barrier to a C-H or C-C double bond and lead to growth. In some cases the C_2 addition can result in new nucleation sites, which may explain crystallite sizes obtained under low hydrogen conditions.

The effect on growth mechanisms of addition of different chemical species such as oxygen- and sulfur-containing species has also been reviewed in this chapter.

Accurate quantum chemical studies can provide much of the information necessary for kinetic modeling of diamond growth processes. The future holds great promise for more accurate studies of reaction pathways, reaction barriers and energies of large number of different possible reactions of growth species with diamond surfaces. Development of new methods and models will make the goal of complete understanding of surface chemistry and growth mechanisms feasible. Coupling quantum chemistry with kinetic modeling and further with plasma reactor modeling will lead to full simulation of diamond growth processes. This will advance our understanding of growth mechanisms and tunability of parameters of diamond synthesis.

ACKNOWLEDGEMENTS

This work is supported in part by the U.S. Department of Energy, BES-Materials Sciences, under Contract W-31-109-ENG-38 (PZ and LAC). This work is supported in part by a grant from the Air Force Office of Scientific Research (MSG) and a grant from the Japanese Society for Physical Science (HT).

REFERENCES

1. *Properties of Diamond*, Ed. J. E. Field, Academic Press, London, 1979.
2. F. P. Bundy, H. T. Hal, H. M. Strong, and R. H. Wentorf, *Nature* **176**, 51 (1955).
3. J. C. Angus and C. C. Hayman, *Science* 241, 913 (1988).
4. *Synthetic Diamond*, Eds. K. E. Spears and J. P. Dismukes, John Wiley and Sons, New York, 1994.
5. D.M. Gruen. *Annu. Rev. Mater. Sci.* **29**, 211-59, (1999).
6. D. Zhou, D. M. Gruen, L.-C. Qin, T. G. McCauley, and A. R. Krauss, *J. Appl. Phys.* **84**, 1981 (1998).
7. (a) D. Huang, M. Frenklach, and M. Maroncelli, *J. Phys. Chem.* **92**, 6379 (1988); (b) M. Frenklach and K. E. Spear *J. Mater. Res.* **3**, 133, (1988).
8. K. E. Spears and M. Frenklach, in *Synthetic Diamonds.* K.E. Spear, J.P.Dismukes, Eds. J. Wiley, New York.. (1994).
9. (a) M. C. Zerner, in *Reviews in Computational Chemistry*, K. B. Lipkowitz and D. B. Boyd, Eds., VCH Publishers, New York, 1991, Vol. 2, pp. 313-365; (b) J. J. P. Stewart, in *Reviews in Computational Chemistry*, K. B. Lipkowitz and D. B. Boyd, Eds., VCH Publishers, New York, 1990, Vol. 1, pp. 45-81.
10. (a) W. J. Hehre, L. Radom, J. A. Pople, and P. v. R. Schleyer, *Ab Initio Molecular Orbital Theory*; Wiley, New York, 1987; (b) F. Jenson, *Introduction to Computational Chemistry*, Wiley, New York, 1999.
11. (a) L. J. Bartolotti and K. Flurchick, in *Reviews in Computational Chemistry*, K. B. Lipkowitz and D. B. Boyd, Eds., VCH Publishers, New York, 1995, Vol. 7, pp. 187-216; (b) W. Kohn, A. D. Becke, and R. G. Parr, *J. Phys. Chem.*, **100**, 12974 (1996).
12. (a) J. C. Slater and G. F. Koster, *Phys. Rev.* **94**, 1498 (1954); (b) I. Lefebvre, M.Lannoo, and G. Allan, *Phys. Rev. B* **39**, 13518 (1989) .
13. (a) L. A. Curtiss, K. Raghavachari, G. W. Trucks, and J. A. Pople, *J. Chem. Phys.*, **94**, 7221 (1991); (b) L. A. Curtiss, K. Raghavachari, P. C. Redfern, V. Rassolov, and J. A. Pople, *J. Chem. Phys.*, **109**, 7764 (1998).
14. R.J. Buenker, S.D. Peyerimhoff, W. Butscher, *Mol. Phys.*, **35**, 771 (1978).
15. R. J. Bartlett and J. F. Stanton, in *Reviews in Computational Chemistry*, K. B. Lipkowitz and D. B. Boyd, Eds., VCH Publishers, New York, 1994, Vol. 5, pp. 65-169.
16. A. D. Becke, *J. Chem. Phys.* **98**, 5648 (1993).

17. P. J. Stephens, F. J. Devlin, C. F. Chabalowski, and M. J. Frisch, *J. Phys. Chem.,* **98**, 11623 (1994).
18. B. G. Johnson, P. M. W. Gill, and J. A. Pople, *J. Chem. Phys*., **98**, 5612 (1993).
19. L. A. Curtiss, K. Raghavachari, P. C. Redfern, and J. A. Pople, *J. Chem. Phys.* **106**, 1063 (1997).
20. M. J. S. Dewar and W. Thiel, *J. Am. Chem. Soc.* **99**, 4899 (1977).
21. M. J. S. Dewar, E. Zoebisch, E. F. Healy, and J. J. P. Stewart, *J. Am. Chem. Soc.* **107**, 3902 (1985).
22. J. J. P. Stewart, *J. Comput. Chem.* **10**, 209 (1989).
23. J. Shoemaker, L. W. Burggarf and M. S. Gordon *J. Phys. Chem. A* **103**, 3245 (1999).
24. F. Maseras and K. Morokuma, *J. Comput. Chem.* **16**, 1170 (1995).
25. D. Porezag, Th. Frauenheim, Th. Kohler, G. Seifert, and R. Kaschner, *Phys. Rev.* ***B*** **51**, 12947 (1995).
26. T. Frauenheim, G. Seifert, M. Elstner, Z. Hajnal, G. Jungnickel, D. Porezag, S. Suhai, R. Scholz, *Physica Status Solidi B: Basic Research* **217**, 41 (2000).
27. M. Elstner, Q. Cui, P. Munih, E. Kaxiras, T. Frauenheim, and M. Karplus, *J Comput Chem* **24**, 565 (2003).
28. S. Zygmunt, M. Sternberg, P. Zapol, P. Redfern, and L. Curtiss, unpublished results.
29. M. P. D'Evelyn, *Handbook of Industrial Diamonds and Diamond Films*, edited by M. A. Prelas, G. Popovici, and L. K. Bigelow, (Marcel Dekker Inc., 1998) pp. 89-146.
30. A. V. Hamza, G. D. Kubiak, R. H. Stulen, *Surf. Sci.* **237**, 35 (1990).
31. B. D. Thoms, J. E. Butler, *Surf. Sci.* **328**, 291 (1995).
32. C. Su, J. C. Lin, *Surf. Sci.* **406**, 149 (1998).
33. D. W. Brenner, *Phys. Rev. B* **42**, 9458, (1990).
34. (a) Y. L. Yang, M. P. D'Evelyn, *J. Am. Chem. Soc.* **114**, 2796 (1992); (b) Y. L. Yang, M. P. D'Evelyn, *J. Vac. Sci. Technol. A* **10**, 978 (1992).
35. (a) W. S. Verwoerd, *Surf. Sci.* **103**, 404 (1981); (b) W. S. Verwoerd, *Surf. Sci.* **108**, 153 (1981).
36. S. P. Mehandru, A. B. Anderson, *Surf. Sci.* **248**, 369 (1991).
37. X. M. Zheng, P. V. Smith, *Surf. Sci.* **256**, 1 (1991).
38. J. L. Whitten, P. Cremaschi, *Appl. Surf. Sci.* **75**, 45 (1994).
39. T. I. Hukka, T. A. Pakkanen, M. P. D'Evelyn, *J. Phys. Chem.* **98**, 12420 (1994).
40. S. Skokov, C. S. Carmer, B. Weiner, M. Frenklach, *Phys. Rev. B* **49**, 5662, (1994)
41. S. H. Yang, D. A. Drabold, J. B. Adams, *Phys. Rev. B* **48**, 5261, (1993).
42. C. Kress, M. Fiedler, W. G. Schmidt, F. Bechstedt, *Phys. Rev. B* **50**, 17697, (1994).
43. J. Furthmüller, J. Hafner, G. Kresse, *Phys. Rev. B* **50**, 15606 (1994).; *Phys. Rev. B* **53**, 7334 (1996).
44. M. J. Rutter, J. Robertson, *Phys. Rev. B* **57**, 9241 (1998).
45. P. Krüger, J. Pollmann, *Phys. Rev. Lett.* **74**, 1155 (1995).
46. T. Ogitsu, T. Miyazaki, M. Fujita, M. Okazaki, *Phys. Rev. Lett.* **75**, 4226 (1995).

47. S. Hong, *Phys. Rev. B* **65**, 153408 (2002).
48. H. Tamura, H. Zhou, K. Sugisako, Y. Yokoi, S. Takami, M. Kubo, K. Teraishi, A. Miyamoto, A. Imamura, and M. N. Gamo, T. Ando, *Phys. Rev. B* **61**, 11025 (2000).
49. H. Tamura, H. Zhou, Y. Hirano, S. Takami, M. Kubo, R. V. Belosludov, A. Miyamoto, A. Imamura, M. N. Gamo, and T. Ando, *Phys. Rev. B* **62**, 16995 (2000).
50. J. A. Steckel, G. Kresse, J. Hafner, *Phys. Rev. B* **66**, 155406 (2002).
51. B. N. Davidson, W. E. Pickett, *Phys. Rev. B* **49**, 11253 (1994).
52. M. D. Winn, M. Rassinger, J. Hafner, *Phys. Rev. B* **55**, 5364 (1997).
53. Th. Frauenheim, U. Stephan, P. Blaudeck, D. Porezag, H. -G. Busmann, W. Zimmermann-Edling, S. Lauer, *Phys. Rev. B* **48**, 18189, (1993).
54. M. W. Schmidt and M. S. Gordon, *Annu. Rev. Phys. Chem.* 49, 233 (1998).
55. B. N. Davidson and W. E. Pickett, *Phys. Rev.* **B49** 11253 (1994)
56. D. R. Alfonso, D. A. Drabold, and S. E. Ulloa, *Phys. Rev.* **B51**, 14669 (1995).
57. G. Kern, J. Hafner, *Phys. Rev.* **B56**, 4203 (1997).
58. A. Scholze, W. G. Schmidt, F. Bechstedt, *Phys Rev* **B 53** 13725 (1996).
59. Hafner, A. Scholze, W. G. Schmidt, F. Bechstedt, *Thin Solid Films*, **281**, 256 (1996).
60. M. Saito, Y. Miyamoto, A. Oshiyama, *Surf. Sci.* **427-428**, 53 (1999).
61. M. Saito, A. Oshiyama, Y. Miyamoto, *Phys. Rev.* **B 57** R9412 (1998).
62. G. Kern, J. Hafner, *Surf. Sci.* **384**, 94 (1997).
63. G. Kern, J. Hafner, G. Kresse, *Surf. Sci*, **366**, 445 (1996).
64. G. Kern, J. Hafner, J. Furthmueller, G. Kresse, *Surf. Sci.* **357**, 422 (1996).
65. G. Kern, J. Hafner, G. Kresse, *Surf. Sci*, **396**, 431 (1998).
66. R. C. Brown, C. J. Cramer, and J. T. Roberts, *J. Phys. Chem.* **B 101**, 9574 (1997).
67. M. Frenklach, *J. Appl. Phys.* **65**, 5142 (1989).
68. M. Frenklach, H. Wang, *Phys. Rev.* **B 43**, 1520 (1991).
69. D. W. Brenner, and M. Page, *J. Am. Chem. Soc.* **113**, 3270 (1991).
70. P. de Sainte Claire, P. Barbarat, W. L. Hase *J. Chem. Phys.* **101** 2476 (1994).
71. K. Larsson, J. Carlsson, *Phys. Rev. B* **59**, 8315 (1999).
72. M. Heggie, G. Jungnickel, C. Latham, *Diamond Related Mater.* **5**, 236 (1996).
73. X. Y. Chang, D. L. Thompson, L. M. Raff, *J. Chem. Phys.* **100**, 1765 (1994).
74. M. Frenklach and S. Skokov, *J. Phys. Chem.*, **B101**, 3025 (1997).
75. E.J. Dawnkaski, D. Srivasta, and B. J. Garrison, *J. Chem. Phys.* **102**, 9401 (1995).
76. S. J. Harris, D. N. Belton, and R. J. Blint, *J. Appl. Phys.* **70**, 2654 (1991).
77. D. N. Belton and S. J. Harris, *J. Chem. Phys.* **96**, 2371 (1992).
78. S. J. Harris and D. N. Belton, Jpn. *J. Appl. Phys.* **30**, 2615 (1991).
79. S. J. Harris and D. G. Goodwin, *J. Phys. Chem.* **97**, 23 (1993).
80. D. Huang, and M.Frenklach, *J. Phys. Chem.* **96**, 1868 (1992).
81. S. Skokov, B. Weiner, and M.Frenklach, *J. Phys. Chem.* **94**, 7073 (1994)
82. S. Skokov, B. Weiner, and M.Frenklach, *J. Phys. Chem.* **99,** 5616 (1995).
83. S. Skokov, B. Weiner, and M. Frenklach, *J. Phys. Chem.* **98**, 8 **(1994)**.
84. M. Frenklach, S. Skokov, and B. Weiner, *Nature* **372** 535 (1994).
85. B.H. Besler, W.L. Hase, and K.C. Hass, *J. Phys. Chem.* **96**, 9369 (1992).

86. C. D. Latham, M. I. Heggie, R. Jones, *Diamond and Related Materials* **2**, 1493 (1993).
87. J. Peploski, D. L. Thompson, and L. M. Raff, *J. Phys. Chem.* **96,** 8538 (1992).
88. M. Frenklach, *J. Chem. Phys.* **97**, 5794 (**1992**).
89. S. Skokov, B. Weiner, and M. Frenklach, *J. Phys. Chem.* **98**, 8 (1994).
90. S. W. Yang, X. Xie, P. Wu, K. P. Loh, *J. Phys. Chem. B* **107**, 985 (2003).
91. H. Kawarada, H. Sasaki, and A. Sato, *Phys. Rev. B* **52**, 11351 (1995).
92. T. Tsuno, T. Tomikawa, S. Shikata, T. Imai, and N. Fujimori, *Appl. Phys. Lett.* **64**, 572 (1994).
93. S. Skokov, B. Weiner, M. Frenklach, Th. Frauenheim, and M. Sternberg, *Phys. Rev. B* **52**, 5426 (1995).
94. T. I. Hukka, T. A. Pakkanen, and M. P. D'Evelyn, *Surf. Sci.* **359**, 213 (1996).
95. J. K. Kang and C. B. Musgrave, *J. Chem. Phys.* **113**, 7582 (2000).
96. H. Tamura, H. Zhou, S. Takami, M. Kubo, A. Miyamoto, M. N.-Gamo, and T. Ando, *J. Chem. Phys.* **115**, 5284 (2001).
97. C. C. Battaile, D. J. Srolovitz, I. I. Oleinik, D. G. Pettifor, A. P. Sutton, S. J. Harris, and J. E. Butler, *J. Chem. Phys.* **111**, 4291 (1999).
98. E. J. Dawnkaski, D. Srivastava, and B. J. Garrison, *J. Chem. Phys.* **104**, 5997 (1996).
99. M. Kaukonen, P. K. Sitch, G. Jungnickel, R. M. Nieminen, S. Poykko, D. Porezag, and Th. Frauenheim, *Phys. Rev. B* **57**, 9965 (1998).
100. J. A. Pople, J. S. Binkley and R. Seeger, *Int. J. Quantum Chem.* **S10**, 1 (1976).
101. (a) H. Nakano, *J. Chem. Phys.* **99**, 7983 (1993); (b) H. Nakano, *Chem. Phys. Lett.* **207**, 372 (1993).
102. *Synthetic Diamonds,* K.E. Spear, J.P.Dismukes, Eds. New York. John Wiley. (1994).
103. D. M. Gruen, S. Liu, A. R. Krauss, and X. Pan, *J. Appl. Phys.* **75**, 1758 (1994).
104. D. M. Gruen, S. Liu, A. R. Krauss, J. Luo, and X. Pan, *Appl. Phys. Lett.* **64**, 1502 (1994).
105. D. A. Horner, L. A. Curtiss, D. M. Gruen, *Chem. Phys. Lett.* **233**, 243 (1995).
106. P. C. Redfern, D. A. Horner, L. A. Curtiss, D. M. Gruen, *J. Phys. Chem.* 100, 11654 (1996).
107. M. Sternberg, M. Kaukonen, R. M. Niemimen, Th. Frauenheim, *Phys. Rev. B* **63**, 1655414 (2000).
108. D. M. Gruen, P. C. Redfern, D. A. Horner, P. Zapol, L. Curtiss, *J. Phys. Chem. B* **103**, 5459 (1999).
109. M. Sternberg, P. Zapol, L. Curtiss, *Phys. Rev. B,* in press.
110. M. Sternberg, P. Zapol, T. Frauenheim, J. Carlisle, D. M. Gruen, and L. A. Curtiss, *Mat. Res. Soc. Symp. Proc.* **675**, W12.11.1 (2001).
111. H. G. Busmann, I. V. Hertel, *Carbon,* **36**, 391 (1998).
112. R. E. Thomas, R. A. Rudder, R. J. Markunas, *J. Vac. Sci. Technol. A* **10**, 2451 (1992).
113. T. Ando, K. Yamamoto, M. Ishii, M. Kamo, Y. Sato, J. *Chem. Soc. Faraday Trans.* **89**, 3635 (1993).
114. P. E. Pehrsson, *Electrochem. Soc. Proc.* **95-4**, 436 (1995).
115. K.P.Loh, X.N.Xie, Y.H.Lim, E.J.Teo, J.C.Zheng, T.Ando, *Surf. Sci.* **505**, 93 (2002).
116. J. L. Whitten and P. Cremaschi, *Appl. Surf. Sci.* **75**, 45 (1994).

117. P. Badziag and W. S. Verwoerd, *Surf. Sci.* **183**, 469 (1987).
118. X. M. Zheng, P. V. Smith, *Surf. Sci.* **262**, 219 (1992).
119. S. Skokov, B. Weiner, and M. Frenklach, *Phys. Rev.* B **49**, 11374 (1994).
120. S. Skokov, B. Weiner, and M. Frenklach, *Phys. Rev.* B **55**, 1895 (1997).
121. N. Eimori, Y. Mori, A. Hatta, T. Ito, A. Hiraki, *Jpn. J. Appl. Phys.* **33**, 6312 (1994).
122. S. Yamanaka, H. Watanabe, S. Masai, D. Takeuchi, H. Okushi and K. Kajimura, *Jpn. J. Appl. Phys.* **37**, L1129 (1998).
123. A. T. Collins, in *Properties and Growth of Diamond* ed. by G. Davies, INSPEC, the Institution of Electronical Engineers, London, UK, **263** (1993), A. T. Collins, *Mat. Res. Soc. Symp. Proc.* **162**, 3 (1990).
124. (a) A. E. Alexenko and B. V. Spitsyn, *Diam. Relat. Mater.* **1**, 705 (1992); (b) S. Koizumi, M. Kamo, Y. Sato, H. Ozaki, and T. Inuzuka, *Appl. Phys. Lett.* **71**, 1065 (1997);. (c) M. Nesladek, K. Meykens, K. Haenen, L. M. Stals, T. Teraji, and S. Koizumi, *Phys. Rev. B* **59**, 14852 (1999).
125. (a) I. Sakaguchi, M. N.-Gamo, Y. Kikuchi, E. Yasu, H. Haneda, T. Suzuki, and T. Ando, *Phys. Rev. B,* **60**, 2139 (1999); (b) M. N.-Gamo, E. Yasu, C. Xiao, Y. Kikuchi, K. Ushizawa, I. Sakaguchi, T. Suzuki and T. Ando, *Diamond Relat. Mater.* **9**, 941 (2000); (c) M. N-Gamo, C. Xiao, Y. Zhang, E. Yasu, Y. Kikuchi, I. Sakaguchi, T. Suzuki, Y. Sato, and T. Ando, *Thin Solid Films* **382**, 113 (2001).
126. J. A. Garrido, C. E. Nebel, M. Stutzmann, E. Gheeraert, N. Casanova, and E. Bustarret, Phys. Rev. B **65**, 165409 (2002).
127. M. Hasegawa, D. Takeuchi, S. Yamanaka, M. Ogura, H. Watanabe, N. Kobayashi, H. Okushi, and K. Kajimura, Jpn. J. Appl. Phys., **38**, L1519 (1999).
128. J. F. Prins, Phys. Rev. B, **61**, 7191 (2000).
129. Y. Liou, R. Weimer, D. Knight, R. Messier, Appl. Phys. Lett. **56**, 437 (1990); Y. Saito, K. Sato, H. Tanaka, K. Fujita, S. Matuda, J. Mater. Sci. **23**, 842 (1988); Y. Muranaka, H. Yamashita, and H. Miyadera, J. Appl. Phys. **69**, 8145 (1991).
130. O. H. Nielsen and R. M. Martin, Phys. Rev. B **32**, 3780 (1985).
131. H, Zhou, Y. Yokoi, H. Tamura, S. Takami, M. Kubo, A. Miyamoto, M. N-Gamo, and T. Ando, Jpn. J. Appl. Phys., **40**, 2830 (2001).

Chapter 8

CHARGE INJECTION IN MOLECULAR DEVICES – ORDER EFFECTS

A. L. Burin[a,b] and M. A. Ratner[a]
a Department of Chemistry & Materials Research Center, Northwestern University, Evanston, IL 60208
b Department of Chemistry, Tulane University, New Orleans, LA 70118

1. INTRODUCTION

Modern molecular devices are constructed using ultra-thin molecular layers of thickness ranging from nanometers to microns incorporated between metal or semiconductor electrodes.[1] Unique properties of charge transport through molecular layers make these systems attractive for several technological applications including molecular wires,[1-3] light emitting devices,[4] light absorbers used in solar cells[5] (see e.g. Ref. 1 for the review). Injection and transport of charge in the insulating molecular levels are very sensitive to various imperfections in the system. For example, the presence of charge traps is known to lead to qualitative changes in the conduction properties when typical trap depths exceed the scale of thermal energy (0.025eV at room temperature).[6] This is a very small scale compared to electronic energies near 1eV and even vibration energies near 0.1eV, so it is not surprising that various structure defects modify the system behavior qualitatively and quantitatively.

It is hard and expensive to make perfect molecular structures for molecular electronic devices, intended for room temperature performance. Most experimental techniques used to make molecular layers of several tens of nanometer thickness, including for instance vacuum deposition,[7] vapor deposition[8] and spin coating,[9] lead to the formation of an essentially amorphous organic layer with extensive disordering. Recently developed self-assembly techniques, where the molecular structure is prepared by careful layer by layer covalent bonding[9,10] can not yet be conveniently made thicker

L.A. Curtiss and M.S. Gordon (eds.),
Computational Materials Chemistry: Methods and Applications, 308–367.

than a few nano-meters. Even in the self assembled layers the transport of charge can be strongly affected by the presence of defects.[11] In addition one recent experimental comparison[10] of charge injection through spin-coated and self-assembled layers made from identical molecules show much better performance of the disordered spin-coated layer. Thus the effect of disordering is not always destructive. On the one hand it reduces charge mobility by making traps within the molecular layer.[6,12,13] This is certainly a destructive effect for molecular device performance. On the other hand the fluctuations in molecular energies can promote efficient injection of charge, thus increasing the injection current.[14-18]

How significant is the effect of disordering? To characterize it we can use the electronic (or hole) energies within the molecular layer. For an isolated molecule the relevant energy is the electron affinity or ionization potential of the molecule. If the molecule is within an amorphous structure, the electronic energies fluctuate from molecule to molecule due to fluctuations in neighboring molecule positions, surface defects, etc. Then we can introduce a characteristic average energy E_M and its typical fluctuation w, describing the effect of disordering.[14-16] If this fluctuation exceeds the thermal energy, molecules with energies below the average one by w become charge traps. The energy of the trapped states has been determined experimentally for various materials. In the widely used organic emitting layer composed of Alq_3 (tris-(8-hydroxyquinoline) aluminum) molecules (e.g. Ref. 9,10), these energies are found to be widely distributed within the range from 0.06eV to 0.5eV.[19] In the standard hole transport layer TPD (N,N'-bis(3-methylphenyl)N,N'-diphenylbenzidine) these energies range from 0.1eV to 0.55eV.[19] These are very large energies compared to the thermal energy k_BT~0.025eV. A similarly wide distribution of defect energies has been found for another light emitting layer, PPV (poly-phenylene-vinylene) molecules.[20]

To get some feeling for these numbers, one can estimate the difference in the population probabilities of two molecules with energies different by $\delta E \sim 0.25$ eV that can be less than the typical defect energy.[19,20] At low concentrations the population ratio can be estimated as $\exp(\delta E/(k_BT)) \sim e^{10} \sim 20000$. This is a dominant effect for the injection process. Fluctuations of the molecular energies obviously lead to fluctuations of the injection energetics and if some electrons see injection barriers below the average by 0.25eV compared to others, their injection will dominate if their fraction exceeds 0.0001.

Thus the disordering is significant for understanding electronic injection and transport in molecular layers. It affects both the charge bulk transport (mobility) and the charge injection. The effect of disordering on the charge transport is described by the theory of traps.[6] Disorder generally suppresses the current because a large fraction of carriers is localized in traps and thus the overall moving charge density is smaller than in an ordered system. For specially correlated trap potentials due to the static polarization of molecules, disorder leads to strong mobility-voltage dependence.[21]

The effect of energy fluctuations on the injection can be quite different [14-16] because of the possibility to reduce the injection barrier. The rare injection channels composed of molecules with fluctuationally reduced energy can conduct charge from the metal into the bulk of molecular material much more efficiently than injection through the average barrier caused by energy mismatch of metal and molecular layer.[16]

Bulk transport or surface injection regimes dominate the electronic properties of molecular layers under different conditions in either the injection-limited regime or the space charge limited regime[6,17,18,22-28] (non Ohmic or Ohmic metal-insulator contact). The injection-limited regime is realized at low applied voltage (and internal current), when the density of carriers inside the molecular layer is small enough to ignore its effect on the potential distribution inside the material. In this case the current is defined by the probability for the charge to enter from the metal to the molecular layer; this requires overcoming the injection barrier. The space-charge limited regime is established when the current is large enough so it partially blocks the injection, thus controlling the charge balance and forming an Ohmic contact. It is clear that the perfect injecting contact for any semiconductor device is an Ohmic contact.[22] Practically it is often hard to reach the space-charge limited regime because it requires very high current, which can lead to damage of the material. It is easier to become space-charge limited when the injection potential barrier formed by the difference between the metal work function and the molecular electronic level is relatively low,[22-23] but it is more difficult with highly stable Al or Au electrodes possessing a large work function. It is interesting that the recent investigation of charge transport through the significant emitting material Alq_3[25] gave the evidence that the performance regime there is always injection-limited. Thus both regimes are interesting from the practical point of view and require theoretical analysis.

The effect of disordering (traps) is relatively well understood in the space-charge limited regime,[6] while the injection into the disordered medium is not so clear. Several studies based on the Monte-Carlo methods and a perturbative analytical approach (see Refs. 14,15,18) show a remarkable increase of the current, arising from fluctuations. An analytical one-dimensional hopping model developed in our previous work[16] predicts strong field and temperature dependence caused by the disorder. The main purpose of this paper is to extend that model[16] to account for disorder effects in real materials. We will combine an analytical treatment with a numerical analysis of the one-dimensional hopping model to compute the charge current dependence on the voltage, temperature and disorder in the injection limited regime. Other important effects will be discussed, including the tunneling injection that can be seen at low temperature[2,24,29] and sometimes takes place even at room temperature[2,11,30] and the influence of disorder on device damage. The most attention will be paid to the current injection into amorphous organic layers of the thickness about a 100 nm placed between metal electrodes. However under certain conditions the results can also be applied to inorganic semiconducting junctions and molecular wires. We will discuss them briefly.

The paper is organized as follows. In section 2 we introduce and discuss the limiting regimes for the charge injection including thermal activation vs. tunneling injection and the injection limited regime vs. the space-charge limited current. Criteria for crossovers between different regimes will be established. The significance of the interaction of mobile electrons with the conducting electrons from the metal, leading to the formation of the image charge and defining the absolute value of the current and current / voltage characteristics, will be pointed out.

Section 3 is devoted to the theory of charge injection into the disordered medium. Following our previous work[16] we will consider the injection path optimization problem and derive the current as a function of the applied voltage, temperature and energy fluctuation (disordering). The average current and its fluctuations will be considered as a possible source for the damage caused by the high local current.

A brief discussion of the experimental data characterizing the electronic properties of the molecular layer in the light of our theory will be made in Section 4. The results will be summarized in Section 5.

2. CHARGE INJECTION AND TRANSPORT

2.1. Energy Levels and Injection Barriers

In this section we consider the most fundamental aspects of the charge transport through the molecular layer in the absence of disordering. The two qualitatively different regimes for charge injection from the metal are charge tunneling and thermally activated hopping inside the layer. One might also expect an intermediate regime of charge injection having features of the thermal activation and tunneling, when part of the barrier is passed by tunneling and part of the energy is taken from the thermal bath. To describe these transport regimes we start with the consideration of the injection barrier. Usually, the metal Fermi level is located between the molecule lowest unoccupied orbital (LUMO) and highest occupied (HOMO) level (see Fig. 1). This is because the typical value of the metal work function W, describing the position of the Fermi energy with respect to the vacuum, scales in the range 2-6eV and is usually larger than the molecular electron affinity EA

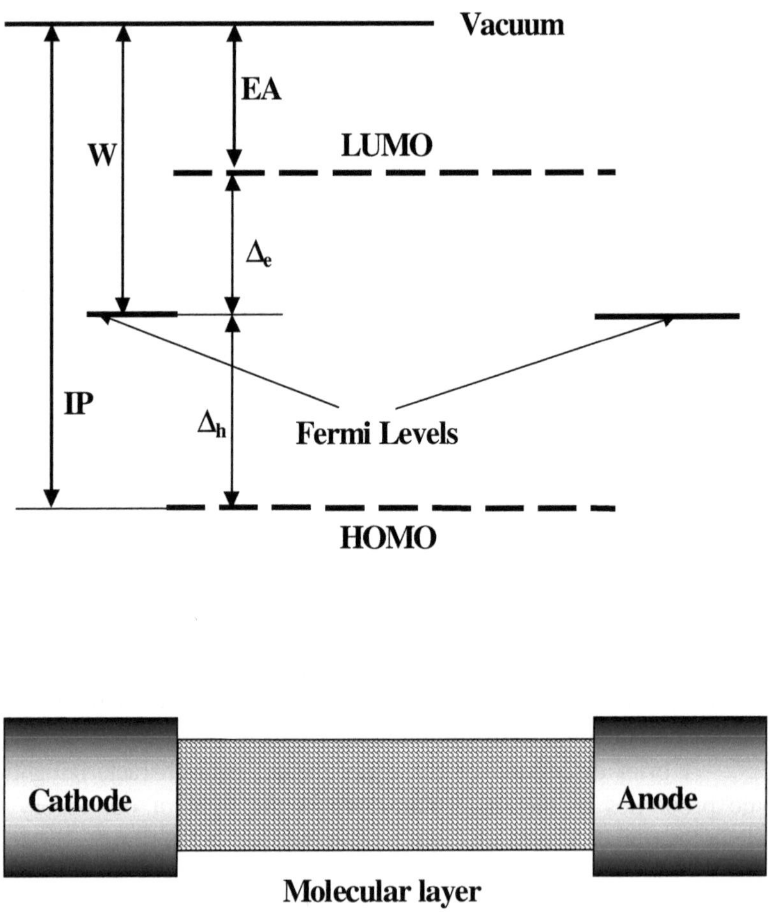

Figure 1. Energetics of metal-molecular layer junction.

<2eV and smaller than the ionization potential IP > 6eV (see, however, Ref. 31,32). Thus the potential barrier Δ_e for the electron to enter from the metal to the molecular layer (electron injection barrier), and the potential barrier Δ_h for the electron to leave the molecular layer for the metal (hole injection barrier) are found as

$$\Delta_e = W - EA, \quad \Delta_h = IP - W. \quad (2.1)$$

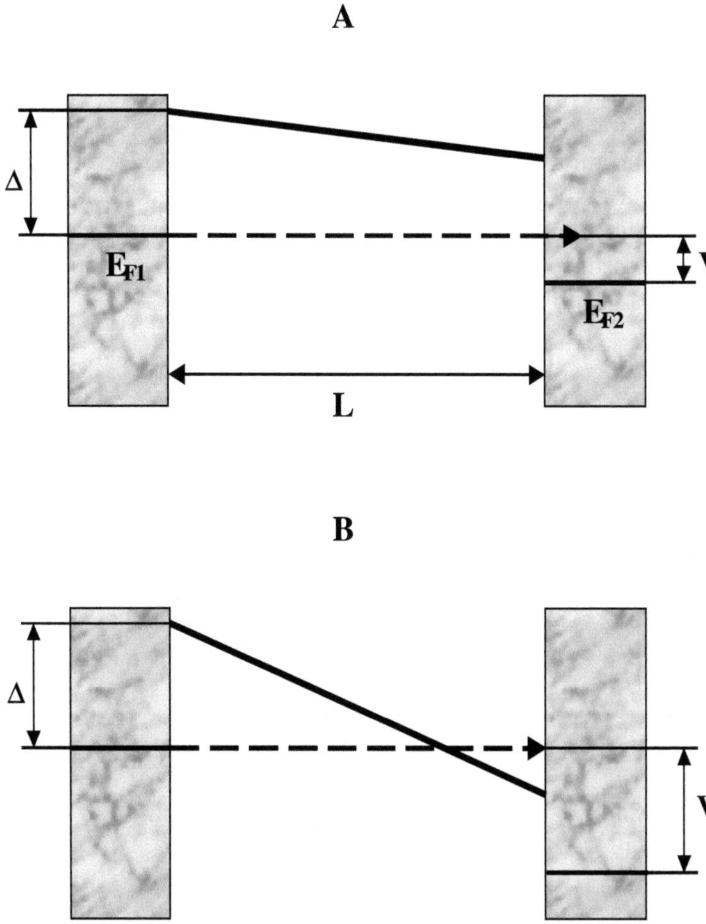

Figure 2. Potential change through the molecular layer in the injection limited regime. In case A the bias voltage is smaller than the injection barrier. In case B it is larger than the injection barrier.

The charge current through a molecular layer is caused by the applied bias voltage V. This voltage should be defined carefully taking into account the built in contribution due to the mismatch of electrodes Fermi levels, surface dipoles, potential drops at contact.[15,17,33,34] This built in potential usually does not exceed 1eV, for instance it is taken to be V_b=0.7V for Alq_3 – Al contact[15] (see, however Ref. 31). Since the performance voltage is within the range 5-20V it gives the small correction to the main applied voltage. We assume everywhere below that it has been taken into account and the bias voltage is defined as the applied bias with subtracted built in potential.

In the injection limited regime there is no effective screening inside the molecular material and we can consider the electric field in the layer to be constant. This also assumes no sharp drop of the potential in the surface and is justified for the thick layers, while for thin layers of about nanometer thickness the surface drop can be very significant (see e.g. Ref. 3). We focus on the thicker layers in this paper. Note that this regime will not be optimal for molecular junctions (see e.g. Ref. 1-3) where the charge can tunnel directly between metals. Thus, in the regime of interest the potential energy of carriers changes linearly inside the material (Fig. 2)

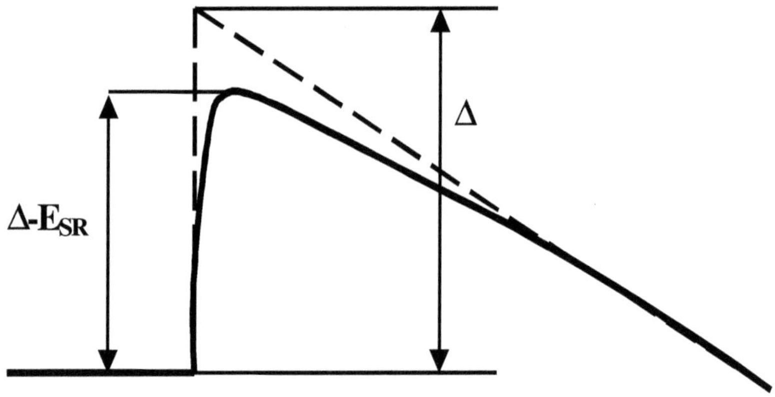

Figure 3. Injection barrier change due to image potential.

$$U(x) = \Delta - \frac{eV}{L}x \tag{2.2}$$

where L is the thickness of the layer and x is the transverse coordinate (x=0 indicates the interface between cathode and layer, while x=L is the interface between layer and anode). Depending on the strength of the applied potential compared to the injection barrier Δ the potential energy can be always higher than the Fermi level at $eV<\Delta$ (Fig. 2A) or can cross the Fermi level in the opposite limit $eV>\Delta$ (Fig. 2B). If the dielectric layer is sufficiently thick ($L >$ 5nm) significant current can be observed only in the second case and we will restrict our consideration to this regime. For example some light emitting diodes become efficient near 10V, that is much greater than any characteristic barrier existing in the system.

Thus the potential barrier for the charge injection has a triangular shape Fig. 3. The charge moving over (or under) the barrier interacts with the metal. If the motion of charge is not very fast compared to the surface plasmon frequency (see e.g. Ref. 35) we can assume that the conducting electrons follow adiabatically the motion of charge. Classically this will result in an image potential, thus changing the potential energy Eq. (2.2) to the form (see e.g. Refs. 6,14)

$$U(x) = \Delta - \frac{eV}{L}x - \frac{e^2}{16\pi\varepsilon\varepsilon_0 x}, \tag{2.3}$$

shown in Fig. 3. Here, ε is the dielectric constant of the molecular layer. For organic molecules it is often about 3.2 and we will use this value in further estimates. Of course, the classical form of Eq. (2.3) does not work when the charge is extremely close to the surface where the image potential formally diverges. This happens only in a very narrow domain near the surface and the effect of this domain on the thermally activated and tunneling transport, defined mostly by the change of the maximum barrier height, can be ignored.

It is clear that the image potential reduces the injection barrier. An estimate of the barrier reduction can be obtained maximizing Eq. (2.3) with respect to x. This leads to the standard Schottky-Richardson expression for the thermionic emission barrier (see e.g. Ref. 6).

$$\Delta_{SR} = \Delta - E_{SR}, \qquad E_{SR} = \sqrt{\frac{V}{L}\frac{e^3}{4\pi\varepsilon_0\varepsilon}}. \tag{2.4}$$

2.1.1 Thermally Activated Injection

The thermally activated mechanism of charge injection is realized when the electron enters the dielectric by passing over the injection barrier (solid line in Fig. 4). In this regime the probability of charge injection into the dielectric medium is defined by the bottleneck of the process, realized at the top of the barrier Δ_{SR}. The thermodynamic injection probability is defined by the Boltzmann factor, giving the Schottky – Richardson expression for the thermionic emission current –voltage dependence

$$j \sim \exp\left(-\frac{\Delta_{SR}}{k_B T}\right) = \exp\left(-\frac{\Delta}{k_B T} + \frac{1}{k_B T}\sqrt{\frac{V}{L}\frac{e^3}{4\pi\varepsilon_0\varepsilon}}\right). \tag{2.5}$$

This exponentially strong and field dependent factor defines the rapid growth of the current with applied voltage in a wide variety of organic materials. With typical parameters for organic layers V~10V, L~100nm, ε~3.2 at room temperature k_BT~0.025eV yields a value near 15 for the second term in the exponent of Eq. (2.5). Thus the image potential can significantly reduce the barrier for the thermally activated injection. If we compare the first and

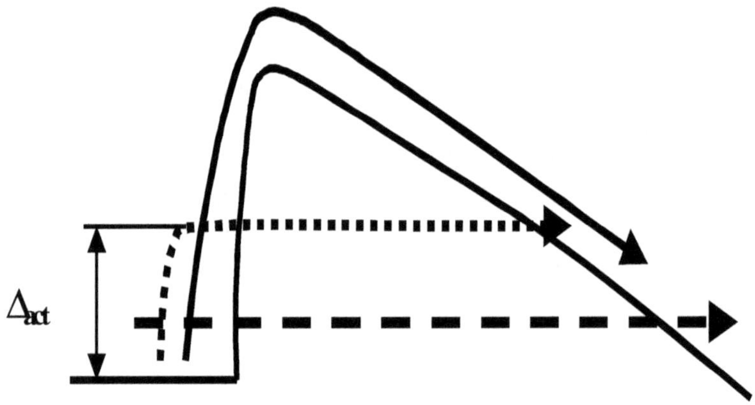

Figure 4. Injection mechanisms including thermal activation (solid line), tunneling (dashed line) and combined mechanism (dotted line)

the second term in the exponent in the right hand side of Eq. (2.5) at the typical barrier taken as Δ~1eV then the image effect contribution is near one third of the main first term.

2.1.2 Tunneling Injection

Tunneling injection occurs when the charge tunnels under the barrier (solid line in Fig. 4). The main parametric dependence is defined by the semi-classical tunneling exponent. The usual Fowler-Nordheim expression for the tunneling current is derived using a continuous medium approach. It defines the tunneling rate in terms of squared exponent of the classical action in the inverted barrier, and can be written with the exponential accuracy as

$$j \sim \exp\left(-\frac{2}{\hbar}\sqrt{2m}\int_{x_1}^{x_2} dx \sqrt{\Delta - eV\frac{x}{L} - \frac{e^2}{16\pi\varepsilon_0\varepsilon x}}\right) \qquad (2.6)$$

where x_1 and x_2 are the turning points of the tunneling path, defined as

$$x_1 = \frac{L}{2eV}\left(\Delta - \sqrt{\Delta^2 - \frac{e^3V}{4\pi\varepsilon_0\varepsilon L}}\right) = \frac{L}{2eV}\left(\Delta - \sqrt{\Delta^2 - E_{SR}^2}\right),$$
$$x_2 = \frac{L}{2eV}\left(\Delta + \sqrt{\Delta^2 - \frac{e^3V}{4\pi\varepsilon_0\varepsilon L}}\right) = \frac{L}{2eV}\left(\Delta + \sqrt{\Delta^2 - E_{SR}^2}\right). \tag{2.7}$$

The integral (2.6) can be generally expressed using the elliptic integrals, but the result is not transparent. It is more helpful to use an asymptotic expansion over the relatively weak image charge term under the square root; this permits us to estimate the image contribution with logarithmic accuracy as

$$\int_{x_1}^{x_2} dx \sqrt{\Delta - eV\frac{x}{L} - \frac{e^2}{16\pi\varepsilon_0\varepsilon x}} \approx$$
$$\approx \int_0^{\Delta L/(eV)} dx \sqrt{\Delta - eV\frac{x}{L}} - \int_{x_1}^{x_2} dx \frac{e^2}{32\pi\varepsilon_0\varepsilon x\sqrt{\Delta - eV\frac{x}{L}}} \approx \tag{2.8}$$
$$\approx \frac{2}{3}\frac{L}{eV}\Delta^{3/2} - \frac{e^2}{32\pi\varepsilon_0\varepsilon\sqrt{\Delta}}\ln\left(\frac{x_2}{x_1}\right).$$

With the same logarithmic accuracy one can replace the integral limits x_1 and x_2 by their limiting values obtained in the absence of the applied bias or the absence of image charge respectively, given by

$$x_1 \approx \frac{e^2}{16\pi\varepsilon_0\varepsilon\Delta},$$
$$x_2 \approx \frac{L\Delta}{eV}. \tag{2.9}$$

Then one can get the final estimate of Eq. (2.8) as

$$\int_{x_1}^{x_2} dx \sqrt{\Delta - eV\frac{x}{L} - \frac{e^2}{16\pi\varepsilon_0\varepsilon x}} \approx \frac{2}{3}\frac{L}{eV}\Delta^{3/2} - \frac{e^2}{32\pi\varepsilon_0\varepsilon\sqrt{\Delta}}\ln\left(\frac{16\pi\varepsilon\varepsilon_0\Delta^2 L}{e^3 V}\right). \quad (2.10)$$

The above expansion is possible when the correction due to the image is much smaller than the main term due to the gap Δ. Taking as before the typical values V~10V, L~100nm, ε~3.2, and the injection barrier Δ~1eV we get the ratio of the first and second terms to be 0.0125. Comparing this term with the image charge effect on the thermal activation problem we can see that it is 20 times less efficient. The reason for this difference is that for thermally activated hopping the effect of the change in the potential barrier comes from the maximum barrier point located relatively close to the metal where the image interaction is still strong. For the tunneling injection this effect comes from the whole tunneling path which extends far from the metal surface.

With exponential accuracy the tunneling current can be expressed using Eqs. (2.6), (2.10) as

$$j \sim \exp\left(-\frac{4}{3}\frac{\sqrt{2mL}}{\hbar eV}\Delta^{3/2} - \frac{e^2\sqrt{2m}}{16\pi\varepsilon_0\varepsilon\hbar\sqrt{\Delta}}\ln\left(\frac{16\pi\varepsilon\varepsilon_0\Delta^2 L}{e^3 V}\right)\right). \quad (2.11)$$

The first term in the exponent is the classical Fowler-Nordheim expression, while the second represents the image charge contribution, which is relatively small and can usually be neglected.

Note that the continuum approach for tunneling can become invalid because of the discrete molecular character of the medium (e.g. Ref. [36,37]). We will discuss this in more detail later. Here we note that the discreteness can reduce the current compared to Eq. (2.11) because of the small tunneling coupling within the molecular layer; this can result in a large effective mass m that needs to be entered into Eq. (2.11). In fact the discrete problem can be approached very accurately by changing the mass definition in Eq. (2.11).

The main limitation on the currents (2.5) and (2.11) comes from the large negative exponents describing the probabilities to overcome the injection barrier. Therefore we will restrict our consideration to comparison of the exponents. The first term (unrelated to the image charge) dominates over the second terms in both Eqs. (2.5) and (2.11) so we can ignore the image charge contributions for comparison purposes. Then one can easily find that the thermally activated injection dominates over tunneling when the temperature is sufficiently large

$$\frac{3}{4}\frac{V}{L}\frac{\hbar e}{\sqrt{2m\Delta}} < k_B T. \tag{2.12}$$

If we take the same parameters as before (V ~ 10 Volts, L ~ 100nm, Δ ~ 1eV) and use the standard mass of the electron then we find that the crossover temperature separating thermal activation and the tunneling regime is about T_C ~ 220K. This estimate suggests that thermally activated injection into the molecular layer might occur in organic light emitting devices at room temperature T_R=300K. In fact, despite the relative nearness of the crossover temperature to room temperature, a large difference emerges in the exponential factors (2.5) and (2.11), describing the injection probabilities. In

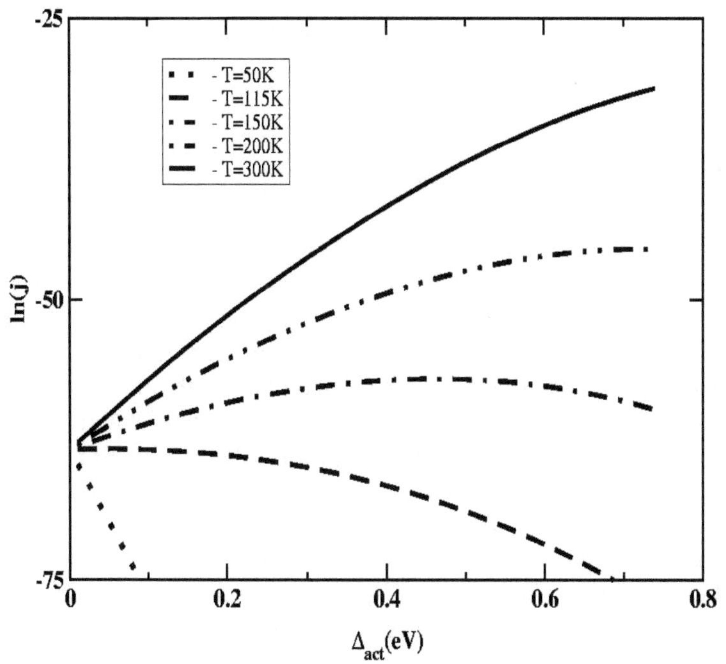

Figure 5. Current density for thermally-activated tunneling regime (ln(j) is taken ignoring the preexponential factor)

addition, the image charge correction is larger in the thermally activated regime, giving additional support to the classical injection mechanism.

Weak electronic coupling between the organic molecules in the amorphous layer should increase the effective mass m of the tunneling electron, leading to greater suppression of tunneling. This explains why the majority of organic light emitting diodes shows strong temperature dependence of the injection current. On the other hand Eq. (2.12) predicts that for bias voltage of 20V and above corresponding to the electric field E~V/L $>2\cdot10^7$V/m the injection mechanism can change back to the tunneling.

Our crude estimates can be compared with the measurements of Ref. 24 for a single layered device composed of a 90nm amorphous layer of TPD placed between an Al cathode and an ITO anode. These measurements show that at bias voltage below 10V the current becomes independent of the temperature at T<100K, so the crossover to tunneling injection occurs even below our estimate. On the hand at V>20V the temperature dependence of current becomes weak even up to room temperature; this agrees with the expected increase of the crossover temperature at higher voltage. Measurements of the conducting properties of self-assembled molecular layers of phenyl and terphenyl placed between carbon and mercury electrodes show tunneling above room temperature. However, the electric field used in these measurements[30] is about $5\text{-}10\cdot10^7$V/m, a great deal higher than in organic light emitting devices.

2.1.3. Tunneling plus Activation

We have compared two limiting regimes of simple tunneling (dashed arrow in Fig. 4) and full thermally activated injection (solid arrow in Fig. 4) (ignoring the image charge). It is reasonable to ask whether an intermediate regime of the thermal activation assisted tunneling (dotted line in Fig. 4) can be realized near the crossover. This can take place if the tunneling is due to the highly excited electrons. There is experimental and theoretical evidence for thermally assisted tunneling in inorganic insulators e.g. Ref. 38. Below we give the results of calculations using the accurate estimates of the tunneling and thermal activated currents Eqs. (2.5), (2.6).

The probability of the combined process involving both thermal activation and tunneling can be expressed with exponential accuracy as the product of the Boltzmann exponent describing first activation by some intermediate energy Δ_{act} (see Fig. 4) and second then tunneling under the reduced barrier (see [38]):

$$j \sim \exp\left(-\frac{\Delta_{act}}{k_B T} - \frac{2}{\hbar}\sqrt{2m}\int_{x_1}^{x_2} dx \sqrt{\Delta - \Delta_{act} - eV\frac{x}{L} - \frac{e^2}{16\pi\varepsilon_0 \varepsilon x}} \right). \quad (2.13)$$

Here the redefined start and end points for tunneling are (cf. Eq.(2.7))

$$x_1 = \frac{L}{2eV}\left(\Delta - \Delta_{act} - \sqrt{(\Delta - \Delta_{act})^2 - \frac{e^3V}{4\pi\varepsilon_0\varepsilon L}}\right),$$
$$x_2 = \frac{L}{2eV}\left(\Delta - \Delta_{act} + \sqrt{(\Delta - \Delta_{act})^2 - \frac{e^3V}{4\pi\varepsilon_0\varepsilon L}}\right). \quad (2.14)$$

Using the standard set of parameters given above we computed the intermediate exponents at various temperatures T=50K, 100K, 150K, 200K, 300K and the dependence of the current in the combined regime on the intermediate activation energy Δ_{act}. The results are shown in Fig. 5. The large negative absolute value of the exponent is compensated by a large pre-exponential factor. For instance, thermally activated transport at room temperature is characterized by a negative exponent 30 describing the probability to overcome the injection barrier. This corresponds to a factor 10^{-13} reduction of the total current, that is compensated by the large charge density in the metal, that in turn can make the pre-exponential factor as high as 10^{10}Acm^{-2}.

Data in Fig. 5 can be summarized as follows: Under standard conditions V~10 Volts, L~100nm and barrier height Δ~1eV the fully tunneling regime is realized at T<100K. At higher temperatures 115K<T<200K, the intermediate regime of thermally assisted tunneling takes place. The current increases by several orders of magnitude in this intermediate regime, which can be realized experimentally and has some real interest. At T>200K the charge injection occurs only through the thermal activation. Thus one can expect that at room temperature the tunneling effect can be neglected. The intermediate regime can take place for a triangular barrier.[38] If the shape of the barrier is rectangular or parabolic the transition between thermal activation and tunneling is instantaneous (see e. g. Ref. 39,40) because the intermediate regime always gives the maximum of the negative tunneling exponent.

We have used the large injection barrier Δ~1eV suitable for instance for the contact Al – Alq_3. However the effective barrier can be lower in other materials. For instance Alq_3-Mg contact has the barrier Δ~0.5eV (see e.g. Ref. 15). For this barrier the tunneling injection should manifest itself at the temperature close to 300K. Experimentally the temperature dependence becomes weak only at T~100K and below.[17] Perhaps this difference is due to the large effective mass of the electron that can remarkably increase the tunneling exponent (2.13). The large effective mass can be due to the small electron overlap integral between different molecules that is not surprising in amorphous organic material. The analysis of the effect in more detail is outside of the scope of this paper.

2.2. Injection-limited and space-charge limited regimes

Thus the model of thermally activated injection is well justified at room temperature, when the electric field is not very high ($E \leq 10^8$V/m). Another significant restriction is the applicability of the injected-limited regime to the performance of the molecular devices. Below we address this question based on the experimentally accessible parameters including current j, voltage V, charge mobility μ and system size L. First we derive a criterion for the injection limited regime. We suppose that if the charge has overcome the injection barrier it has no chance to return back to the electrode but will proceed to the anode. This assumption is based on the large height of the injection barrier, leading to an exponentially small probability to return. On the other hand, the motion out of the cathode is dominated by the applied bias.

Then our analysis is correct if the charge density n within the bulk molecular layer is too small to affect the potential distribution inside the material. Accordingly the current through the system will be defined entirely by the probability to overcome the injection barrier at the given electric field E=V/L.

The restrictions for the charge density n can be found from the solution of the Maxwell equation

$$\frac{d^2U}{dx^2} = -\frac{n}{\varepsilon_0 \varepsilon}, \tag{2.15}$$

where x is the transverse coordinate directed from the anode to the cathode (see Eq. (2.2)). Without the volume charge (n=0) this solution corresponds to the constant electric field (2.2). The correction to that solution can be estimated as

$$\delta U \sim \frac{n}{2\varepsilon_0 \varepsilon} L^2. \tag{2.16}$$

The expression (2.2) is applicable until the correction becomes comparable with the applied bias V. Thus the condition for the injection-limited regime reads

$$\frac{n}{2\varepsilon_0 \varepsilon} L^2 < V. \tag{2.17}$$

It is convenient to express the charge density through the experimentally known parameters of the electric current density j and the charge mobility μ

$$n = \frac{jL}{V\mu}. \tag{2.18}$$

Combining Eqs. (2.17), (2.18) one can derive the restriction for the bias voltage V, defining whether the charge transport through the system is injection limited

$$\sqrt{\frac{j}{2\varepsilon_0 \varepsilon \mu e} L^3} < V. \tag{2.19}$$

The typical performance value for current density in organic light emitting devices is within the range 10-100A/m^2. Mobilities of charge depend on the material, preparation, temperature and electric field.[21-24] For the amorphous organic TPD layer often used for the hole transport, the hole mobility is $10^{-7}m^2V^{-1}s^{-1}$.[24] According to Eq. (2.19) the bias voltage should be greater than 0.3V. This is certainly satisfied in the performance regime V~10-20V. The mobility of electrons in the electron transport or emissive layer composed of Alq_3 or PPV are smaller, and strongly depend on the system parameters. The mobility of electrons in the PPV layer is found[7] to be $10^{-8}m^2V^{-1}s^{-1}$. This gives the crossover voltage ~1V, certainly less than in the standard performance regime, whiles in Alq_3 it changes from 10^{-10}-$10^{-7}m^2V^{-1}s^{-1}$ with increasing the applied voltage.[20,23,24] Only at the lowest possible mobility value $10^{-10}m^2V^{-1}s^{-1}$ is charge density large enough to break the injection limited regime. However, this low mobility corresponds to small applied bias while at operational voltage V~10V it is certainly higher. Thus typical devices operate in the injection-limited regime. This is not the optimum regime (see e.g. Ref. 22), but it is still very hard to reach Ohmic contact conditions. Our conclusion agrees with the experimental evidence of injection limited regime for Alq_3.[18] It is hard to make the careful comparison of our analysis with theoretical and experimental studies of Refs. 26-28,33,34, where the criterion of the space charge limited regime has been established through the value of the injection barrier that must be less than 0.4-0.5eV for reaching the Ohmic injection. This crossover value of barrier is too close to the actual injection barrier (see e.g. Sect. 4) so both regimes have a chance to be realized.

Summarizing the consideration of this section we can conclude under room temperature for the standard experimental setup, molecular devices based on an amorphous layer placed between two electrodes operate in the injection limited-regime, and injection is caused by the thermally activated hopping over the injection barrier. We can not and do not wish to claim that the above regime shows the optimum performance. However under usual conditions it is significant, and requires careful study. As we will see in the following section many significant properties of the system can be strongly modified by disorder.

3. EFFECT OF DISORDER ON THERMALLY ACTIVATED INJECTION

3.1 Models of disordering

Based on the discussion of the previous section we can conclude that transport through amorphous molecular layers can be dominated by thermally activated hopping at room temperature. This assumption agrees with both theoretical expectations and experimental data. A tunneling regime at room temperature can occur at a high applied bias in specially prepared self-assembled structures (see e.g. Ref. 2,30,41). We will discuss the effect of disorder in that regime in the next section.

In addition we assume that injection limited regime dominates. As we discussed in the previous section, this condition is realized in many materials. Although it is not clear whether this regime is best from the point of view of performance, stability and low preparation cost it requires understanding because of its wide realization.

To describe the charge injection properly, one has to take into account

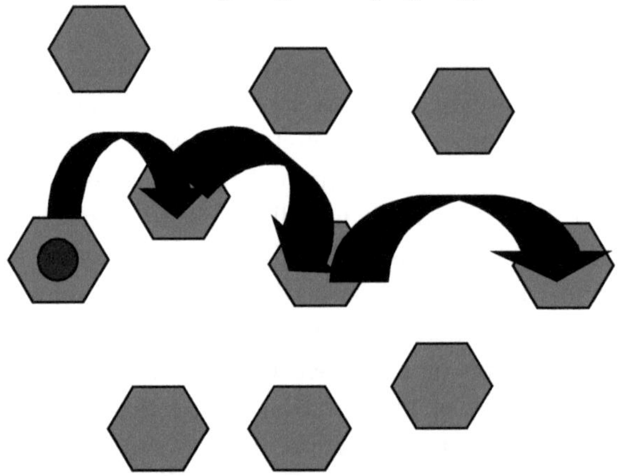

Fig. 6. Charge hopping transport between molecules

the nature of the states of excess charge within the insulating layer. We believe that these states are strongly localized inside the amorphous organic material because of the disorder in molecular energies. This requires that energy disorder w be greater than the intermolecular charge hopping amplitude b (see e.g. Ref. 42 and references therein). Additional localization effects come from the large vibrational reorganization of relatively small molecules in charged states. The only alternative option to localization is a band-like charge conduction. To the best of our knowledge this has never be seen within amorphous organic layers (see, however Ref. 43 for regular doped organic materials). In this limit of weak intermolecular coupling we can attribute each localized state to a single molecule. Then the kinetics of the population of each molecule within the bulk obeys the rate equation

$$\frac{dn_i}{dt} = \sum_j (W_{ji} n_j - W_{ij} n_i). \tag{3.1}$$

Here the rates W_{ij} describes the jump rates from the molecule j to the molecule i.

Equation (3.1) describes random walk hopping motion of the charge between the molecules within the bulk of the molecular layer (see Fig. 6). The hopping amplitudes W_{ij} (from the molecule j to the molecule i) depend on the electronic energies E_i and E_j at the molecules i and j, respectively. They obey the detailed balance relationship

$$W_{ji} = W_{ij} \exp\left(-\frac{E_j - E_i}{k_B T} \right). \tag{3.2}$$

There is expected to be general energy and distance dependence of the hopping amplitudes W_{ij} on the intermolecular distance $r_{ij} = |\mathbf{r}_i - \mathbf{r}_j|$ and molecular energies E_i and E_j. One can conventionally separate the distance and energy dependence in the form

$$\begin{aligned} W_{ij} &= \varsigma(r_{ij}) f(E_i - E_j), \\ \varsigma(r_{ij}) &\sim \exp(-\kappa r_{ij}), \end{aligned} \tag{3.3}$$

where the distance dependence reflects exponential decrease of the electron transfer integral with the intermolecular separation, and the decrement κ is usually close to 1 $Å^{-1}$.[1,3,44,45] One should note that in addition to the intermolecular distance dependence the charge transfer rate can be very sensitive to the orientation of molecules due to for instance the anisotropy of the pi-bond usually responsible for the electron transfer (see e.g. the first principles analysis of the orientation dependence of the electronic overlap integral for DNA bases,[46,47] where that anisotropy substantially affects the electronic coupling). The energy dependence of the factor f, responsible for detailed balance is sensitive to the energy change within the scale of thermal energy k_BT~0.025eV.

Thus disordering both in position and in temperature can be significant for charge transport. Fluctuations in the intermolecular distances and orientations leads to disorder in potential barriers, while the energy fluctuations leads to the disorder in potential well depths (see e.g. Refs. 48, 49). It is usually believed that for system dimensionality greater than one the potential well disorder is more significant than the disordering in the barriers, since it forms traps, while the electron can avoid the high potential barrier by choosing the different path around it.[48,49] The dominating effect of energy disordering on the hopping transport has been demonstrated in the Monte-Carlo simulations of Refs 14,21. One should also note that the fluctuations in orientation do not change the electronic integral by more than one order of magnitude, while a fluctuation in energy by the relatively small value δE~0.1eV can change the hopping rate by the Boltzmann factor $\exp(\delta E/(k_BT))>50$. Therefore we will ignore the position disorder within our approximate treatment and concentrate on the fluctuations in energy. A simplest model for this regime is a lattice model for the molecules within the layer, assuming that the molecules form a simple cubic lattice with the period a. The hopping takes place between neighboring molecules only (according to simulations of Ref. 14, hopping by more than one intermolecular distance is not significant at room temperature).

We will approximate the dependence of the hopping amplitude on the difference in molecular energies in two ways. First is the minimalist model [14,16,21,50] assuming a single parameter decay

$$W_{ji} = \frac{W_0}{1+\exp((E_j - E_i)/(k_BT))}. \tag{3.4}$$

This model suggests the inverted gap law regime for the energy change $|E_i-E_j|$ modified by the vibration quantum overlap effect.[50,51]

The alternative approach can be made using the classical Marcus formula for the electron transfer rate[44]

$$W_{ji} = W_0 \exp\left(-\frac{(E_i - E_j + \lambda)^2}{4\lambda k_BT}\right), \tag{3.5}$$

that takes into account the interaction with the polarization of the medium.[50] Here λ is the energy of the molecular reorganization caused by its charging. Modification of Eq. (3.5) taking into account the presence of high frequency polarized modes[45,51] can make the resulting energy dependence similar to Eq. (3.4).[50] We will study the charge injection efficiency with the both definitions of charge hopping rates and show that the result does not change significantly. On the other hand the consideration of charge transport mobility following Ref. 21 for the hopping rate defined by Eq. (3.5) will lead to the qualitatively different behavior from the case of Eq. (3.4).

The disordering in energy can be introduced as a random correction φ_i to the energy of each lattice site k describing the single molecule

$$\varepsilon_k = \Delta + \varphi_k, \tag{3.6}$$

where Δ is the injection barrier (See Figs. 1, 2) and φ_i is a random correction to the energy caused by the fluctuations in the neighboring molecules positions and orientations. It is assumed that the zero of energy corresponds to the cathode Fermi level; φ_i is assumed to have zero average and is characterized by the distribution function $f(\varphi)$. The simplest approach is to ignore the correlation of the potential fluctuations at different molecules. Then we will consider two cases of the distribution $f(\varphi)$. Most attention will be paid to the gaussian distribution

$$f(\varphi) = \frac{1}{\sqrt{2\pi}w} \exp\left(-\frac{\varphi^2}{2w^2} \right). \tag{3.7}$$

The gaussian distribution can be expected because it comes from the field of all neighboring molecules in the closely packed structure. The number of neighbors in the three dimensional closely packed structure is about 10, so we can use the law of large numbers for the energy fluctuations.[14,21] The significant parameter w characterizes the strength of disordering, which is significant when it exceeds the thermal energy. As was mentioned in the introduction an experimental estimate of w can be made from the energy of trap states;[19,20,52] it can reach the scale of tenths of eV.

An alternative distribution of random energies is exponential

$$f(\varphi) = \frac{1}{2w} \exp\left(-\frac{|\varphi|}{w} \right). \tag{3.8}$$

It is based on the properties of the density of defect states below the band edge of disordered semiconductors.[6,53]. In particular it has been used in Ref.

[6] to describe the charge mobility in the space charge limited regime for systems containing traps. Note that, the distribution (3.8) is more relevant for the inorganic semiconductors, where the band-like conduction can take place at sufficient doping,[53] while for the organic amorphous systems under study the gaussian distribution Eq. (3.7) is more justified.[14]

Correlations of random potentials φ at different molecules can arise for many reasons. In particular finite polarization of molecules in the ground state leads to the long-range correlation[21]

$$< \delta\varphi_i \delta\varphi_j > \sim \frac{1}{r_{ij}}. \tag{3.9}$$

As was pointed out in Ref. 21 these correlations can be responsible for the very strong dependence of mobility on the bias voltage

$$\mu \propto \exp(\beta\sqrt{U}). \tag{3.10}$$

We will take them into account for the injection problem within the generalized Gaussian distribution (3.7).

Significant disordering can take place if the molecular layer is doped by the molecules having larger electron affinity than the host molecule. For example, incorporation of a small concentration of quinacridone molecules into an Alq_3 layer remarkably increases the conducting and light emitting properties of the device.[31,54] The distribution of molecular energies in the presence of the dopant of mole fraction x can be approximated as

$$f(\varphi) = (1-x)\delta(\varphi) + x\delta(\varphi - \varphi_0), \tag{3.11}$$

where φ_0 is the difference of the host molecule and dopant molecule energies. This significant case will receive special consideration.

If charge concentrations are small, so that no significant screening occurs, we are not in the space-charge limited regime[6] and the presence of the finite bias voltage V (electric field E=V/L) modifies the potential energy in each molecule adding the linear correction to Eq. (3.6) (see Eq. (2.2))

$$E_k = \Delta + \varphi_k - eEx_k, \tag{3.12}$$

where x_k is the distance from the molecule k to the cathode. As we have seen in the previous section, the image interaction can also be significant for charge injection and transport. It can be included into the energy definition (3.12) similar to Eq. (2.3)

$$E_k = \Delta + \varphi_k - eEx_k - \frac{e^2}{16\pi\varepsilon_0 \varepsilon x_k} \tag{3.13}$$

Now all necessary parameters are defined and we can proceed to study the injection of charge into the random medium.

3.2 Fluctuational paths

3.2.1 General Approach

We start with the most simple model, ignoring the image charge (3.13) and treating the energy fluctuations in the simple gaussian approach (3.7). To illustrate the strength of the disordering effect, consider the thermal populations of molecules located near the cathode at zero applied bias (see Refs. 17,52,55). In the ordered system this population is defined by the Boltzmann factor

$$P \propto \exp\left(-\frac{\Delta}{k_B T}\right). \tag{3.14}$$

In the disordered system the population of each site k is different because of the energy difference from site to site (3.6) and the average population should be used

$$\begin{aligned} < P > &\propto \exp\left(-\frac{\Delta}{k_B T}\right)\left\langle \exp\left(-\frac{\varphi_i}{k_B T}\right)\right\rangle \\ &= \exp\left(-\frac{\Delta}{k_B T}\right)\int d\varphi \exp\left(-\frac{\varphi}{k_B T}\right) f(\varphi). \end{aligned} \tag{3.15}$$

For gaussian distribution $f(\varphi)$ (3.7) the integral in Eq. (3.15) can be evaluated as

$$< P > \propto \exp\left(-\frac{\Delta}{k_B T}\right)\exp\left(\frac{w^2}{4(k_B T)^2}\right). \tag{3.16}$$

If one takes the random energy w~0.1eV, then the average population of molecular levels increases by two orders of magnitude compared to the regular case. This estimate tells us that disordering can strongly enhance injection by randomly reducing the injection barrier. This is demonstrated schematically in Fig. 7. The fluctuationally-induced reduction of the energies of the group of molecules providing the path from the metal to the dielectric layer over the injection barrier can reduce the effective injection energy barrier from its average value Δ to some smaller value Δ_*. This low barrier injection path increases the injection current by the exponential factor $\exp((\Delta-\Delta_*)/(k_BT))$. On the other hand the probability $P(\Delta_*)$ to find a path having the smaller injection barrier decreases with increasing reduction of the barrier. The theoretical problem is to find the optimum path, maximizing the product of the entropy loss factor $P(\Delta_*)$ and the energy gain factor $\exp((\Delta-\Delta_*)/(k_BT))$.

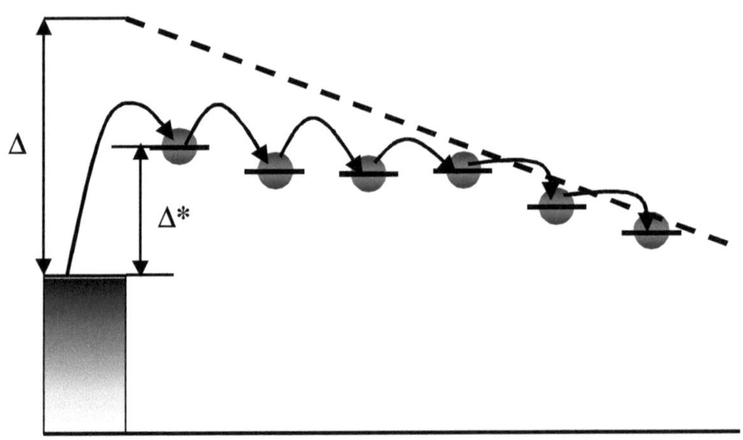

Figure 7. Formation of low barrier path for charge injection into the molecular layer.

One should note that Eq. (3.16) does not describe the overall reduction of the injection barrier for the charge as it was discussed in Ref. 17,52,55. It is insufficient to reduce fluctuationally the energy of one molecule nearby the metal to insure the faster charge injection into the bulk because the charge entering the molecule near the interface will return back to the metal instead of going further into the bulk. The only configuration, which consists of many molecules, like that in Fig. 7 can promote the charge injection.

The situation simplifies if we restrict our consideration to one-dimensional injection paths only. The one-dimensional model is easier for numerical study,[56] and permits exact analytical evaluation of the current through the system.[21] As shown in Fig. 8 we then ignore the complicated paths of charge from the cathode to the dielectric layer containing steps parallel to the cathode shown by the dashed line, and restrict our considera-

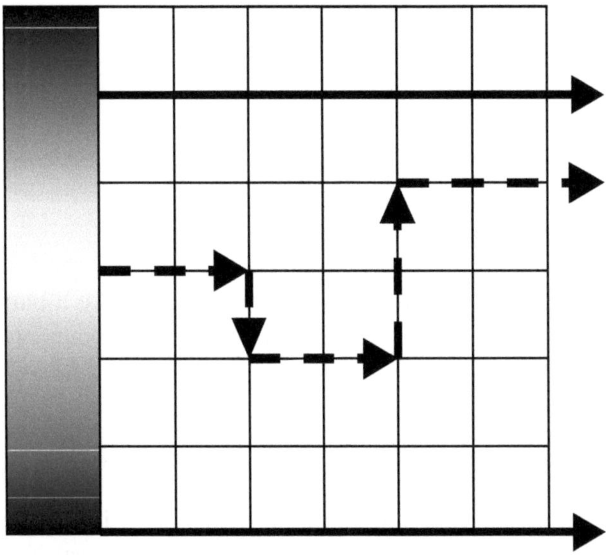

Figure 8. Injection paths from the electrode into the medium.

tion to the paths shown by the solid lines directed out from the cathode. Formally deviations from the straight path requires more molecules to have a given fluctuation of the injection barrier $\varphi_* \sim \Delta\text{-}\Delta_*$ (see Fig. 7). The exponentially small probability of such a fluctuation at $\varphi_* > w$ is a sufficient condition to ignore non-straight paths. As we will see from the final analysis this condition is satisfied when the electric field is strong enough so that the voltage drop over one intermolecular distance a exceeds the thermal energy. Comparison of one-dimensional modeling results with numerical simulations of three-dimensional hopping shows that the one-dimensional approach is good enough to characterize the charge transport generally.[14,21]

For a sequence of one-dimensional chains conducting charge from the cathode to the bulk, each chain can be described as a sequence of sites k characterized by the distance a·k from the cathode (recall that a is the average intermolecular distance). Each site can be described by the potential energy

$$E_k = \Delta + \varphi_k - eEak, \qquad (3.17)$$

and the neighboring sites k, k+1 are connected by the hopping amplitudes $W_{k,k+1}$. Assume that each chain has N sites. The populations of non-boundary sites P_k $(1<k<N)$ obey the rate equations

$$\frac{dP_k}{dt} = -W_{k,k+1}P_k - W_{k,k-1}P_k + W_{k+1,k}P_{k+1} + W_{k-1,k}P_{k-1}. \qquad (3.18)$$

The population at the first site is pumped from the cathode over the first highest step of injection barrier

$$\frac{dP_1}{dt} = W_{01}n_0 - W_{10}P_1 + W_{21}P_2 - W_{12}P_1, \qquad (3.19)$$

where n_0 is the density of carriers at the cathode. For the last site the final step (to anode or different transport layer) does not have a barrier and can be considered very fast. Therefore we can set the zero boundary condition for the right end site

$$P_N = 0. \qquad (3.20)$$

We will see that the final answer is not really sensitive to the choice of the right boundary condition (3.20) at sufficiently large N ($eaEN > \Delta$ that is satisfied in the system under study, see Fig. 2B).

The steady state solution of equations (3.18), (3.19) describes the stationary regime of interest. The current can be expressed analytically through the zero site density n_0 using methods developed in Ref. 21. The steady state solution of Eq. (3.18) can be expressed at the constant current j through the chain

$$j = W_{k,k+1}P_k - W_{k+1,k}P_{k+1} = -W_{k,k-1}P_k + W_{k-1,k}P_{k-1}. \qquad (3.21)$$

Then we can derive the recursion relationship for the local densities

$$P_k = P_{k-1}\frac{W_{k-1,k}}{W_{k,k-1}} - \frac{j}{W_{k,k-1}}. \qquad (3.22)$$

Making use of the detail balance equation (3.2) we can rewrite Eq. (3.22)

$$P_k = P_{k-1}\exp\left(-\frac{E_k - E_{k-1}}{k_B T}\right) - \frac{j}{W_{k,k-1}}. \qquad (3.23)$$

The equation (3.23) can be applied k times to relate the population P_k and the electron density n_0 (see Ref. 21)

$$P_k = n_0 \exp\left(-\frac{E_k}{k_B T}\right) - j\sum_{p=1}^{k} \frac{\exp\left(-\frac{E_k - E_p}{k_B T}\right)}{W_{p,p-1}}, \qquad (3.24)$$

we have used the definition of the cathode Fermi energy $E_0=0$. Then the current through the chain can be expressed using the zero boundary condition (3.20) at k=N

$$j = \frac{n_0}{\left(\sum_{p=1}^{N} \frac{\exp\left(E_p/(k_B T)\right)}{W_{p,p-1}}\right)}. \qquad (3.25)$$

The injection-limited regime should be treated at constant electronic density of states n_0 at the cathode layer, since it has much higher conductivity than the dielectric and each loss of charge is immediately compensated. The total current can be found summing the contributions (3.25) over all conducting chains (see Fig. 8); this is equivalent to averaging of the result (3.25) over all realizations of random energies E_p

$$j = \left\langle \frac{n_0}{\left(\sum_{p=1}^{k} \frac{\exp\left(E_p/(k_B T)\right)}{W_{p,p-1}}\right)} \right\rangle. \qquad (3.26)$$

This is different from the definition of the charge mobility, made in Ref. 19, that should be considered at the given average current j for the infinite chain without distinguishing of the first and last steps. Then the mobility is defined as the ratio of the current and the average charge density multiplied by the electric field

$$\mu \propto \frac{j}{<n> E} = \frac{1}{E\left\langle \sum_{p=1}^{N} \frac{\exp(E_p/(k_B T))}{W_{p,p-1}} \right\rangle}. \tag{3.27}$$

The different averaging in Eq. (3.27) compared to Eq. (3.26) leads to qualitatively different behavior of mobility, compared to the conduction in the injection limited regime. In particular the mobility decreases exponentially with increasing disorder, opposite to the injection current behavior.

We will ignore the possible sample inhomogeneity in our consideration. This can be not always the case because of the specifics of the interface molecular layer nearby the metal. If the molecules are polarized the strong image forces can affect the orientation of all molecules in the layer that will influence their energies, built in voltage and charge injection through them.[17,57] If the length l of the effective injection barrier ($l \sim \Delta/(eE)$, Fig. 2.B) is longer than the single molecular layer (that normally takes place) then the above effect will lead to the small corrections. Note that our consideration can be extended to treat this effect. However the lack of the knowledge of the surface layer properties does not permit us to do that in this work.

Let us use Eq. (3.26) to calculate the current through the system in the injection limited regime. First consider the hopping rates defined in the simple way (3.4). Then Eq. (3.26) can be rewritten as

$$j \approx \left\langle \frac{1}{2} \frac{n_0 W_0}{\left(\sum_{p=1}^{N} \exp(E_p/(k_B T)) \right)} \right\rangle. \tag{3.28}$$

Using the definition of energies Eq. (3.17) (still ignoring the image potential) one can rewrite Eq. (3.28) as

$$j \approx \left\langle \frac{1}{2} \frac{n_0 W_0 \exp(-\Delta/(k_B T)) \exp(eEa/(k_B T))}{\left(\sum_{p=0}^{N} \exp((\varphi_p - eEap)/(k_B T)) \right)} \right\rangle. \tag{3.29}$$

The exponential decrease of the contribution of large p terms in the denominator of Eq. (3.29) guarantees weak sensitivity to the boundary condition at large p=N. The exponentially small term describing the first step has been ignored in the denominator of Eq. (3.29) (cf. Ref. 16).

One can evaluate Eq. (3.29) using the standard integral transform.[16]

$$j \approx \frac{n_0 W_0 \exp(-\Delta/(k_B T) + eEa/(k_B T))}{2}$$
$$\times \int_0^{+\infty} ds \prod_{p=1}^{N} \int d\varphi f(\varphi) \exp(-s \exp((\varphi - eEap)/(k_B T))). \tag{3.30}$$

It is convenient to make the replacement of the variable $s=\exp(-\varphi_0/(k_B T))$. The expression for the current can be rewritten as

$$j \propto \exp(-\Delta/(k_B T) + eEa/(k_B T))$$
$$\times \int_0^{+\infty} d\varphi_0 \exp(\varphi_0/(k_B T)) \prod_{p=1}^{N} \int d\varphi f(\varphi) \exp(-\exp((-\varphi + \varphi_0 - eEap)/(k_B T))). \tag{3.31}$$

It is impossible to evaluate the integral (3.31) explicitly. In the case of interest when the disordering exceeds the thermal energy $w > k_B T$ one can replace the exponent in the integrals in Eq. (3.31) by the Θ-function. Then each factor in the product (3.31) can be approximated as

$$I_p(\varphi_0 - eEp) =$$
$$\int d\varphi f(\varphi) \exp(-\exp((-\varphi + \varphi_0 - eEap)/(k_B T))) \approx \int_{\varphi_0 - eEap}^{+\infty} d\varphi f(\varphi). \tag{3.32}$$

3.2.2 Gaussian Disorder

Now we turn to the case of the gaussian random potential Eq. (3.7). In that case we need the asymptotic behavior of the error function Eq. (3.32) that is different for different lower limits of the integral

$$I_p(\lambda) \approx \frac{w}{\sqrt{2\pi}a} e^{-\frac{\lambda^2}{2w^2}}, \ \lambda > w,$$
$$I_p(\lambda) \approx \frac{1}{2} - \frac{\lambda}{\sqrt{2\pi w}}, \ w > \lambda > -w, \tag{3.33}$$
$$I_p(\lambda) \approx 1 - \frac{w}{\sqrt{2\pi} \mid \lambda \mid} e^{-\frac{\lambda^2}{2w^2}} \approx 1, \ \lambda < -w.$$

In the regime of strong fluctuations, corresponding to the effective values of φ_0>w the major contribution to the integral comes from the small p, corresponding to φ_0-eEap>w. The product of asymptotic factors (3.33) for this part of the whole product in Eq. (3.31) can be expressed as

$$P_1 \sim \exp\left[-\sum_{p=0}^{\frac{\varphi_0-\eta w}{eEa}} \frac{(\varphi_0 - eEap)^2}{2w^2}\right]\left(\frac{w}{\sqrt{2\pi}}\right)^{\frac{\varphi_0-\eta w}{eEa}+1} \prod_{p=0}^{\frac{\varphi_0-\eta w}{eEa}} \frac{1}{\varphi_0 - eEap}, \tag{3.34}$$

Here η is the unknown constant of order of 1 that reflects the constraint a>w (3.33). The first exponent in Eq. (3.34) gives the main contribution that can be evaluated after replacement the summation over p with the integration as

$$\exp\left[-\sum_{p=0}^{\frac{\varphi_0-\eta w}{eEa}} \frac{(\varphi_0 - eEap)^2}{2w^2}\right] \approx \exp\left[-\frac{(\varphi_0 + eEa/2)^3}{6w^2 eEa} + \eta' \frac{w}{eEa}\right], \tag{3.35}$$

$\eta' \sim 1.$

Eq. (3.35) defines the largest contribution in φ_0. As we will see from the comparison of analytical theory with numerical evaluation of Eq. (3.29) this approximation is insufficient in the domain of interest and corrections should be taken into account. The remaining part of Eq. (3.34) can be evaluated as

$$\left(\frac{w}{\sqrt{2\pi}}\right)^{\frac{\varphi_0-\eta w}{eEa}+1}\prod_{p=0}^{\frac{\varphi_0-\eta w}{eEa}}\frac{1}{\varphi_0-eEap}\approx$$
$$\approx\exp\left[-\left(\frac{\varphi_0-\eta w}{eEa}\right)\left(\ln\left(\frac{\sqrt{2\pi}\varphi_0}{w}\right)+1\right)\right]. \tag{3.35}$$

Finally the contribution of the domain $-w<a<w$ (see Eq. (3.33)) is composed of the product of w/(eEa) integrals, all being of order of 1/2. The product can be approximated as

$$\exp\left(-\varsigma\frac{w}{eEa}\right),\ \varsigma\sim 1. \tag{3.36}$$

Thus, the whole product of integrals from Eq. (3.31) can be approximated as

$$\prod_{p=1}^{N}\int d\varphi f(\varphi)\exp(-\exp((-\varphi+\varphi_0-eEap)/(k_BT))\sim$$
$$\sim\exp\left[-\frac{(\varphi_0+eEa/2)^3}{6w^2eEa}-\frac{\varphi_0}{eEa}\left(\ln\left(\frac{\sqrt{2\pi}\varphi_0}{w}\right)+1\right)-\alpha\frac{w}{eEa}\right], \tag{3.37}$$
$$\alpha\sim 1.$$

Our estimate is defined with the accuracy to one unknown parameter α of order of unity that is significant for the corrections to the main dependence. As we will see from the answer $w<\varphi_0$, when the one-dimensional model is applicable (see Eq. (3.42) and the third term in Eq. (3.37) is the smallest. These parameters will be defined from the comparison of analytical and numerical results. Within the approximation (3.37) the total current Eq. (3.31) can be expressed through the integral

$$j\propto\exp(-\Delta/(k_BT)+eEa/(k_BT))$$
$$\times\int_0^{+\infty}d\varphi_0\exp\left(\frac{\varphi_0}{k_BT}-\frac{(\varphi_0+eEa/2)^3}{6w^2eEa}-\frac{\varphi_0}{eEa}\left(\ln\left(\frac{\sqrt{2\pi}\varphi_0}{w}\right)+1\right)-\alpha\frac{w}{eEa}\right). \tag{3.38}$$

This integral contains large exponents. It is defined by the parameter $\varphi_0=\varphi_*$ that makes the integrand maximum. The total integral then can be estimated as the integrand taken at $\varphi_0=\varphi_*$. The exponent reaches maximum where its derivative with respect to φ_0 is zero

$$0=\frac{1}{k_BT}-\frac{3(\varphi_*+eEa/2)^2}{6w^2eEa}-\frac{1}{eEa}\ln\left(\frac{\sqrt{2\pi\varphi_*}}{w}\right). \tag{3.39}$$

Assuming $eEa > k_BT$ (see the criterion (3.42)) we can approximate the solution of Eq. (3.39) as

$$\varphi_* = \sqrt{2}w\sqrt{\frac{eEa}{k_BT}-\ln\left(2\sqrt{\frac{\pi eEa}{k_BT}}\right)}-\frac{eEa}{2}. \tag{3.40}$$

Then the current in Eq. (3.38) can be expressed as

$$j \propto \exp\left(-\frac{\Delta}{k_BT}-\frac{2\sqrt{2}}{3}\frac{w}{k_BT}\sqrt{\frac{eEa}{k_BT}}\left(1-\frac{k_BT}{eEa}\ln\left(2\sqrt{\frac{\pi eEa}{k_BT}}\right)\right)^{3/2}+\frac{eEa}{2k_BT}-\alpha\frac{w}{eEa}\right). \tag{3.41}$$

Consider the applicability of Eq. (3.41). We should note that our derivation perfectly maps the fluctuation of the potential shown in Fig. 7. In fact the integration variable $\varphi_0=\varphi_*$ describes the reduction of the potential barrier Δ-Δ_*, leading to the gain factor $\exp(\varphi_0/(k_BT))$ (see Eq. (3.38)), while the product of integrals (3.31) within the approximation (3.32) describes the probability of such a fluctuation. Thus Fig. 7 describes the optimum injection path, and the optimum reduction of the potential barrier φ_* is given by Eq. (3.40). The one-dimensional approach (Fig. 8) requires that reduction to be greater than the typical fluctuation of potential energy w to ignore the deviations from the straight path. This takes place when the electric field is sufficiently large

$$eEa > k_BT. \tag{3.42}$$

This condition also justifies the restriction of our consideration to one-dimensional straight paths (Fig. 8) since deviations from the direct hopping occurs with exponentially small probability factor $\exp(-(\varphi_*/w)^2) \ll 1$. The condition (3.42) is satisfied in the performance regime of light emitting diodes at the bias voltage V~10V, system size L~100 nm and the intermolecular distance a~ 5Å. We will discuss the relationship to the experiment in detail later in Section 4.

If the field is so large that the number of sites involved into the sum in the exponent in Eq. (3.35)

$$n_* \sim \frac{\varphi_*}{eEa} \sim \frac{w}{\sqrt{eEak_B T}} \tag{3.43}$$

is smaller or equal to unity then the only one term is significant in the denominator of Eq. (3.29). The calculation of the gaussian integral yields

$$j \propto \exp\left(\frac{eEa}{k_B T} + \frac{w^2}{2(k_B T)^2}\right) \tag{3.44}$$

for the current dependence on the voltage and the temperature. This takes place when the voltage drop between the nearest molecules exceeds the trap energy, in the case of gaussian distribution given by $w^2/(k_B T)$. The result Eq. (3.41) is applicable if

$$eEa < \frac{w^2}{k_B T}. \tag{3.45}$$

If both constraints (3.42) and (3.45) are satisfied then the most significant voltage and temperature dependence in Eq. (3.41) is defined as [16]

$$\begin{aligned} j &\propto \exp(-\Delta/(k_B T))\exp\left[\frac{\varphi_*}{k_B T} - \frac{\varphi_*^3}{6w^2 eEa}\right] = \\ &= \exp(-\Delta/(k_B T))\exp\left[\frac{2\sqrt{2}}{3}\frac{w}{k_B T}\sqrt{\frac{eEa}{k_B T}}\right]. \end{aligned} \tag{3.46}$$

Thus, for the gaussian distribution of random potentials we have obtained the voltage and temperature dependence of the injection current. The voltage dependence coincides with that for the Schottky-Richardson law Eq. (2.5), discussed in the previous section. The difference is seen from the temperature dependence of the exponent, that is stronger in the case (3.46), when disordering defines the injection. The exponent increases with decreasing temperature as $T^{-3/2}$, contrary to 1/T dependence for the Schottky Richardson law.

As we found the optimum fluctuation conducting the charge into the dielectric layer is that where the effective potentials of n_* (3.43) centers are reduced to the lower values, allowing the reduction of the overall potential barrier by the energy φ_* Eq. (3.40). The probability of such a fluctuation is given by

$$\Omega_S \sim \exp\left(-\frac{(\varphi_* + eEa/2)^3}{6w^2 eEa} - \frac{\varphi_*}{eEa}\left(\ln\left(\frac{\sqrt{2\pi\varphi_*}}{w}\right) + 1\right) - \alpha\frac{w}{eEa}\right). \quad (3.47)$$

This probability describes the difference between the average current j and maximum local current j_{max}. In fact in the fluctuational picture of the charge injection the current is almost zero everywhere excluding the fraction (3.47) of the total junction area. Therefore the current through the optimum fluctuation will be remarkably larger than the average one and its typical value can be estimated as

$$j_{max} \sim \frac{j}{\Omega_S(\varphi_*)} \propto j\exp\left[\frac{\sqrt{2eEaw}}{3(k_B T)^{3/2}}\right]. \quad (3.48)$$

The local high current can be significant because it can lead to strong heating and damage of the metal-insulator junction. We will discuss this effect in detail later in Section 4.

In the case of a small electric field, opposite to Eq. (3.42), the most significant dependence is due to the last term in exponent in Eq. (3.41)

$$j \sim \exp\left(-\alpha\frac{w}{eEa}\right). \quad (3.49)$$

Although this dependence is derived for the one-dimensional case we believe it should be valid qualitatively in the three-dimensional regime of interest, based on the similarity of hopping transport in systems with energy disordering in all dimensions.[48,49] The dependence (3.49) on the electric field is even stronger than the Schottky-Richardson behavior Eq. (3.46). It has the same functional form as the tunneling regime Fowler-Nordheim dependence (2.11). However the remarkable temperature dependence in Eq. (3.41) caused by the thermal activation factor $\exp(-\Delta/(k_B T))$ describing the average injection barrier can distinguish between the two regimes. It is interesting that when the electric field is relatively large (3.42) the disordering assists the charge injection Eq. (3.46), while in the opposite limit it suppresses the overall current Eq. (3.49).

The interesting question is how small should be the electric field E to reach the Ohmic injection regime

$$j \propto \left\langle \frac{1}{\left(\sum_{p=0}^{N} \exp\left((\varphi_p - eEap)/(k_B T)\right)\right)} \right\rangle \propto$$

$$\propto \frac{1}{\sum_{p=0}^{N} < \exp\left(\varphi_p /(k_B T)\right) > \exp\left(-eEap/(k_B T)\right)} \propto \tag{3.50}$$

$$\propto \frac{eEa}{k_B T} \exp\left(-\frac{w^2}{2(k_B T)^2}\right),$$

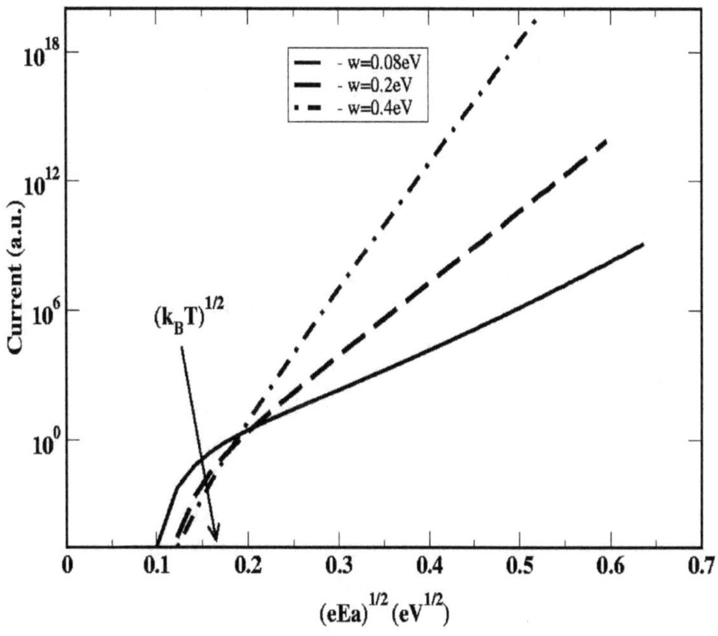

Figure 9. Current – Voltage dependence at different disordering and room temperature for intermolecular distance a=6Å

where we can perform the averaging of the denominator based on the law of large numbers (many terms there). This regime requires sufficiently long injection path containing the number of steps n~w/(eEa) to enable the equilibrium energy fluctuation $\varphi \sim w^2/(k_BT)$[48,49] to be realized. The probability of such a fluctuation is given by $\exp(-w^2/(2(k_BT)^2))$ (see the definition (3.7) of the potential distribution). Thus we get the condition

$$eEa < w \exp\left(-\frac{w^2}{2(k_BT)^2}\right), \tag{3.51}$$

for the Ohmic injection regime. We should note that it is very sensitive to the parameters of the system. In the case of strong disordering and low temperature it requires an extremely small bias voltage. It is clear that in the Ohmic injection regime disordering strongly suppresses the charge transport.

We have used the exponential approach to derive the temperature and field dependence of the current in the injection limited regime Eq. (3.41). Still we have the undefined parameter α. To compute it and analyze the accuracy of the analytical approach Eq. (3.41) we have completed a numerical evaluation of Eq. (3.31) for a gaussian distribution, to compare with the theory predictions. The results are given below. One-dimensional model is relatively easy for numerical analysis and should give a reasonable description of the system. This follows from Ref. 21, where the predictions made in one-dimensional system were supported by three-dimensional Monte-Carlo simulations.[48,49]

The results of the computation of current voltage dependence are shown in Fig. 9 for different extents of disorder. It clearly demonstrates the exponential dependence on the square root of the applied voltage as suggested by the analytical theory (3.46) when the potential drop between the subsequent molecules exceeds the thermal energy (3.42). In that case increasing disorder leads to current increase. The opposite behavior similar to Eq. (3.49) is seen in the opposite regime.

Analytical theory (3.41) that takes into account the most significant corrections to the main exponential form (3.46) satisfactorily describes the current dependence on the voltage and the temperature in the regime (3.42). We demonstrate this comparing the logarithmic derivative of the current with respect to the square root of electric field at room temperature in Fig. 10. The free parameter α (3.41) should be taken

$$\alpha = 1.2 \tag{3.52}$$

to get essentially a perfect fit.

The simple exponential approach[16] based on Eq. (3.46) is insufficient to describe the current voltage dependence with enough accuracy. The relative accuracy of the exponential approach increases with increasing disordering

w. Our simulations at other temperatures have demonstrated that the analyti-

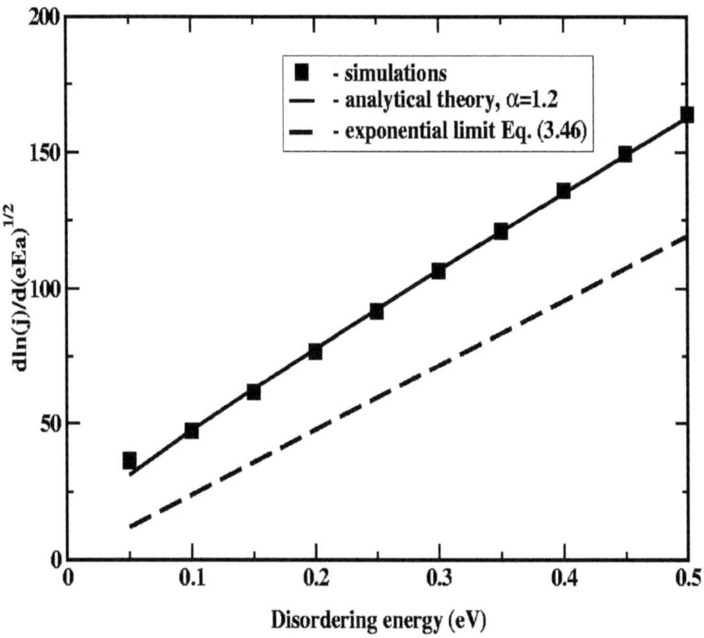

Figure 10. Logarithmic slope of Current – Voltage dependence vs. disordering at room temperature and intermolecular distance a = 6 Å.

cal approach (3.41) has sufficient accuracy when the condition (3.42) is satisfied, and the free parameter α is chosen appropriately Eq. (3.52). Thus one can describe the current in the injection limited regime using the equation

$$j \propto \exp\left(\begin{array}{l} -\dfrac{\Delta}{k_B T} - \dfrac{2\sqrt{2}}{3} \dfrac{w}{k_B T} \sqrt{\dfrac{eEa}{k_B T}} \left(1 - \dfrac{k_B T}{eEa} \ln\left(2\sqrt{\dfrac{\pi eEa}{k_B T}} \right) \right)^{3/2} \\ + \dfrac{eEa}{2k_B T} - 1.2 \dfrac{w}{eEa} \end{array} \right). \tag{3.53}$$

3.2.3 Image Potential Effect

We have ignored the image charge in the above consideration. The image effect should be significant at relatively small disordering, when the characteristic change of the potential energy due to the image potential Eq. (2.4) exceeds the disordering fluctuation φ_* (3.40)

$$w < \sqrt{\frac{e^2 k_B T}{8\pi\varepsilon_0 \varepsilon a}} \sim 0.15\text{eV}. \tag{3.54}$$

The estimate is made at room temperature $k_B T = 0.025$eV and for the interatomic distance a~5 Å. Thus both image charge and disordering can be significant at room temperature and more careful study of them simultaneously needs to be performed. Decreasing temperature leads to increasing significance of disordering.

As in the previous case we have approached the problem both analytically and numerically. The exponential approach similar to Eq. (3.46) has been derived in our previous work.[16] It shows that the exponential dependence on the square root of the applied voltage is retained. This is not surprising since both the Schottky Richardson mechanism of injection and the disorder induced injection lead to this dependence. The accuracy of the simple exponential approximation is not very good, just as for the exponential ap-

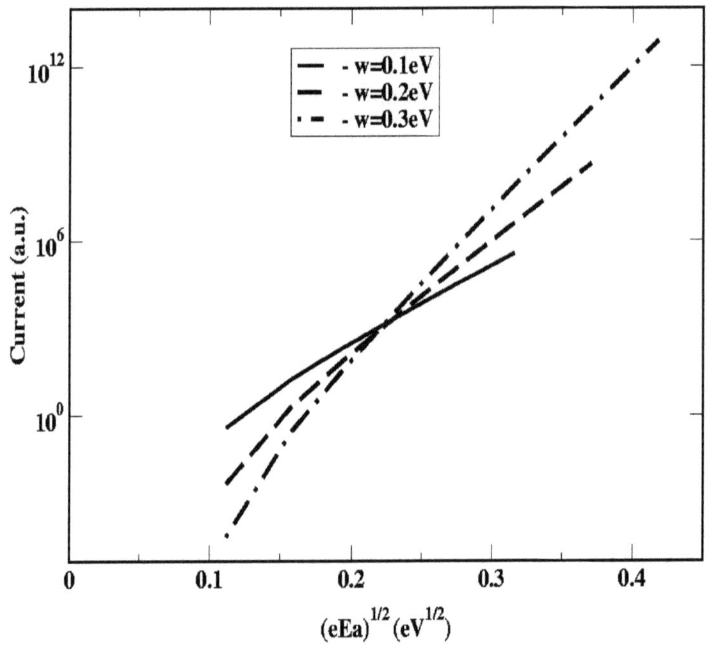

Figure 11. Current – Voltage dependence at different disordering and room temperature for intermolecular distance a=6Å. Image charge is taken into account

proach without image charge (dashed line in Fig. 10). Instead we propose an easy interpolation formula that contains all effects and fit experiment very well for reasonable fields. This approach expresses the current behavior using the sum of all contributing exponents from the image potential Eq. (2.4), disorder Eq. (3.46) and direct effect of applied field (cf. Eq. (3.31))

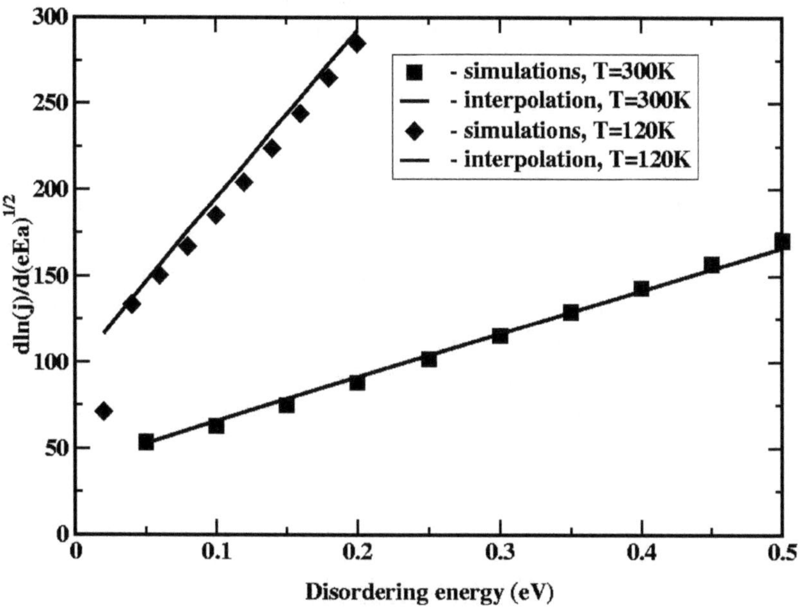

Figure 12. Logarithmic slope of Current – Voltage dependence vs. disordering at intermolecular distance a=6Å. Image charge is taken into account.

$$\ln(j) \approx -\frac{\Delta}{k_B T} + \sqrt{\frac{e^3 E}{4\pi\varepsilon_0\varepsilon}}\frac{1}{k_B T} + \frac{2\sqrt{2}}{3}\frac{w}{k_B T}\sqrt{\frac{eEa}{k_B T}} + \frac{eEa}{k_B T} \qquad (3.55)$$

We show the computed current voltage dependence in Fig. 11 for the intermolecular distance 5 Å. The results are qualitatively similar to those shown in Fig. 9. When the voltage drop at the intermolecular distance exceeds the thermal energy Eq. (3.42), disordering assists the charge injection. To compare the simple interpolation Eq.(3.55) with the simulations we computed the slope of the current voltage dependence at two temperatures T=300K and T=120K and compare them with Eq. (3.55). The results are shown in Fig. 12.

In both cases the interpolation (3.55) works very well excluding the case of the very small disorder w~0.05eV where strong deviations can be expected because of the inapplicability of the exponential approach that requires $w \gg k_B T$. The combined exponential approach (3.55) shows reasonable accuracy when both image potential and disordering are included, possibly due to the compensation of pre-exponential terms, which become significant in the absence of an image potential (see Fig.10). Based on this we can expect that our estimates for the properties of the optimum fluctuation of the injection path in the system without the image potential can be used to characterize the system even in the presence of the image.

3.2.4 Injection in case of Marcus law charge transfer

The results above form the qualitative and semi-quantitative basis for discussion of the experimental data. Now we consider the effect of possible changes of our model, including the change in the transfer rate from the gap law Eq. (3.4) to the Marcus classical formula Eq. (3.5) and the change in the distribution of the molecular levels from the gaussian Eq. (3.7) to exponential Eq. (3.8) or bimodal Eq. (3.11). Also the possible correlation effect in the potentials of different molecules Eq. (3.9) will be discussed.

Use of Eq. (3.5) for the charge transfer rate modifies our expression for the injection current (3.29) to become

$$j \propto \left\langle \frac{1}{\left(\sum_{p=0}^{N} \exp\left(\frac{\varphi_p - eEap}{k_B T} + \frac{(\varphi_p - \varphi_{p+1} - eEa + \lambda)^2}{4\lambda k_B T}\right)\right)} \right\rangle, \qquad (3.56)$$

where λ is the reorganization energy. Using the same concept of the optimum path as before (see Fig. 7) we obtain a similar derivation with an identical (in the exponential approach) optimization problem for the optimum fluctuation of the potentials φ_* with a typical value of the site potential φ_p=-$|\varphi_*|$+eEp. The only difference comes from the exponent in the transfer rate expression, leading to the current reduction by the factor

$$f_M = \exp\left(-\frac{\lambda^2}{4\lambda k_B T}\right), \qquad (3.57)$$

Thus in this regime the expression for the total current (3.55) should be modified to

$$\ln(j) \approx \sqrt{\frac{e^3 E}{4\pi\varepsilon_0 \varepsilon}\frac{1}{k_B T}} + \frac{2\sqrt{2}}{3}\frac{w}{k_B T}\sqrt{\frac{eEa}{k_B T}} + \frac{eEa}{k_B T} - \frac{\Delta}{k_B T} - \frac{\lambda}{4k_B T}. \qquad (3.58)$$

Thus the effective barrier height increases by the quarter of the reorganization energy. Analysis of possible changes in the shape of optimum fluctuation shows that the contribution of the reorganization effects can not significantly affect the current voltage dependence at small electric fields

$$eEa < \lambda. \tag{3.59}$$

(Eq. (3.45) is also supposed to be satisfied). For the reasonable values of λ~0.5eV [1-3,6] this effect will show up at very high electric fields ~ 1V/nm that are not sustainable in the molecular layers. The correction to the injection barrier λ~0.1eV is small compared to the typical barrier 0.5-1eV. Thus the overall effect of the change of the definition of the charge transfer rate to the Marcus classical law leads to the relatively small corrections to the gap law behavior but possibly large reduction of the overall current.

It is interesting that the effect of the change of the charge transfer rate from the gap law (3.4) to the Marcus law (3.5) crucially affects the mobility temperature dependence (3.27). The evaluation with the gap law leads to the known super Arrhenius dependence

$$\mu \propto \exp\left(-\frac{w^2}{2(k_B T)^2}\right). \tag{3.60}$$

As was already mentioned, the effect of disordering on mobility and on injection is quite opposite. Disordering strongly reduces the steady state mobility because of the deep traps Eq. (3.60) while the it supports injection Eq. (3.55) creating low barrier paths for electrons from the metal to the molecular layer. If we evaluate the expression for the mobility with the Marcus charge transfer rates (3.5) we get behavior qualitatively different from Eq. (3.60)

$$\mu \propto \frac{1}{\sqrt{1-\dfrac{T}{T_g}}} \exp\left(-\frac{w^2}{2(k_B T)^2} - \frac{w^2 (eEa)^2}{2k_B^2 T \lambda^2 (T - T_g)}\right), \tag{3.61}$$

$$T_g = \frac{w^2}{\lambda k_B}. \tag{3.62}$$

The problem has a critical temperature T_g defined in Eq. (3.62). Below this temperature the steady state mobility becomes zero. The mobility dependence on the field resembles the Tamann-Vogel-Fulcher law (e.g. Ref. 58) describing very strong reduction of the finite field conduction in the vicinity of the critical temperature. The result (3.62) can be possibly applied to describe other systems (e. g spin glasses[58] and/or super-cooled liquids[53]).

3.2.5 Exponential Disorder

Consider now the regime of the exponential distribution of the random potential, Eq. (3.8). In this case the calculations can be repeated similarly to the gaussian distribution. The calculations are easier because, all integrals are exponential. In the case of the exponential distribution the general expression (3.55) should be replaced with

$$\varphi_* \approx \frac{weEa}{k_B T},$$
$$\ln(j) \approx -\frac{\Delta}{k_B T} + \frac{1}{k_B T}\sqrt{\frac{e^3 E}{4\pi\varepsilon_0\varepsilon} + \frac{w^2 (eEa)^2}{4(k_B T)^2}} + \frac{eEa}{k_B T}. \tag{3.63}$$

At small disordering and/or electric field Eq. (3.63) yields the standard Schottky-Richardson behavior Eq. (2.5). At high disorder and /or applied voltage this dependence should change to the exponential as the function of the electric field

$$j \propto \exp\left(\frac{weEa}{2(k_B T)^2}\right). \tag{3.64}$$

To the best of our knowledge the dependence (3.64) has never been reported in organic molecular layers while the square root voltage dependence is widespread. Perhaps this is because the gaussian distribution is more relevant in these systems.

3.2.6 Dopant Induced Disorder

The situation with dopant induced disorder Eq. (3.11) is very special. The problem of the optimum bridge can be defined as whether the bridges containing only low energy dopants are favorable compared to the bridges containing the host molecules. The gain in the potential barrier is given by the exponential factor

$$\exp(\varphi_0 /(k_B T)), \tag{3.65}$$

where φ_0 is the energy difference between the host and dopant molecules. The path should involve $n=\varphi_0/(eEa)$ dopant molecules. Therefore the probability to make this dopant bridge between the electrode and the bulk is given by

$$\Omega_S \sim x^{-\varphi_0/(eEa)}, \tag{3.66}$$

where x is the fraction of dopant per the molecule. The product of factors (3.65) and (3.66) should be compared with the contribution of the host molecules, that is 1. This leads to the conclusion that at sufficiently large electric field

$$eEa > k_B T \ln(1/x). \tag{3.67}$$

the injection will be through the dopants while at lower field it occurs through the host. If, for instance x~0.01, that is a reasonable value for the dopant concentration,[9] the voltage applied to a 100nm thickness molecular layer should be about 20V, which is also reasonable. It is interesting that the result does not depend on the energy difference between the guest and host molecules. This is because the gain in energy is compensated by the increase of the length of the required dopant bridge. The dependence of the current on the field and the temperature can be summarized as following

$$j \propto \exp\left(-\frac{\Delta}{k_B T}\right)\begin{pmatrix} \Theta(eEa - k_B T\ln(1/x))\exp\left(\frac{\varphi_0}{k_B T} - \frac{\varphi_0}{eEa}\ln(1/x)\right) \\ + \Theta(eEa - k_B T\ln(1/x)) \end{pmatrix} \tag{3.68}$$

It shows the sharp change of field and temperature dependencies at the crossover point (3.67) and the exponential increase with the field at large electric fields.

3.2.7 Correlated Disorder

Finally consider the effect of correlations in random potentials (3.9) on the injection current. We concentrate on the case with the long-range correlations caused by the dipolar field

$$<\varphi_i \varphi_j> = w^2(\delta_{ij} + (1-\delta_{ij})\eta / r_{ij}),\ \eta < 1. \tag{3.69}$$

(see Eq.(3.9) and Ref. [21]). Then the probability of any specific fluctuation of site potentials (φ_1,.. φ_n) along the straight path from the electrode to the bulk of the molecular layer (see Fig. 8) can be estimated as

$$\exp\left(-\frac{1}{2}\sum_{i=1,j=1}^{n} Q_{ij}\varphi_i\varphi_j\right), \tag{3.70}$$

where Q_{ij} is the matrix inverse with respect to the correlation matrix (3.69)

$$\sum_k Q_{ik} < \varphi_k\varphi_j > = \delta_{ij}. \tag{3.71}$$

The probability of a fluctuation similar to Fig. 7 is studied (the image potential effect does not modify our answer significantly). It can be described as

$$\varphi_k = eEa(n-k),\ n = \varphi_* /(eEa), \tag{3.72}$$

where φ_* is the optimum reduction of the potential barrier.

We can not resolve the problem for the correlation function (3.69) analytically. However to predict the correct parametric dependence we can study the similar but solvable problem of the fluctuation φ_k=φ_* for k=1,..n, at the circular circuit, where the site 1 follows the site n. For this fluctuation we find using Eq. (3.70)

$$P_{sim} \sim \exp\left(-\frac{n\varphi_*^2}{2}\sum_{j=1}^{n} Q_{1j}\right) = \exp\left(-\frac{nQ_0\varphi_*^2}{2}\right),$$
$$Q_0 = \sum_{j=1}^{n} Q_{1j}. \tag{3.73}$$

Since in a periodic system the direct and inverse matrixes have an eigen mode of all unity components, the corresponding eigen values Q_0 Eq. (3.73) and $K_0=\Sigma<\varphi_k\varphi_1>\approx w^2(1+\eta\ln(n))$ should give a unity product. Thus the probability of the fluctuation can be expressed as

$$P_{sim} \sim \exp\left(-\frac{n\varphi_*^2}{2w^2(1+\eta\ln(n))}\right). \tag{3.74}$$

We are interested in the different fluctuation (3.72). Its probability should have the form similar to Eq. (3.74) with different numerical parameters

$$P_* \sim \exp\left(-\frac{n\varphi_*^2}{6w^2(1+c\eta\ln(n))}\right). \tag{3.75}$$

The factor 2 is replaced with 6 in the denominator of (3.75) because of the known limit at $\eta=0$ (Eq. (3.46)). The parameter c has been found to be equal 1.47 from the comparison of Eq. (3.75) with the numerical evaluation of the problem.

The logarithmic correction (3.75) of correlations (3.69) modifies the current behavior (3.46) as

$$j \propto \exp(-\Delta/(k_B T))\exp\left[\frac{2\sqrt{2}}{3}\frac{w}{k_B T}\sqrt{\frac{eEa(1+1.67\eta\ln(w/(eEak_B T)^{1/2}))}{k_B T}}\right]. \tag{3.76}$$

The logarithmic factor in Eq. (3.76) does not affect our conclusions qualitatively since it depends weakly on the parameters. It can however be significant quantitatively since any change in the exponent strongly modifies the total result. One should note that if the correlations are of short range, i.e. they decrease with distance faster than Eq. (3.69), then the expected correction is the additional constant factor in exponent (3.76) without logarithm form that holds in the case of Eq. (3.69). This difference is due to the logarithmic divergence of the sum of matrix elements (3.69) at long distance.

3.3 Disorder Effect on Tunneling Injection

The predicted current voltage dependencies Eqs. (3.53), (3.55), (3.63) contains the stronger temperature dependence of the increasing component $T^{-3/2}$ for the gaussian distribution of potential fluctuations or T^{-2} in the case of exponential distribution than that for the main decreasing contribution $-\Delta/(k_B T)$ caused by the injection barrier. At sufficiently low temperature, where the characteristic fluctuation of the potential barrier φ_* for the optimum injection path reaches the scale of the injection barrier Δ the optimum path becomes non-thermally activated. In fact the reduction with the subsequent vanishing of the temperature dependence of the current with decreasing the temperature has been seen in numerical simulations[14,15] at high disordering. However even earlier the barrier will become small enough and the thermal activation should be replaced by tunneling in accordance with the discussion in Sect. 2. It is interesting to consider the disordering effect on the charge injection in the tunneling regime. We will restrict our consideration to the continuous model and the tunneling current for the given disorder configuration $\varphi(x)$ can be described within the semi-classical expression similar to Eq. (2.6)

$$j \sim \left\langle \exp\left(-\frac{2}{\hbar}\sqrt{2m} \int_0^{\Delta/(eE)} dx \sqrt{\Delta - eEx - \varphi(x)} \right) \right\rangle, \tag{3.77}$$

image charge is ignored as the weak correction in the tunneling regime. Averaging in Eq. (3.77) should be performed with respect to all realization of the random potential $\varphi(x)$ having the white noise correlation properties

$$< \varphi(x)\varphi(x') >= w^2 a\delta(x - x'). \tag{3.78}$$

This correlation property approximately corresponds to the gaussian distribution of site energies with the width w and the characteristic separation of sites a.

Then the averaging problem is equivalent to the evaluation of the functional integral

$$j \sim \exp\left(-\frac{1}{2w^2 a} \int_0^{\Delta/(eE)} dx\varphi^2(x) - \frac{2}{\hbar}\sqrt{2m} \int_0^{\Delta/(eE)-\varphi(L)} dx \sqrt{\Delta - eEx - \varphi(x)} \right).$$

(3.79)

Assuming that the disordering gives the weak correction to the main exponent one can make linear expansion of the integral (3.79) with respect to $\varphi(x)$ and then perform the explicit gaussian integration leading to the following result

$$j \sim \exp\left(-S_0 + \frac{2m(aw^2)^2}{\hbar^2} \int_0^{(\Delta-w)/(eE)} dx \frac{1}{\Delta - eEx} \right),$$
$$S_0 = \frac{2}{\hbar}\sqrt{2m} \int_0^{\Delta/(eE)} dx \sqrt{\Delta - eEx} = \frac{4}{3\hbar} \frac{(2m\Delta)^{1/2}\Delta}{eE}. \tag{3.80}$$

Here S_0 is the classical action leading to the Fowler-Nordheim expression for the injection current. The upper integral limit for the disorder contribution is reduced by w/(eE) to avoid the irrelevant logarithmic divergence at the distances, where the expansion of the square root in Eq. (3.79) with respect to $\varphi(x)$~w becomes invalid. Easy integration of Eq. (3.80) leads to the following expression for the tunneling current increased by the random fluctuations

$$j \sim \exp\left(-S_0 + \frac{2maw^2}{\hbar^2 eE}\ln\left(\frac{\Delta}{w}\right)\right). \qquad (3.81)$$

The correction factor in Eq. (3.81) possesses the same electric field dependence as the main factor S_0. It is interesting to consider the relative effect of disordering compared to the main contribution. The ratio of the second and first terms in Eq. (3.81) can be expressed as

$$\delta_{tun} \sim \frac{w^2}{\Delta^{3/2}(\hbar^2/(ma^2))^{1/2}}\ln\left(\frac{\Delta}{w}\right). \qquad (3.82)$$

Assuming w~0.1eV, Δ~0.5eV, a~0.5nm, we get δ_{tun}~0.15. Thus disordering leads to the 15% reduction of the tunneling exponent without any changes in the field dependence. Although this effect increases the tunneling current at the performance electric field 10^8V/m by 2-3 orders of magnitude it does not strongly affect the transition temperature from thermal activation to the tunneling. The disordering effect at the same field and room temperature leads to much stronger changes in the thermally activated regime that can reach up to 10 orders of magnitude (see Fig. 11).

Recapping this (mostly theoretical) section, we can conclude that disordering remarkably affects the efficiency of charge injection. When the electric field is sufficiently large that the voltage drop between nearest-neighbor molecules exceeds the thermal energy, disorder strongly enhances the injection of charge, creating fluctuational low barrier paths and promoting efficient charge transport over the injection barrier. Various assumptions for system disorder lead to exponentially strong dependence of the injection current on the temperature and field caused by the random potential fluctuations. The effect of disordering in the tunneling injection regime is similar but does not change qualitative behavior and is smaller quantitatively. In the following section we will use the theoretical approach developed here to discuss some experimental data.

4. DISCUSSION OF EXPERIMENT

The interpretation of experimental data in terms of the effect of disordering is a hard problem. First, the most significant current/voltage (electric field) dependence $j \sim \exp(\beta E^{1/2})$ is expected both for thermionic emission in a regular system Eq. (2.5) and for the gaussian disorder stimulated charge injection (see Eq. (3.46), Figs. 9, 11). Therefore, it is hard to distinguish between these two behaviors, which can lead to different interpretations for the same data. For instance, measurements of current voltage characteristics in one of the most popular light emitting materials Alq_3, leads to a measured slope $d\ln(j)/dE^{1/2}$ for the current voltage dependence.[14,15,59,60] For the Schottky Richardson thermionic emission injection Eq. (2.5) this slope can be evaluated as

$$\beta_{SR} = \frac{d\ln(j)}{dE^{1/2}} = \frac{1}{k_B T}\sqrt{\frac{e^3}{4\pi\varepsilon_0\varepsilon}}. \qquad (4.1)$$

This leads to the slope value $\beta = 0.017/\varepsilon^{1/2}\ (\mathrm{cm/V})^{1/2}$ at room temperature. To interpret the experimentally observed value $\beta \approx 0.014\text{-}0.017$ it was assumed in Ref. 60 that the dielectric constant of Alq_3 is anomalously small $\varepsilon \sim 1 - 1.7$ compared to the characteristic value 3.2-3.5 for organic materials, while the interpretation of the similar observations in later studies[14,15] is based on the effect of disorder. Since the slope Eq. (4.1) possesses different temperature dependence for the disordered system (3.46)

$$\beta = \frac{d\ln(j)}{dE^{1/2}} = \frac{2\sqrt{2}w}{k_B T}\sqrt{\frac{ea}{k_B T}} \qquad (4.2)$$

the joint analysis of current/electric field and current temperature dependence has been used in Refs. 14,15 to choose the right injection mechanism.

In addition the effect of disordering is different at strong and weak electric fields Eq. (3.49). When the characteristic potential change over one intermolecular distance exceeds the thermal energy, disordering supports charge injection while at the lower fields it acts in the opposite way (Figs. 9,11). The main disorder-induced field dependence at high fields is given by the strongly temperature dependent effect (4.2) while at low field the temperature independent term $-1.2w/(eEa)$ of Eq. (3.49) will dominate in the exponent describing the current voltage dependence in the system. The experimental data are often taken on the different sides of the inequality Eq. (3.42) (e.g. Ref. 15) and therefore it is hard to interpret them from the same point of view.

An additional problem is the proper definition of the electric field E that causes the injection of charge. In multi-layered devices the electric fields can be different in different layers even in injection limited regime due to the surface charge density that can be formed in the interface separating the layers (see e.g. Ref. 61). One can expect, based on Ref. 61, that in the double layer device composed of an Alq_3 emissive layer and a TPD hole transport layer the electric field will be larger in Alq_3 and smaller in TPD to provide the balance between charge currents, i.e. reduce the injection efficiency of majority carriers (holes) and increase that for the minority carriers (electrons). Therefore the quantitative application of theory to the multi-layered devices requires the solution of the balance problem, that lies outside of the scope of this paper.

Thus the direct application of theory, developed in Sect. 3, can be made to single layer devices. Experiments studying both current/voltage and current/temperature dependence are preferable because the temperature dependence enables us to distinguish between Schottky-Richardson and disorder induced injection. Finally the injection limited regime is realized in the case of a relatively high injection barrier. This seem to take place in Alq_3 (see the recent work[17]).

Measurements clarifying the charge current in Alq_3 for the wide range of fields and temperatures have been reported in Ref. 15. Here we interpret the results of Ref. 15 related to the current dependence on the voltage and temperature in terms of theory developed in Sect. 2 for a gaussian distribution of random potentials, and discuss qualitatively the more recent work

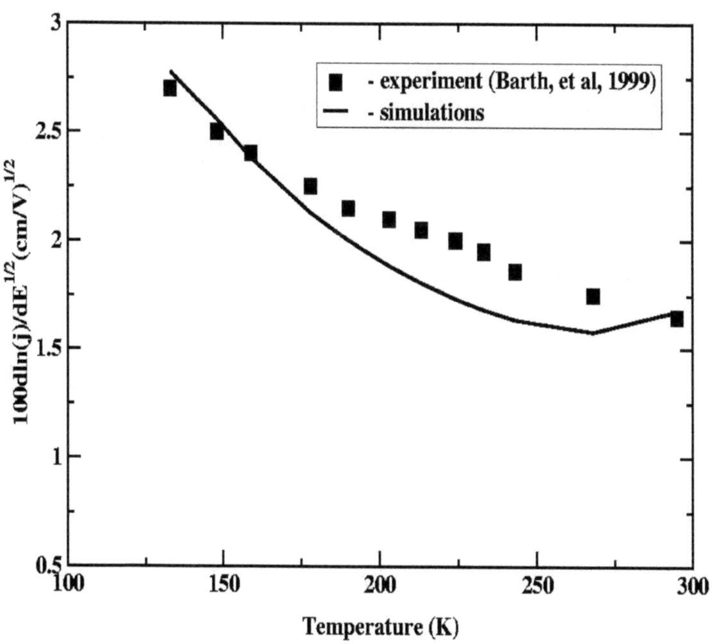

Figure 13. Logarithmic slope of Current – Voltage dependence vs. disordering and intermolecular distance a=6Å. Simulations vs. experiment

(see Ref. 17).

The slope of the logarithmic field dependence Eq. (4.1) at different temperatures has been systematically studied in Ref. 15. The results of measurements for the electron injection current from Mg:Ag electrode to Alq_3 molecular layer, taken from Ref. 15, are shown by the filled squares in Fig. 13. Deviations from the thermionic emission model take place (the similar results have been also obtained for TPD hole transport layer[24]). First, the absolute value of the derivative exceeds the expected value Eq. (4.1) by roughly a factor of 2, using the proper dielectric constant of Alq_3, ε~3.2 – 3.5. Second, the temperature dependence of the slope is weaker than the 1/T expected for thermionic emission (4.1). Note that disordering at strong fields Eq. (3.42) leads to the opposite behavior, Eq. (4.2). To interpret this anomalous behavior, note that the slope data in Fig. 13 were taken at different applied voltages (larger at lower temperatures and smaller at high temperatures) to have reasonable current through the system.[15] The slope at the highest temperature T~290K has been defined at the lowest field E~$2.5\cdot10^5$V/cm, while the slope at the lowest temperature T~133K was measured at the highest field E~$7\cdot10^5$V/cm. This leads to different values of the voltage drop energy eEa, for an inter-molecular distance that can be taken to be equal a=6 Å according to Ref. 14. At the lowest field corresponding to the highest temperature T=290K, this product is 0.015eV smaller than the thermal energy k_BT~0.025eV. At highest field and lowest temperature T~133K it is about 0.05eV, certainly larger than the thermal energy k_BT~0.01eV.

To interpret the data we thus can not use any analytical interpolation and need to integrate numerically the definition Eq. (3.29). The electric field for different temperatures has been chosen by linear interpolation between the maximum value E_1= $7\cdot10^5$V/cm at T_1 =133K and minimum value E_2=$2.5\cdot10^5$V/cm at room temperature T_2=290K

$$E(T) = E_1 + (T - T_1)\frac{E_2 - E_1}{T_2 - T_1}. \qquad (4.3)$$

To characterize the system disorder introduced by the random potential φ, we need to approximate the width w of the potential distribution (3.7). This has been defined experimentally for Alq_3 in Ref. 18 as w~0.06eV.

Using all these available parameters (no adjustable parameters) we have computed the slopes within the one-dimensional model (3.29). The results are shown by the solid line in Fig. 13. The agreement is reasonable taking into account the one-dimensional approach. Possibly a better fit can be made with modified parameters, but we do not think that this makes sense because the one-dimensional theory itself can lose validity at small electric fields (see Sect. 3).

We believe that the slope behavior found for the TPD hole transport layer in Ref. 24 can be interpreted using similar considerations, because of the qualitative similarity of the studies in Ref. 15 and Ref. 24 including increasing measuring field with decreasing temperature. However the lack of the necessary data (slopes) in Ref. 24 and the strong possibility for space charge contribution to current/voltage dependence in TPD does not permit careful analysis of the data.

Our theory can also be applied to the temperature dependence of the current measured in Ref. 15 at fixed electric field $E=9.5\cdot10^5$ V/cm used in Ref. 15 (Fig. 7 there). To interpret the current dependence on the temperature, one more parameter was needed: the injection barrier Δ at zero field and disorder. This value has been chosen as an adjustable parameter, and the optimum fit of the experimental data (Fig. 14) has been obtained at Δ=0.47eV. This value is in the reasonable agreement with the estimate of Ref. 15 and the theory seems to be in the good agreement with the experiment (see Fig. 14).

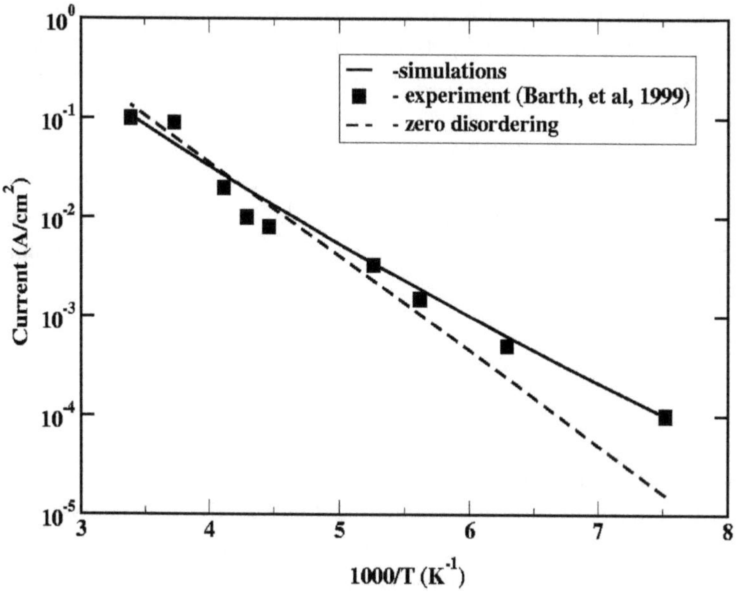

Figure 14. Temperature dependence of current at the bias voltage $9.1\cdot10^5$V/cm: Experiment vs. theory

Monte-Carlo study of the charge injection within a three dimensional hopping model in Ref. 14 is certainly a better approach than our one-dimensional model. We believe, however, that the character of the parametric dependencies should not differ in different dimensionalities based on the analytical studies[21,48,49] and numerical study in Ref. 21. Computationally, three-dimensional Monte-Carlo is much more expensive, which may account for the problems in the interpretation of the slope in Ref. 15 at low temperatures. This is because the injection is defined by the very rare fluctuations that makes hard to collect the sufficient statistics for proper averaging. Our numerical temperature dependence of the current agrees with the experimental data shown in Fig. 14.

The measurements of Ref. 15 on Alq_3 have been extended in Ref. 17 to lower temperatures, T~ 20K. The temperature dependence of the current was shown to saturate at T<100K, which was interpreted as the effect of the fluctuations within two molecular layers close to the cathode reducing the injection barrier to almost zero. The hopping rate expression Eq. (3.5) from the Marcus classical theory was used and the temperature where the fluctuations suppress the injection barrier has been estimated close to our estimate for T_g Eq. (3.62). In our interpretation this temperature corresponds to the loss of microscopic mobility. The analysis of Ref. 17 considers only a couple of hopping steps from the electrode and does not consider the whole path to the bulk (Fig. 7). Therefore the energy barrier for the injection is underestimated since even if the two neighboring molecules adjacent to the electrode have the low potentials; this may be insufficient to conduct the charge to the bulk. Actually one needs the strong fluctuations in about $n_* \gg 1$ molecules (see Eq. (3.43)); this has much lower probability than that estimated in Ref. 18 (a similar problem exists in the analytical theory suggested in Ref. 14).

Our theoretical approach Eq. (3.55) for the overall temperature dependence of the injection current also predicts saturation of the temperature dependence at very low temperature

$$k_B T \approx \frac{9}{4} eEa \frac{w^2}{\Delta^2}. \tag{4.4}$$

At this temperature the optimum fluctuation of the energy barrier reaches the value of the barrier. We do not think that there is the real chance to reach this regime since lowering the barrier by the fluctuations will first lead to change of the injection mechanism to tunneling. We believe that the temperature independence of the injection current at low temperatures observed in Ref. 17 is the consequence of the change of the injection mechanism to tunneling, that is always expected when the temperature is small enough. The analysis of this injection mechanism change in detail is outside of the scope of this paper.

It is interesting that the slope temperature dependence $T^{-3/2}$ predicted by Eq. (4.2) has been observed in the recent experiments [30] in the narrow temperature range from 0 to 30 C for charge transport through self-assembled multi-layer nitroazobenzene (NAB) films of several nano-meters thickness placed between carbon and mercury electrodes. However the absolute value of the slope is smaller than that expected for the thermionic emission (4.1). Therefore we can not apply our theory for quantitative interpretation of Ref. 24]. The low value of the slope is possibly caused by the potential drop at contacts between electrodes and molecules that can be significant in this very thin system (e.g. Ref. 3). Accordingly the sensitivity of the current to the applied voltage can be weaker. The interpretation of all changes in detail is outside of the scope of this paper.

The assumption of a gaussian distribution for the random potentials used to describe the injection process in disordered systems leads to current/voltage dependence similar to the classical theory of the thermionic emission. One can distinguish between the two regimes by studying the temperature dependence of the injection current. The disorder then is significant for the charge injection into, for example, an Alq_3 amorphous layer and the theory of Sect. 3 can be used to interpret the experimental data. Our fits of the experimental data (Figs. 13, 14) have been based on the experimentally available parameters for the disorder, and can be modified once these parameters are defined more accurately. It would be particularly valuable to complete experimental study of the current/voltage dependence in more detail at fixed large field to compare the slope of the observed current voltage Eqs. (4.1), (4.2) dependence with the prediction (3.55). The full quantitative interpretation of the charge injection in multi-layered devices requires consideration of the charge balance between electrodes.

Another effect of disordering on the current voltage characteristics can be significant in both injection limited and space charge limited regimes. This is the mobility voltage dependence that also shows an exponential dependence on the square root of applied voltage

$$\mu \sim \exp(\beta_\mu \sqrt{E}) \tag{4.5}$$

[7,17,21,25]. The equilibrium charge mobility for the infinite system is defined by Eq. (3.27). According to Ref. 21, it leads to the dependence (4.5) only when the random potentials at different sites have the long-range correlations Eq. (3.69). The question arises whether the equilibrium mobilities can be used to describe the experimental data. In fact, if we consider the exponential disordering Eq. (3.8) or the charge transfer rates defined by the Marcus law (3.5), at some temperature the macroscopic equilibrium mobility simply approaches zero. In addition the charge should visit $\exp(w^2/(2(k_BT)^2))$ sites to be equilibrated. This is not necessarily satisfied at strong disordering $w>>k_BT$, because the number of hops between two electrodes is about 200 at high enough field (the number of hops is given by the ratio of the device thickness ~100nm over the intermolecular distance

~0.5nm, and the hopping is unidirectional at high fields), while the exponential factor can be much larger. Then the mobility will be defined by the deepest trap and the process of leaving the trap can be considered within an optimization procedure similar to the analysis of the previous section. The largest contribution to the current will be defined by those paths, where the escape from the deepest trap is fluctuationally reduced. Then a dependence similar to Eq. (4.5) can be derived. Additional arguments supporting this expectation can be made based on the recent mobility measurements.[25] It was shown in Ref. 25 that at steady state the mobility and injection current, (defined entirely by the injection barrier) are proportional to each other. As we have seen in the previous section, the correlations of potential responsible for the behavior (4.5) of the equilibrium mobility do not strongly modify the injection current Eq. (3.76). On the other hand disordering can lead to the similar behaviors. Thus the effect of disorder on the mobility should be carefully tested.

Similarly one can expect that the exciton quenching will be affected by disorder. It is the common conclusion of many experimental studies that at sufficiently high applied voltage the luminance increase saturates due to the field induced quenching processes (see e.g. Ref. 62). The non-radiative exciton decay can be induced by the external field. It is similar to the transition over the injection barrier (see Fig. 7) with the only difference that the barrier height is given by the exciton binding energy that is several tenths of electron Volt (see e.g. Ref. 16). In fact the exciton diffuses over tens of nanometers (e.g. Ref. 63) before decaying, thus visiting hundreds of molecules. Some of them might have the barriers for exciton dissociation reduced by the fluctuations, and the density of reduced barriers will strongly increase with the bias field. When the field exceeds the certain limiting value the quenching rate in some visited site will become larger than the exciton radiative decay rate. Then the light emission efficiency will decrease with further increase of the voltage. This process requires special study. It is clear however that the disordering reduction will increase the emission efficiency at high fields.

Disorder can affect the device stability. As we discussed in Sect. 3 the fluctuational injection mechanism leads to the strong differences between the average current through the surface layer and the local current occurring at the places of optimum fluctuation. This difference is given by the exponential factor Eq.(3.48) that can become very large at strong disordering $w \gg k_B T$. In fact the experimental analysis of damage shows (see Refs. 64 and references therein) that it occurs at isolated spots on the ITO surface, and is strongly sensitive to the surface contact. Experimental analysis of a copper phthalocyanine (CuPc) layer, inserted between the ITO electrode and N,N'-diphenyl-N,N'-(2-naphthyl)-(1,1'-phenyl)-4,4'-diamine (NPB) hole transport layer shows rapid increase in the device stability simultaneously with the reduction of the hole injection efficiency.[65] If we assume that the disordering is weaker in the CuPc layer, the reduction of the hole injection and the increase of the stability can be explained by the suppression of fluctuations in the energies of molecules nearby the ITO electrode. The fluctuations near the interface are of most significance since they compensate the high injection barrier at the interface. The hole injection is reduced because no fluctuational pathways (see Fig. 7) can be formed and the stability increases due to the homogeneity of the current through the molecular layer.

As has been mentioned in Sect. 3 the gaussian distribution (3.7) of the potential fluctuations is most suitable for the amorphous organic molecular layers, while the current behavior (3.63) for the exponential distribution Eq. (3.8) have been never seen in the organic materials. Exponential decay of the density of states with energy is a characteristic property of the density of states below the band edge in disordered semiconductors (see e.g. Ref. 53 and references therein). Therefore it is more reasonable to expect the manifestations of the behavior (3.63) in doped inorganic materials. Such behavior has been observed experimentally in Au/n-GaAs Schottky diodes,[66,67] PtSi/Si diodes and many others (see Refs. 55,56,67 and references therein). In these reports the current voltage dependence at high voltages has the form Eq. (3.63). Theoretical explanation for this behavior[55] uses the assumption of the gaussian distribution of the overall injection barrier φ_*.

We can easily demonstrate that the assumption of an exponential distribution of the potential fluctuations will lead to the gaussian distribution of the barrier heights. In fact the barrier height φ requires the potentials at $n=\varphi/(eEa)$ adjacent sites (Fig. 7) to be reduced by φ. For the molecule separated from the electrode by k periods a, the potential fluctuation is given by φ-eEak. The probability of such a fluctuation is given by the product of n exponents $\exp(-(\varphi\text{-}eEak)/w)$ leading to the total barrier fluctuation probability

$$P(\varphi) \sim \exp\left(-\frac{\varphi^2}{2weEa}\right). \tag{4.5}$$

If the sample thickness L is so small that $eEL < \varphi$ then the distribution becomes exponential

$$P(\varphi) \sim \exp\left(-\frac{\varphi L}{2wa}\right). \tag{4.6}$$

The distribution (4.5) gives the same result as the gaussian distribution suggested in Ref. 55 to explain the experimental data. Certainly the parameters of the simple hopping model used in Sect. 3 should be modified since for inorganic materials we can not consider the overall electronic state as belonging to the single molecule. In particular the intermolecular distance a should be replaced by the localization radius l>>a. Thus our model can be applied to the fluctuational charge injection in inorganic materials, after proper modification of random potentials distribution.

5. DISCUSSION AND CONCLUSIONS

In this paper we have discussed the general problem of charge injection from a metal electrode into an amorphous organic molecular layer. Attention has been paid to the effect of disordering on the charge current through the layer. This problem is significant for fundamental understanding of the charge transport in organic materials and for technical applications including modern light emitting devices and molecular electronics.

We have briefly discussed the general problem of the charge injection mechanism, which can be caused by the quantum mechanical tunneling or thermally activated hopping over the injection barrier. Our estimate agrees with the experimental fact that at room temperature the injection of charge occurs in the thermally activated regime, but the crossover temperature between the thermally-activated and tunneling regimes is overestimated (~200K versus the observed 50-100K[17]). This is possibly because of the assumed continuous model for tunneling that should perhaps be replaced with the tight binding picture. The corresponding increase of the effective mass by a factor of 4 should explain the existing relationships. It is interesting that in addition to the limiting thermally activated and tunneling regimes, an intermediate regime of thermally activated tunneling should be realized in the vicinity of the crossover temperature.

Thermally activated injection is strongly sensitive to various changes of the injection barrier including the image charge and disordering, while these effects lead to relatively small corrections in the tunneling regime. At sufficiently strong bias field (corresponding, however, to the performance regime of light emitting devices), such that the potential change at the intermolecular distance exceeds the thermal energy, disordering enhances the injection current. This enhancement is due to formation of low-barrier paths leading the charge from the electrode into the organic layer. Theoretical analysis of these paths has been performed in the gaussian model of disordering, most reasonable for amorphous materials. In this case we found an analytical solution for the current – temperature – voltage dependence. It has the same exponential dependence on the square root of the electric field as that for the Schottky Richardson thermionic emission, but with more complicated temperature and disordering dependence. Theory can be used to interpret the complicated temperature and voltage dependence of current density in the most popular emitting material Alq_3.[14-15] The results do not change much for different definitions of the energy dependence of the charge transfer rate or if we take into account the correlations in the fluctuations of molecular energies. Note that these changes crucially affect the charge mobility.

Our model can also be extended to systems with exponential density of random electronic state energies. This is more suitable for metal - semiconductor contacts. In that case the theory predicts an exponential increase of the current with the applied voltage. This behavior is in fact known in semiconducting materials[66] and our modeling predictions can be directly applied to interpretation of that data.

We have briefly discussed other effects related to disordering. It can affect the device performance negatively by reducing the device stability and supporting exciton quenching. In fact the fluctuational paths promoting the charge injection into the medium are very rare. Therefore the current through them is much higher than the average current. Thus strong heating and subsequent device damage should be expected at the metal-molecular layer contact, as is seen in the experiment.[64]

Since the exciton dissociation process is similar to the charge injection (exciton binding energy acts like the injection barrier, remaining charge Coulomb field acts like the image potential) the optimum configurations have significantly increased probability of exciton quenching by its dissociation. Therefore disorder can reduce the emission efficiency.

Thus, materials with high extent of disordering can be used to promote the efficient injection of charge with the effective barrier reduced remarkably compared to its average value. On the other hand disordering can reduce the emission efficiency. To avoid the latter problem one can use multi-layered structures, where the emissive layer is more ordered compared to the random electron transport layer. To optimize the disorder contribution to the device performance it is significant to enable control of it. In the recent works[68] it was demonstrated that disordering in Alq_3 is very sensitive to the presence of oxygen. Thus the annealing process in different oxygen atmospheres should give the certain opportunities to alter the device properties in the desirable direction.

ACKNOWLEDGEMENT

The authors are grateful to the Chemistry Division of the ONR, to the DARPA Moletronics Program, to the DoD/MURI program and to the NSF-MRSEC program for support of this research.

REFERENCES

1. *Molecular Electronics: Science and Technology*, edited by A. Aviram and M. A. Ratner (New York Academy of Sciences, New York, 1998); M. D. Newton, *Adv. Chem. Phys.* **106**, 303 (1999).
2. J. Chen, M. A. Reed, A. M. Rawlett, and J. M. Tour, *Science*, **286**, 1550 (1999).
3. V. Mujica, A. E. Roitberg, and M. A. Ratner, *J. Chem. Phys.* **112**, 6834 (2000); S. N. Yaliraki, A. E. Roitberg, C. Gonzalez, V. Mujica, and M. A. Ratner, *J. Chem. Phys.* **111**, 6997 (1999).
4. N. R. Armstrong, R. M. Wightman, and E. M. Gross, *Ann. Rev. Phys. Chem.* **52**, 391 (2001); J. R. Sheats, H. Antoniadis, M Hueschen, W. Leonard, J. Miller, R. Moon, D. Roitman, and A. Stocking, *Science* **273**, 884 (1996); H. Bassler, *Polym. and Adv. Techn.* **9**, 402 (1998)
5. D. Vuillaume, B. Chen and R.M. Metzger, *Langmuir* **15**, 4011 (1999); E. Punkka and R.F. Rubner, *J. Elect. Mat.* **21**, 1057 (1992).
6. M. A. Lampert and P. Mark, *Current Injection in Solids* ~Academic, (New York, 1970); Electronic processes in organic crystals and polymers, Martin Pope, Charles E. Swenberg. 2nd ed.: New York : Oxford University Press, 1999.
7. L. Bozano, S. A. Carter, J. C. Scott, G. G. Malliaras, and P. J. Brock, *Appl. Phys. Lett.*, **74**, 1132 (1999); W. Geens, D. Tsamouras, J. Poortmans, and G. Hadziioannou, *Synth. Met.* 8, 191 (2001).
8. M. K. Lee, R. H. Horng, and L. C. Haung, *J. Appl. Phys.* **72**, 5420 (1992).
9. W. J. Li, Q. W. Wang, J. Cui, H. Chou, S. E. Shaheen, G. E. Jabbour, J. Anderson, P. Lee, B. Kippelen, N. Peyghambarian, N. R. Armstrong, and T. J. Marks, *Adv. Mater.* **11**, 730 (1999); M. L. Swiggers, G. Xia, J. D. Slinker, A. A. Gorodetsky, G. G. Malliaras, R. L. Headrick, B. T. Weslowski, R. N. Shashidhar, and C. S. Dulcey, *Appl. Phys. Lett.* **79**, 1300 (2001).
10. J. Cui, Q. Huang, J. C. G. Veinot, H. Van, Q. Wang, G. Hutchinson, A. G. Richter, G. Evmenenko, P. Dutta, and T. J. Marks, Langmuir **18**, 9958 (2002).
11. L. E. Calvet, R. G. Wheeler, and M. A. Reed, *Appl. Phys. Lett.* **80**, 1761 (2002).
12. P. E. Burrows and S. R. Forrest, *Appl. Phys. Lett.* **64**, 2285 (1994).
13. H. Tang, F. Li and J. Shinar, *Appl. Phys. Lett.* **71**, 2560 (1997).
14. U. Wolf, V. I. Arkhipov and H. Bässler, *Phys. Rev. B*, **59**, 7507, 1999; V. I. Arkhipov, E. V. Emelianova, Y. H. Tak, and H. Bassler, *J. Appl. Phys.* **84**, 848, (1998).
15. S. Barth, U. Wolf, H. Bässler, P. Muller, H. Riel, H. Vestweber, P. F. Seidler, and W. Rieβ, *Phys. Rev. B*, **60**, 8791 (1999); U. Wolf, S. Barth, H. Bässler, *Appl. Phys. Lett.* **75**, 2035 (1999).
16. A. L. Burin and M. A. Ratner, *J. Chem. Phys.* **113**, 3941 (2000).
17. M. A. Baldo and S. R. Forrest, *Phys. Rev. B* **64**, 085201 (2001).
18. M. A. Baldo, Z. G. Soos and S. R. Forrest, *Chem. Phys. Lett.* **347**, 297 (2001).
19. S. Karg, J. Steiger, and H. von Seggern, *Synth. Met.* **111**, 277 (2000).

20. W. Brutting, E. Buchwald, G. Egerer, M. Meier, K. Zuleeg, and M. Schwoerer, *Synth. Met.* **84**, 677 (1997).
21. S. V. Novikov, D. H. Dunlap, V. M. Kenkre, P. E. Parris, A. V. Vannikov, *Phys. Rev. Lett.* **81**, 4472 (1998); D. H. Dunlap, P. E. Parris, V. M. Kenkre, *Phys. Rev. Lett.* **77**, 542 (1996); M. W. Klein, D. H. Dunlap, and G. G. Malliaras, *Phys. Rev. B* **64**, 195332 (2001).
22. A. J. Campbell, D. D. C. Bradley, and H. Antoniadis, *J. Appl. Phys.* **89**, 3343 (2001).
23. J. Staudigel, M. Stossel, F. Steuber, J. Blassing, and J. Simmerer, *Synth. Met.* **111**, 69 (2000).
24. A. J. Campbell, D. D. C. Bradley, J. Laubender, and M. Sokolowski, *J. Appl. Phys.* **86**, 5004 (1999).
25. Y. Shen, M. W. Klein, D. B. Jacobs, J. C. Scott, G. C. Malliaras, *Phys. Rev. Lett.* **86**, 3867 (2001).
26. P. S. Davids, I. H. Campbell, and D. L. Smith, *J. Appl. Phys.* **82**, 6319 (1997).
27. I. H. Campbell, P. S. Davids, D. L. Smith, N. N. Barashkov, and J. P. Ferraris, *Appl. Phys. Lett.* **72**, 1863 (1998).
28. G. G. Malliaras and J. C. Scott, *J. Appl. Phys.* **85**, 7426 (1999).
29. S. Barth, U. Wolf, H. Bassler, P. Muller, H. Riel, H. Vestweber, P. F. Seidler, and W. Riess, *Phys. Rev. B* **60**, 8791 (1999).
30. F. Anariba, R. L. McCreery, *J. Phys. Chem.* B **106**, 10355 (2002).
31. In several cases it is possible to reach the conditions where the Fermi level is located below the highest occupied molecular level. For instance this takes place for Indium Tin Oxide (ITO) electrode coated by the self-assembled siloxane layer (see J. E. Malinsky, J. G.C. Veinot, G. E. Jabbour, S. E. Shaheen, J. D. Anderson, P. Lee, A. G. Richter, A. L. Burin, M. A. Ratner, T. J. Marks, N. R. Armstrong, B. Kippelen, P. Dutta, and N. Peyghambarian, to appear in *Chemistry of Materials*, **14**, 3054 (2002)). In these systems the image potential captures the extra charge and can still create a large injection barrier for conducting holes [Ref. 26].
32. V. Kasperovich, K. Wong, G. Tikhonov, and V. V. Kresin, *Phys. Rev. Lett.* **85**, 2729 (2000).
33. B. K. Crone, P. S. Davids, I. H. Campbell, and D. L. Smith, *J. Appl. Phys.* **87**, 1974 (2000).
34. G. G. Malliaras, J. R. Salem, P. J. Brock, and J. C. Scott, *J. Appl. Phys.* **84**, 1583 (1998).
35. H. Ness and A. J. Fisher, *Appl. Phys. A – Mater. Sci. & Processing* **66**, S919, Part 2 Suppl. S, (1998).
36. E. Etterdgui, H. Razafitrimo, Y. Gao, and B. R. Hsieh, *Appl. Phys. Lett.* **67**, 2705, (1995).
37. A. Onipko, Y. Klymenko, and L. Malysheva, *J. Chem. Phys.* **107**, 7331 (1997).

38. M. Koehler and I. A. Hümmelgen, *Appl. Phys. Lett.* **70**, 3254 (1997); C. Y. Lee, M. Z. Tidrow, K. K. Choi, W. H. Chang, L. F. Eastman, F. J. Towner, and J. S. Ahearn, *Appl. Phys. Lett.* **65**, 442 (1994).
39. S. N. Burmistrov and L. B. Dubovskii, *Zh. Eksp. Teor. Fiz.***100**, 1844 (1991).
40. J. K. Schon, C. Kloc, and B. Batlogg, *Phys. Rev. B* **63**, art. no. 125304 (2001).
41. R. E. Holmlin, R. F. Ismagilov, R. Haag, V. Mujica, M. A. Ratner, M. A. Rampi, and G. M Whitesides, *Angew. Chem. Int. Ed.* **40**, 2316 (2001).
42. J. Klafter and J. Jortner, *Chem. Phys. Lett.* **60**, 5 (1978).
43. E. M. Monberg and R. Kopelman, *Chem. Phys. Lett.* **58**, 497 (1978).
44. R. A. Marcus and N. Sutin, *Biochim. et. Biopys. Acta*, **811**, 265 (1985).
45. N. R. Kestner, J. Logan, and J. Jortner, *J. Phys. Chem.* 78, 2148 (1974).
46. H. Sugiyama, and I. Saito, *J. Amer. Chem. Soc.* **118**, 7063 (1996).
47. A. A. Voityuk, N. Rosch, M. Bixon, and J. Jortner, *J. Phys. Chem. B* **104**, 9740, (2000).
48. Y. A. Berlin and A. L. Burin, *Chem. Phys. Lett.* **257**, 665 (1996).
49. J. P. Bouchaud and A. Georges, *Phys. Rep.* **195**, 127 (1990).
50. Y. N. Garstein and E. M. Conwell, *Chem. Phys. Lett.* **217**, 41 (1994), see also A. L. Burin and M. A. Ratner, *J. Chem. Phys.* **109**, 6092 (1998).
51. R. Englman and J. Jortner, *Mol. Phys.* **18**, 145 (1970).
52. B. K. Crone, I. H. Campbell, P. S. Davids, D. L. Smith, C. J. Neef, and J. P. Ferraris, *J. Appl. Phys.* **86**, 5767 (1999).
53. S. John, M. Y. Chou, M. H. Cohen, and C. M. Soukoulis, *Phys. Rev. B* **37**, 6963 (1988).
54. T. J. Marks, J. G. C. Veinot, J. Cui, H. Yan, A. Wang, N. L. Edleman, J. Ni, Q. Huang, P. Lee, and N. R. Armstrong, *Synth. Met.* **127**, 29 (2002); G. E. Jabbour, B. Kippelen, N. R. Armstrong, and N. Peyghambarian, *Appl. Phys. Lett.* **73**, 1185 (1998).
55. J. H. Werner and H. H. Güttler, *J. Appl. Phys.* **69**, 1522 (1991).
56. J. Staudigel, M. Stößel, F. Steuber, and J. Simmerer, *J. Appl. Phys.* **86**, 3895 (1999).
57. P. Sigaud, J. N. Chzalviel, and F. Ozanam, *J. Appl. Phys.* **89**, 466 (2001).
58. L. Leuzzi and Th. M. Nieuwenhuizen, *J. Phys. – Cond. Matt.* **14**, 1637 (2002).
59. D. Thirumalai and R. D. Mountain, *Phys. Rev. E* **47**, 479 (1993); V. I. Arkhipov, H. Bassler, *J. Phys. Chem.* **98**, 662 (1994).
60. M. Matsumura, T. Akai, M. Saito, and T. Kimura, *J. Appl. Phys.* 79, 264 (1996); M. Matsumura, and Y. Jinde, *Appl. Phys. Lett.* **73**, 2872 (1998).
61. See D. V. Khramtchenkov, V. I. Arkhipov, and H. Bassler, *J. Appl. Phys.* **79**, 9283 (1996); despite of the assumption of the tunneling injection mechanism that was not confirmed by the later studies this work can also be applied to the thermally activated injection after weak modifications.
62. R. H. Young, C. W. Tang, A. P. Marchetti, Appl. Phys. Lett. 80, 874, 2002
63. A. L. Burin, M. A. Ratner, J. Phys. Chem. A, 104, 4704, 2000.

64. J. S. Kim, R. H. Friend, F. Cacialli, Appl. Phys. Lett. 74, 3084, 1999; C. I. Chao, K. R. Chuang, S. A. Chen, Appl. Phys. Lett. 69, 2894, 1996.
65. E. W. Forsythe, M. A. Abkowitz, and Y. Gao, *J. Phys. Chem. B* **104**, 3948 (2000).
66. M. K. Hudait and S. B. Krupanidhi, *Physica B* **307**, 125 (2001).
67. J. P. Sullivan, R. T. Tung, M. R. Pinto, and W. R. Graham, *J. Appl. Phys.* **70**, 7403 (1991).
68. G. G. Malliaras, Y. Shen, D. H Dunlap, H. Murata and Z. H. Kafafi, **79**, 2582 (2001).

INDEX

L.A. Curtiss and M.S. Gordon (eds.),
Computational Materials Chemistry: Methods and Applications, 368–372.